Universitext

Universitext is a series of textbooks that presents material from a wide variety of mathematical disciplines at master's level and beyond. The books, often well class-tested by their author, may have an informal, personal, or even experimental approach to their subject matter. Some of the most successful and established books in the series have evolved through several editions, always following the evolution of teaching curricula, into very polished texts.

Thus as research topics trickle down into graduate-level teaching, first textbooks written for new, cutting-edge courses may find their way into *Universitext*.

Claude Zuily

Selected Topics in Partial Differential Equations

Claude Zuily
Institut Mathématique d'Orsay
Université Paris-Saclay
Orsay, France

ISSN 0172-5939 ISSN 2191-6675 (electronic)
Universitext
ISBN 978-3-032-24081-1 ISBN 978-3-032-24082-8 (eBook)
https://doi.org/10.1007/978-3-032-24082-8

Mathematics Subject Classification: 35-01

The original submitted manuscript has been translated into English. The translation was done using artificial intelligence. A subsequent revision was performed by the author(s) to further refine the work and to ensure that the translation is appropriate concerning content and scientific correctness. It may, however, read stylistically different from a conventional translation.

This Springer imprint is published by the registered company Springer Nature Switzerland AG
The registered company address is: Gewerbestrasse 11, 6330 Cham, Switzerland

For my wife Yasmine

Preface

This text was born from the desire to introduce third-year undergraduate and first-year Master's students to some aspects of the theory of partial differential equations (PDEs). These students had already taken, during previous semesters, courses in differential and integral calculus, holomorphic functions, and, for those in the Master's program, distribution theory.

As part of their TER (teaching and research work), which is included in their curriculum, these students were required to read, possibly complete and write a report, then to present orally a mathematical text that uses their knowledge and, if possible, opens up to current research topics. It is in this spirit and for them that, over several years, the following chapters have been written. Most of them are self-contained and require, for their understanding, only concepts already covered in their previous courses. However, in addition to the students described above, I believe that this text could also be read with benefit by Master 2 students and even by young doctoral students, who are often very early focused on their research topics and to whom it will offer other aspects of PDEs.

Given the vastness of the theory of PDEs, it was obvious that it was out of the question to claim or even consider exhaustiveness. This text is therefore clearly not a course book centered on a single question, nor a reference text. It is rather a journey through the countless possible topics in this field.

To begin, we have chosen to present methods for solving nonlinear partial differential equations, first in the real domain (method of characteristics) and then in the complex domain (Cauchy-Kovalevski Theorem). The first method uses classical notions of differential calculus as well as the theory of differential equations. As for the second, it is based on the theory of holomorphic functions, and for the proof, we have chosen to use a recent approach, essentially due to L. Nirenberg[1], which provides a demonstration using a "fixed point" type method.

In the following chapter, we have detailed a method widely used in analysis that allows one to determine the asymptotic behavior of certain integrals depending on a parameter: the method of stationary phase. This method is useful, as shown in

[1] Louis Nirenberg (1925–2020), Abel Prize 2015.

this text, for determining the asymptotic behavior of solutions to certain differential equations. It also plays an essential role in more advanced theories such as that of Fourier integral operators.

In PDEs, it is always crucial to specify the functional spaces in which to seek solutions. In this respect, Sobolev spaces occupy a prominent place. The Littlewood-Paley theory, which is described in this chapter, allows them to be characterized efficiently for applications. It is also important to be able to study functions defined on these spaces and to determine their continuity. This third chapter contains, in particular, a very general continuity result that has been recently published. This result applies especially to the study of the continuity of the map (called "the flow") which, to the initial data, associates the solution (assumed unique) of an evolution PDE.

As an excellent example of a nonlinear PDE, one can mention the Burgers equation. It is an evolution equation that was studied by J. M. Burgers (see the Comments) to model problems in fluid mechanics. Chapter 4 provides a very detailed study of this equation when the initial data is Lipschitz, and then when it belongs to a Sobolev space. In particular, it is shown that its flow is continuous but not uniformly continuous, a characteristic property of quasi-linear equations.

An interesting system, originating from physics, is the one that describes the propagation of waves on the surface of a fluid (waves). In Chap. 5, we begin by introducing some physical notions that describe this propagation: Newton's law, incompressibility, irrotationality, then we show how these notions can be described by mathematical objects. In other words, this chapter describes the mathematical modeling of surface wave propagation. It turns out that this leads to rather complicated PDEs. That is why, in the 18th century, renowned mathematicians such as Cauchy, Stokes, Boussinesq and others proposed simplified versions of these equations and then solved them, while introducing relevant physical concepts that remain pertinent to this day. These simplified equations allow, for example, the calculation of the speeds of different types of waves: wind waves (the most common), tsunamis. All this is explained in this chapter using tools accessible to our students. It turns out that an important object, the Dirichlet-Neumann operator, allows the problem to be simplified. It is then introduced and studied in Chap. 7.

Spectral theory is a major issue in mathematical physics and in partial differential equation. That is why we dedicate five of the following six chapters to its study.

In Chap. 8 we begin with a few special cases namely those of the Laplacian on the sphere and on the torus. This will lead us to the study of the spherical harmonics and to multi-dimensional Fourier series.

In Chap. 9, we remain within the framework of spectral theory and we focus on a problem from quantum mechanics: that of the hydrogen atom. As is well known, the hydrogen atom is composed of an electron and a proton interacting according to Coulomb's law. According to the Bohr model, the electron orbits the nucleus along specific paths, each orbit corresponding to an energy level. The lowest energy level is called the "ground state," while the others are referred to as "excited states." In this chapter, a complete mathematical theory is presented, through which all these objects are described and calculated. The methods used are particularly instructive for students, as they draw on various theories such as spectral theory, differential equations, holomorphic functions, Sobolev spaces, etc.

The next two chapters are also devoted to the same problem, but in a different context: that of compact self-adjoint operators. There is an abstract theory that is presented here, in the general framework of operators in Hilbert spaces. The spectral theory of the Dirichlet problem for the Laplacian operator in a bounded open subset of $\mathbf{R}^d$ fits perfectly into this framework and allows for a complete description of the spectrum: it is then composed of an increasing sequence $(\lambda_n)_{n\in\mathbf{N}}$ of eigenvalues tending to $+\infty$ as n increases. Besides the links with physics, the importance of this theory comes from the fact that, for bounded open sets, it plays the role that the Fourier transform plays in $\mathbf{R}^d$.

The next question is to describe the asymptotic behavior of λ_n as n tends to $+\infty$. It has been shown that the different terms in this expansion are closely related to the geometry of the open set: volume, measure of the boundary, number of holes, etc. For reasons of difficulty, we have limited ourselves here to describing the first term of this expansion, leading to a famous result known as "Weyl's law."

The following chapter is devoted to a central question in PDEs, that of the uniqueness of solutions. The question discussed here is the description of the places where a solution to an elliptic PDE can vanish. This is a problem that, due to its applications, is still very much current. It turns out that, under certain conditions, nontrivial solutions of elliptic equations cannot vanish on an open set (no matter how small) nor even on a subset of strictly positive measure. Several approaches are possible to prove these results. We have chosen here to use an elegant and effective theory known as "Almgren's theory," named after its inventor. To avoid overloading the text with technicalities that might obscure the essentials, we have limited ourselves to the case of the Laplacian, but it is important to note that these methods go far beyond this simple case.

In the last chapter we go back to spectral theory and we detail a beautiful theorem due to Richard Courant, which sheds light on the structure of the zeros of the eigenfunctions of the Laplacian.

I would like to conclude this description with a few remarks.

The main purpose of this book being to lead the reader towards questions leading to research, it does not contain exercises. However, readers who miss them can consult the book [4] for the Master 1 level and the book [6] for the Master 2 level.

On the other hand, I would like to emphasize that several results contained in this text do not appear in the classical literature, and that some others are found in more advanced documents that are often out of reach for the audience to whom this book is intended.

Next, for the convenience of the beginning reader, we have included, at the beginning of each chapter, a short section specifying the prerequisites necessary for its reading, while at the end of some chapters an appendix gathers some basic results used.

Furthermore, at the end of the book, the reader will find a short bibliography dealing with the prerequisites and another, accessible one, allowing them, if they wish, to continue exploring this vast field of partial differential equations, as well as an index.

Finally, it goes without saying that any possible error contained in this book would be the sole and exclusive responsibility of the author.

Orsay, France
February 2026

Claude Zuily

Acknowledgements I would like to express my gratitude to all the students in mathematics at the University Paris-Saclay who, for their Bachelor's or Master's thesis, chose one of the topics presented here. Their careful readings, relevant comments, and enthusiasm have greatly contributed to enriching the content, both in substance and in form. Thank you all.

I would also like to thank the anonymous reviewers of this book. Their high standards, as well as their suggestions and requests for clarification, have contributed to the final improvement of the text.

I am grateful to Ms. L. Jammet, editor at Dunod, for her kind permission to reproduce Chap. 13 from my earlier work "Elements de Distributions et d'équations aux dérivées partielles", in the present book.

My thanks also go to Springer, and in particular to Ms. Fairle T., for their warm welcome and their assistance in the production of this book.

During the preparation of this book, I had numerous exchanges with Mr. Remi Lodh. I would like to extend my warm thanks to him for his support, kindness, and efficiency. Thank you, dear Sir, for making the publication process so pleasant.

Finally, I would like to express my deepest thanks to my colleagues and friends, Thomas Alazard, Nicolas Burq, Patrick Gérard, Luc Robbiano, and Nicola Visciglia with whom, for so many years, I have had fruitful discussions on various topics of this book.

Competing Interests The author has no competing interests to declare that are relevant to the content of this manuscript.

Contents

Chapter 1
The Method of Characteristics

1.1 Prerequisites

Elements of differential calculus: implicit function theorems and local inversion; notions of hypersurface, of vector field tangent to a hypersurface, of integral curves of a vector field. Differential equations; Cauchy-Lipschitz theorem.

1.2 The Context

In the field of partial differential equations, a fundamental question is that of the existence of solutions. In the overwhelming majority of cases it is illusory, even pointless, to try to establish explicit formulas for these solutions. It is therefore necessary to develop various methods, depending on the nature of the equation, to answer this question.

This chapter aims to describe a method that allows us to solve certain nonlinear partial differential equations called evolution equations (that is, equations that depend on a "time" variable for which the value of the solution at the initial time, for example $t = 0$, is known).

In this method, two things are important: the equations must be **of first order** and all the **data** must be **real** (see Remark 1.5.4 (ii) on this point). As we will see, it is based on classical ingredients of differential calculus (implicit function theorem, notion of submanifold, tangent space) as well as on the theory of differential equations.

Concerning nonlinear evolution partial differential equations, unlike linear equations, the solution does not necessarily exist for all later times. The method of characteristics also allows, in certain cases, to specify exactly up to what time the solution exists as a C^1 function.

C. Zuily, *Selected Topics in Partial Differential Equations*, Universitext,
https://doi.org/10.1007/978-3-032-24082-8_1

Note that for the purposes of the proof, in Sect. 1.5, we will state and then prove (in a particular case) in the Appendix a beautiful theorem due to J. Hadamard, which characterizes global C^1-diffeomorphisms from $\mathbf{R}^d$ to $\mathbf{R}^d$.

To conclude, a point of terminology. As we will see, the method is based on the fact that there exist curves along which the partial differential equation reduces to a differential equation that is easier to solve. These curves were called as early as the 18th century, in particular by Lagrange and Monge, "characteristic curves," which gives the method its name.

1.3 The Problem

As with differential equations, we will be interested in the Cauchy problem. Let us specify the hypotheses and the problem.

Hypothesis 1.3.1 *Let $t_0 \in \mathbf{R}$, $x_0 \in \mathbf{R}^d$. Let I be an open neighborhood of t_0, Ω_{x_0} be an open neighborhood of x_0. We are given functions $b = b(t, x, z)$ and $a_j = a_j(t, x, z)$ for $j = 1, \dots, d$, of class C^1 on $I \times \Omega_{x_0} \times \mathbf{R}$ with real values and a function $u_0 : \Omega_{x_0} \to \mathbf{R}$ of class C^1.*

The question that interests us is the following: does there exist a function $u = u(t, x)$ of class C^1 on an open neighborhood of (t_0, x_0) of the form $J \times V_{x_0}$ contained in $I \times \Omega_{x_0}$ with real values which is a solution in $J \times V_{x_0}$ of the problem

$$\begin{aligned} &\frac{\partial u}{\partial t}(t, x) + \sum_{j=1}^{d} a_j(t, x, u(t, x)) \frac{\partial u}{\partial x_j}(t, x) = b(t, x, u(t, x)), \\ &u|_{t=t_0} = u_0. \end{aligned} \tag{1.3.1}$$

By setting $t' = t - t_0$ we reduce to the case where $t_0 = 0$, which we will assume from now on.

Remark 1.3.2 There is another important point to emphasize. It will have been noticed that since the initial surface is $\{(t, x) \in \mathbf{R} \times \mathbf{R}^d : t = t_0\}$, the coefficient of $\frac{\partial u}{\partial t}$ in the equation is nonzero (here equal to 1). In that case we say that the initial surface is "non-characteristic" for the equation, which constitutes another essential hypothesis for the development of the method.

1.3.1 A Bit of Terminology

Equations of type (1.3.1) are called **quasi-linear** (of first order) because they are linear with respect to the first order derivatives (but nonlinear in u). If the a_j do not depend on the variable z they are called **semi-linear**, i.e.

$$\frac{\partial u}{\partial t}(t,x) + \sum_{j=1}^{d} a_j(t,x)\frac{\partial u}{\partial x_j}(t,x) = b(t,x,u(t,x)).$$

If moreover b does not depend on the variable z either, they are called **linear**, i.e.

$$\frac{\partial u}{\partial t}(t,x) + \sum_{j=1}^{d} a_j(t,x)\frac{\partial u}{\partial x_j}(t,x) = b(t,x).$$

Finally, equations that are not linear with respect to the first order derivatives are called **fully nonlinear**. Here is an example

$$\frac{\partial u}{\partial t}(t,x) + \sum_{j=1}^{d} \left(\frac{\partial u}{\partial x_j}(t,x)\right)^2 = 0.$$

1.4 The General Method of Characteristics

1.4.1 The Main Result

The purpose of this paragraph is to prove the following result.

Theorem 1.4.1 *Under the Hypothesis 1.3.1 on a_j, b and u_0, there exists an open interval $J \subset \mathbf{R}$ containing t_0, an open neighborhood V_{x_0}, and a unique function u of class C^1 from $J \times V_{x_0}$ into $\mathbf{R}$ that solves the problem* (1.3.1).

The proof will require several steps and will be given at the end of this section. All neighborhoods considered in what follows will be open.

1.4.2 Some Reminders from Differential Calculus

The Implicit Function Theorem

Theorem 1.4.2 *Let $(x_0, y_0) \in \mathbf{R}^p_x \times \mathbf{R}^d_y$, U_{x_0} (resp. V_{y_0}) a neighborhood of x_0 (resp. of y_0), and $F : U_{x_0} \times V_{y_0} \to \mathbf{R}^d$ a C^1 map.*

Suppose that $F(x_0, y_0) = 0$ and that the linear map $\frac{\partial F}{\partial y}(x_0, y_0) \in \mathcal{L}(\mathbf{R}^d, \mathbf{R}^d)$ is invertible. Then there exist neighborhoods $\widetilde{U}_{x_0}$, $\widetilde{V}_{y_0}$ and a C^1 map $Y : \widetilde{U}_{x_0} \to \widetilde{V}_{y_0}$ such that

$$\Big((x,y) \in \widetilde{U}_{x_0} \times \widetilde{V}_{y_0},\ F(x,y) = 0\Big) \Longleftrightarrow y = Y(x).$$

The Local Inversion Theorem

Theorem 1.4.3 *Let F be a C^1 map from an open set U of $\mathbf{R}^d$ into $\mathbf{R}^d$ and x_0 a point in U. If the Jacobian determinant of F is nonzero at x_0, there exist neighborhoods V of x_0 and W of $F(x_0)$ such that F is a C^1 diffeomorphism from V onto W.*

Hypersurfaces

Let us now recall some basics of differential geometry. In this chapter, we will only use the notion of hypersurface. What are called hypersurfaces of $\mathbf{R}^d$ are the submanifolds of dimension $d-1$. There are several equivalent ways to define such objects. We will recall two here: the implicit definition and the parametric definition.

Implicit definition: $S \subset \mathbf{R}^d$ is a C^k hypersurface, $k \geq 1$, if for every point $x_0 \in S$ there exists a neighborhood V_{x_0} and a function $\varphi : V_{x_0} \to \mathbf{R}$ of class C^k such that

$$S \cap V_{x_0} = \{x \in V_{x_0} : \varphi(x) = 0\} \quad \text{and} \quad d\varphi(x) \neq 0 \quad \forall x \in S \cap V_{x_0}.$$

The tangent space to S at a point $x \in S$ is then the $(d-1)$-dimensional plane given by

$$T_x S = \Big\{ v \in \mathbf{R}^d : \langle v, d\varphi(x) \rangle := \sum_{j=1}^{d} v_j \frac{\partial \varphi}{\partial x_j}(x) = 0 \Big\}. \tag{1.4.1}$$

Parametric definition: $S \subset \mathbf{R}^d$ is a C^k hypersurface if for every x_0 there exists a neighborhood V_{x_0}, an open set $U \subset \mathbf{R}^{d-1}$, and a map $\Phi : U \to S \cap V_{x_0}$ such that

(i) Φ is of class C^k from U into $\mathbf{R}^d$,

(ii) Φ is bijective and $\Phi^{-1} : S \cap V_{x_0} \to U$ is the restriction to $S \cap V_{x_0}$ of a C^k map from an open set $W \subset \mathbf{R}^d$ into $\mathbf{R}^{d-1}$ such that $S \cap V_{x_0} \subset W$,

(iii) for every $m \in U$, the map $d\Phi(m) \in \mathcal{L}(\mathbf{R}^{d-1}, \mathbf{R}^d)$ is injective.

In this case, the tangent space to S at a point $x = \Phi(m) \in S \cap V_{x_0}$ is given by

$$T_x S = d\Phi(m)(\mathbf{R}^{d-1}).$$

Example 1.4.4 A simple example is that of the unit sphere in $\mathbf{R}^d$,

$$\mathbf{S}^{d-1} = \{x \in \mathbf{R}^d : |x| = \Big(\sum_{j=1}^{d} x_j^2\Big)^{1/2} = 1\}.$$

For the implicit definition, we take $\varphi(x) = |x|^2 - 1$. Then $d\varphi(x) = 2x \neq 0$ if $x \in \mathbf{S}^{d-1}$. The tangent plane to $\mathbf{S}^{d-1}$ at the point $x \in \mathbf{S}^{d-1}$ is then the plane $\{v \in \mathbf{R}^d : \langle v, x \rangle = 0\}$.

For the parametric definition, we proceed as follows. By rotating the coordinates, we may assume that $x_0 = (0, \ldots, 0, 1)$. For $x = (x_1, \ldots, x_{d-1}, x_d) \in \mathbf{R}^d$, we write $x' = (x_1, \ldots, x_{d-1})$. We set, for $\varepsilon > 0$ small enough,

$$V_{x_0} = \{x = (x', x_d) \in \mathbf{R}^d : |x'| < \varepsilon, |x_d - 1| < \varepsilon^2\}, \quad U = \{x' \in \mathbf{R}^{d-1} : |x'| < \varepsilon\},$$

and we define the map $\Phi : U \to \mathbf{S}^{d-1} \cap V_{x_0}$ by $\Phi(x') = (x', \sqrt{1 - |x'|^2})$. We have $\Phi(x') \in \mathbf{S}^{d-1} \cap V_{x_0}$ because $|x'| < \varepsilon$, $|\sqrt{1 - |x'|^2} - 1| = \frac{|x'|^2}{1+\sqrt{1-|x'|^2}} < \varepsilon^2$ and $|x'|^2 + 1 - |x'|^2 = 1$. Moreover, this mapping is bijective.

For $x' \in U$, the matrix (a_{ij}) of $d\Phi(x')$ satisfies $a_{ij} = \delta_{ij}$ for $1 \le i, j \le d-1$ and $a_{dj} = -\frac{x_j}{\sqrt{1-|x'|^2}}$. Thus,

$$d\Phi(x')(t_1, \dots, t_{d-1}) = \Big(t_1, \dots, t_{d-1}, -\sum_{j=1}^{d-1} \frac{x_j}{\sqrt{1 - |x'|^2}} t_j\Big).$$

Therefore, it is indeed injective.

The tangent plane to $\mathbf{S}^{d-1}$ at $x = \Phi(x') \in \mathbf{S}^{d-1} \cap V_{x_0}$ is the set $d\Phi(x')\mathbf{R}^{d-1}$. It is thus the plane in $\mathbf{R}^d$ defined by

$$\Big\{T = (T_1, \dots, T_d) \in \mathbf{R}^d : T_d = -\sum_{j=1}^{d-1} \frac{x_j}{\sqrt{1 - |x'|^2}} T_j\Big\}.$$

It is indeed orthogonal to the vector $x \in \mathbf{S}^{d-1}$. Indeed, if $x \in \mathbf{S}^{d-1} \cap V_{x_0}$ we have $x_d > 0$ and $x_d^2 = 1 - |x'|^2$. Therefore,

$$\langle x, T\rangle = \sum_{j=1}^{d-1} x_j T_j + x_d T_d = \sum_{j=1}^{d-1} x_j T_j - \sqrt{1 - |x'|^2} \sum_{j=1}^{d-1} \frac{x_j}{\sqrt{1 - |x'|^2}} T_j = 0.$$

How is a hypersurface changed by diffeomorphism?

Let S be a hypersurface given implicitly in a neighborhood U_{x_0} of a point $x_0 \in \mathbf{R}^d$ by a function φ.

Let $\theta : U_{x_0} \to V_{\theta(x_0)}$ be a C^1 diffeomorphism. Set

$$\psi = \theta^{-1}, \quad y_0 = \theta(x_0), \quad \widetilde{\varphi}(y) = \varphi(\psi(y)).$$

Then in the y coordinates S becomes the hypersurface $\widetilde{S}$ given implicitely in V_{y_0} by the function $\widetilde{\varphi}$.

Let $J_\theta(x) = \big(\frac{\partial \theta_i}{\partial x_j}(x)\big)_{1 \le i, j \le d}$ be the Jacobian matrix of θ. Then, denoting by ∇ the gradient, we have

$$\nabla_x \varphi(x) = {}^t J_\theta(x) \nabla_y \widetilde{\varphi}(\theta(x)), \tag{1.4.2}$$

where ${}^t J_\theta$ denotes the transpose matrix of J_θ.

Vector Fields and Integral Curves

A vector field $\mathcal{X}$ on $\mathbf{R}^d$ of class C^1 in a neighborhood of $x_0 \in \mathbf{R}^d$ is a mapping $\mathcal{X} : V_{x_0} \to \mathbf{R}^d$, $x \to (a_1(x), \dots, a_d(x))$, of class C^1.

A vector field $\mathcal{X}$ is said to be tangent to the hypersurface S at $x_0 \in S$ if it belongs to the tangent plane to S at x_0.

According to (1.4.1), this fact is expressed by the equation

$$\langle \mathcal{X}(x_0), \nabla_x \varphi(x_0) \rangle = \sum_{j=1}^{d} a_j(x_0) \frac{\partial \varphi}{\partial x_j}(x_0) = 0,$$

where $\langle \cdot, \cdot \rangle$ denotes the scalar product in $\mathbf{R}^d$.

How do vector fields transform under a diffeomorphism θ?

Keeping the above notations, the vector field $\mathcal{X}(x) = (a_1(x), \dots, a_d(x))$ is transformed to a vector field $\widetilde{\mathcal{X}}$ given by

$$\widetilde{\mathcal{X}}(y) = J_\theta(\psi(y))\mathcal{X}(\psi(y)), \tag{1.4.3}$$

where here $\mathcal{X}$ is considered as a column vector.

This formula guarantees that the concept of tangency remains invariant under diffeomorphism.

Indeed, denoting by $\langle \cdot, \cdot \rangle$ the scalar product in $\mathbf{R}^d$, using (1.4.2) and (1.4.3) and setting $\theta(x) = y$ which is equivalent to $x = \psi(y)$, we can write

$$\begin{aligned} \langle \mathcal{X}(x), \nabla_x \varphi(x) \rangle &= \langle \mathcal{X}(x), {}^t J_\theta(x) \nabla_y \widetilde{\varphi}(y) = \langle J_\theta(\psi(y))\mathcal{X}(\psi(y)), \nabla_y \widetilde{\varphi}(y) \rangle, \\ &= \langle \widetilde{\mathcal{X}}(y), \nabla_y \widetilde{\varphi}(y) \rangle, \end{aligned} \tag{1.4.4}$$

and the right hand side vanishes if and only if the left hand side vanishes.

Let us now introduce the notion of "integral curve" of a vector field.

Let $\mathcal{X}$ be a vector field on $\mathbf{R}^d$ of class C^1. An integral curve of $\mathcal{X}$, starting from the point $x_0 = (x_{0,1}, \dots, x_{0,d})$, is a curve $t \to (x_1(t), \dots, x_d(t))$ of class C^1 such that

$$\begin{cases} \dot{x}_1(t) = a_1(x_1(t), \dots, x_d(t)), \ x_1(0) = x_{0,1}, \\ \dots \\ \dot{x}_d(t) = a_d(x_1(t), \dots, x_d(t)), \ x_d(0) = x_{0,d}. \end{cases} \tag{1.4.5}$$

The Cauchy-Lipschitz theorem ensures the existence (and uniqueness) of the integral curve for some time $|t| \le \delta$.

We then have the following theorem.

Theorem 1.4.5 *Let S be a hypersurface of class C^2 and $\mathcal{X}$ a vector field of class C^1. Suppose that $\mathcal{X}$ is tangent to S at every point, that is, $\mathcal{X}(x) \in T_x S$ for all $x \in S$. Then every integral curve of $\mathcal{X}$ starting from a point x_0 of S remains in S as long as it exists.*

Proof We will use the implicit definition of hypersurfaces, that is,

$$S \cap V_{x_0} = \{x \in V_{x_0} : \varphi(x) = 0\} \text{ with } d\varphi(x) \neq 0 \text{ for } x \in S \cap V_{x_0}.$$

Without loss of generality, we may assume that $\frac{\partial \varphi}{\partial x_d}(x_0) \neq 0$. We will make a change of coordinates.

Set

$$y_j = x_j - x_{0,j}, \quad 1 \leq j \leq d-1, \quad y_d = \varphi(x).$$

Set $y = F(x)$. We have $F(x_0) = 0$ and the determinant, at x_0, of the Jacobian of F is equal to $\frac{\partial \varphi}{\partial x_d}(x_0)$. It is nonzero, by hypothesis. The local inversion theorem then implies that there exist neighborhoods W_{x_0} of x_0 and U_0 of zero in $\mathbf{R}^d$ such that F is a diffeomorphism from W_{x_0} onto U_0. In the new coordinates, the point x_0 becomes $y_0 = 0$, the hypersurface $S \cap V_{x_0}$ becomes

$$\widetilde{S} = \{y \in U_0 : y_d = 0\}, \tag{1.4.6}$$

and $\mathcal{X}$ becomes a vector field $\widetilde{\mathcal{X}} = (b_1(y), \ldots, b_d(y))$ of class C^1 in U_0 which, by (1.4.4), is tangent to $\widetilde{S}$ at every point. Writing $y' = (y_1, \ldots, y_{d-1})$ this last property translates to the fact that

$$b_d(y', 0) = 0. \tag{1.4.7}$$

The integral curve of $\widetilde{\mathcal{X}}$ starting from the point $y = 0$ is the solution of the differential system

$$\dot{y}_j(t) = b_j(y(t)), \quad y_j(0) = 0, \quad 1 < j \leq d. \tag{1.4.8}$$

By the Cauchy-Lipschitz Theorem, this system admits a unique solution for $t \in [0, T_1^*)$. According to the same theorem, letting $z' = (z_1, \ldots z_{d-1})$, the system

$$\dot{z}_j(t) = b_j(z'(t), 0), \quad z_j(0) = 0, \quad 1 \leq j \leq d-1 \tag{1.4.9}$$

admits a unique solution defined for $t \in [0, T_2^*)$. Using (1.4.7) we see that the function $(z(t), 0)$ is also a solution of the system (1.4.8) on $[0, T_2^*)$. By uniqueness, we deduce that $y(t) = (z(t), 0)$; this shows that, as long as it exists, the integral curve of $\widetilde{\mathcal{X}}$ remains in $\widetilde{S}$. □

1.4.3 The Method of Characteristics

The Method

The starting point is the following observation. Let us consider the vector field in $\mathbf{R}^{d+2}$ given in $I \times \Omega_{x_0} \times \mathbf{R}$ by

$$\mathcal{X}(t,x,z) = (1, a_1(t,x,z), \dots, a_d(t,x,z), b(t,x,z)). \tag{1.4.10}$$

Let $J \subset I$, $V_{x_0} \subset \Omega_{x_0}$ and S be a hypersurface of $\mathbf{R}^d$ of the form,

$$S = \{(t,x,z) \in J \times V_{x_0} \times \mathbf{R} : z = u(t,x)\}, \tag{1.4.11}$$

where u is a C^1 function from $J \times V_{x_0}$ into $\mathbf{R}$.

Then,

$$\mathcal{X} \text{ is tangent to } S \text{ at every point point } \iff u \text{ satisfies equation (1.3.1).} \tag{1.4.12}$$

Indeed, S is defined by the function $\varphi(t,x,z) = z - u(t,x)$ so that

$$\frac{\partial \varphi}{\partial t} = -\frac{\partial u}{\partial t}, \qquad \frac{\partial \varphi}{\partial x_j} = -\frac{\partial u}{\partial x_j}, \qquad \frac{\partial \varphi}{\partial z} = 1.$$

We deduce that at every point $(t, x, u(t,x))$ of S we have

$$(\mathcal{X} \cdot d\varphi)(t,x,u(t,x)) = -\frac{\partial u}{\partial t}(t,x) - \sum_{j=1}^{d} a_j(t,x,u(t,x))\frac{\partial u}{\partial x_j}(t,x) + b(t,x,u(t,x)),$$

and the left-hand side is zero if and only if the right-hand side is also zero, that is, if u satisfies the equation.

From this, the strategy for the proof of Theorem 1.4.1 is clear. The vector field $\mathcal{X}$ being given by (1.4.10), it suffices to find a hypersurface S of the form (1.4.11) to which $\mathcal{X}$ is tangent.

Let us consider the integral curve of the vector field $\mathcal{X}$ starting from a point $m = (0, y, u_0(y))$ belonging to $I \times U_{x_0} \times \mathbf{R}$. It is given by

$$\begin{cases} \dot{t}(s) = 1, & t(0) = 0, \\ \dot{X}_j(s) = a_j(t(s), X(s), Z(s)), & X_j(0) = y_j, \quad j = 1, \dots, d, \\ \dot{Z}(s) = b(t(s), X(s), Z(s)), & Z(0) = u_0(y). \end{cases} \tag{1.4.13}$$

We obviously have $t(s) = s$. Next, the classical theorems on differential systems allow us to say that there exists a neighborhood $\mathcal{O}_{x_0}$ and $\delta_0 > 0$ such that for all $y \in \mathcal{O}_{x_0}$ this system admits a unique solution defined for $|s| < \delta_0$ and that if we denote by $(X(s, y), Z(s, y))$ the solution of (1.4.13), the maps $(s, y) \to X(s, y)$ and $(s, y) \to Z(s, y)$ are of class C^1 on $(-\delta_0, \delta_0) \times \mathcal{O}_{x_0}$.

Proposition 1.4.6 *There exists $\delta_1 \in]0, \delta_0]$ and $W_{x_0} \subset \mathcal{O}_{x_0}$ such that if we set*

$$S = \{(s, X(s, y), Z(s, y)) : |s| < \delta_1, y \in W_{x_0}\}, \tag{1.4.14}$$

then,

(i) S is a hypersurface of $\mathbf{R}^{d+2}$,
(ii) the vector field $\mathcal{X}$ is tangent to S at every point,
(iii) S is of the form (1.4.11).

To prove these points we will need the following result.

Lemma 1.4.7 *Let $F : (-\delta_0, \delta_0) \times \mathcal{O}_{x_0} \times \mathbf{R} \to \mathbf{R}^d$ defined by $F(s, x, y) = X(s, y) - x$. There exists $\delta_1 > 0$, neighborhoods V_{x_0}, W_{x_0} of x_0 and a function $Y : (-\delta_1, \delta_1) \times V_{x_0} \to W_{x_0}$ of class C^1 such that*

$$\Big(|s| < \delta_1, x \in V_{x_0}, y \in W_{x_0}, F(s, x, y) = 0\Big) \Longleftrightarrow y = Y(s, x)$$

Proof We use the Implicit Function Theorem 1.4.2 where the variable x is replaced by the variable (s, x) and $y_0 = x_0$. Since $X(0, y) = y$ we have $F(0, x_0, x_0) = 0$ and $\frac{\partial F}{\partial y}(0, x_0, x_0) = \text{Identity}$. The conclusion then follows from the Implicit Function Theorem 1.4.2. □

Remark 1.4.8 (i) Since $X(0, y) = y$ we have $Y(0, x) = x$.
(ii) For $(s, y) \in (-\delta_1, \delta_1) \times V_{x_0}$ we have $Y(s, X(s, y)) = y$. We deduce that

$$\frac{\partial Y}{\partial x}(s, X(s, y))\frac{\partial X}{\partial y}(s, y) = Id,$$

which shows that the matrix $\big(\frac{\partial X}{\partial y}(s, y)\big)$ is invertible in this set.

Proof of Proposition 1.4.6 (i) We use the parametric definition of hypersurfaces. We shall take $U = (-\delta_1, \delta_1) \times V_{x_0}$ where δ_1 and V_{x_0} are those found in Lemma 1.4.7 and

$$\Phi(s, y) = (s, X(s, y), Z(s, y)).$$

The map Φ is of class C^1. By definition it is surjective from U onto S. Let us show that it is injective. If $(s, X(s, y), Z(s, y)) = (s', X(s', y'), Z(s', y'))$ we have $X(s, y) = X(s, y') := x$. According to Lemma 1.4.7 we have $y = Y(s, x)$ and $y' = Y(s, x)$ so that $y = y'$. Thus Φ is bijective. Let us consider the map

$\Psi : (-\delta_1, \delta_1) \times V_{x_0} \times \mathbf{R} \to \mathbf{R} \times \mathbf{R}^d$ defined by $\Psi(s, x, z) = (s, Y(s, x))$. According to Lemma 1.4.7 this is a C^1 map and we have

$$\Psi \circ \Phi(s, y) = (s, Y(s, X(s, y))) = (s, y)$$

which shows that the restriction of Ψ to S coincides with Φ^{-1}.

Now, let us show that for every $(s, y) \in U$, the map $d\Phi(s, y) \in \mathcal{L}(\mathbf{R}^{d+1}, \mathbf{R}^{d+2})$ is injective. We have $\Phi = (\Phi_0, \Phi_1, \dots, \Phi_d, \Phi_{d+1})$ where

$$\Phi_0(s, y) = s, \quad \Phi_j(s, y) = X_j(s, y), \; 1 \le j \le d, \quad \Phi_{d+1}(s, y) = Z(s, y).$$

The matrix of $d\Phi(s, y)$ in the canonical basis is therefore

$$\begin{bmatrix} 1 & 0 & \dots & 0 \\ \frac{\partial X_1}{\partial s} & \frac{\partial X_1}{\partial y_1} & \dots & \frac{\partial X_1}{\partial y_d} \\ \vdots & \vdots & \ddots & \vdots \\ \frac{\partial X_d}{\partial s} & \frac{\partial X_d}{\partial y_1} & \dots & \frac{\partial X_d}{\partial y_d} \\ \frac{\partial Z}{\partial s} & \frac{\partial Z}{\partial y_1} & \dots & \frac{\partial Z}{\partial y_d} \end{bmatrix}$$

Let $v = (v_0, v_1, \dots, v_d) \in \mathbf{R}^{d+1}$ such that $d\Phi(s, y)(v) = 0$. From above, we see that $v_0 = 0$ and $\frac{\partial X}{\partial y}(s, y)v' = 0$ where $v' = (v_1, \dots v_d)$. Since the matrix $\frac{\partial X}{\partial y}(s, y)$ is invertible, we have $v' = 0$.

(ii) Let $e_0 = (1, 0, \dots, 0) \in \mathbf{R}^{d+1}$ be the first basis vector. From above, the system (1.4.13) and (1.4.10) give

$$\begin{aligned} d\Phi(s, y)e_1 &= (1, \dot{X}(s, y), \dots, \dot{Z}(s, y)), \\ &= (1, a(s, X(s, y), Z(s, y)), \dots, b(s, X(s, y), Z(s, y)), \\ &= \mathcal{X}((s, X(s, y), Z(s, y)), \end{aligned}$$

thus $\mathcal{X}(p) \in T_pS$ where $p = (s, X(s, y), Z(s, y)) \in S$.

(iii) According to Lemma 1.4.7, $X(s, y) = x$ is equivalent to $y = Y(s, x)$ so that, using (1.4.14), we can write

$$S = \{(s, x, Z(s, Y(s, x)) : |s| < \delta_1, x \in V_{x_0}\}.$$

Setting

$$u(s, x) = Z(s, Y(s, x)), \tag{1.4.15}$$

we will have

$$S = \{(s, x, z) \in (-\delta_1, \delta_1) \times V_{x_0} \times \mathbf{R} : z = u(s, x)\},$$

which shows that S is indeed of the form (1.4.11). The function u defined in (1.4.15) is therefore a solution of the equation appearing in (1.3.1). Moreover, it follows from Remark 1.4.8 and from the system (1.4.13) that we have

$$u(0,x) = Z(0, Y(0,x)) = Z(0,x) = u_0(x).$$

□

Proof of Theorem 1.4.1 The existence part of Theorem 1.4.1 then follows from Proposition 1.4.6. Indeed, the function u defined in (1.4.15) is a solution of the equation appearing in (1.3.1) and $u(0,x) = u_0(x)$.

Let us prove uniqueness. Let u_1 and u_2 be two solutions of problem (1.3.1) and S_1, S_2 the hypersurfaces (1.4.11) defined by these solutions. Since $u_1(0,x) = u_2(0,x) = u_0(x)$, we have $\{(0,x,u_0(x)) : x \in V_{x_0}\} \subset S_1 \cap S_2$.

According to (1.4.12), the vector field $\mathcal{X}$ is tangent to S_1 and S_2 at every point. By Theorem 1.4.5, the integral curve of $\mathcal{X}$ starting from a point $(0, y, u_0(y))$ is contained in S_1 and S_2, which means that

$$Z(s,y) = u_1(s, X(s,y)) = u_2(s, X(s,y)), \quad \forall x \in V_{x_0}.$$

It follows from Lemma 1.4.7 that $u_1(s,x) = u_2(s,x)$ for all (s,x) in $(-\delta_1, \delta_1) \times V_{x_0}$.

This completes the proof of Theorem 1.4.1. □

1.5 Cauchy Problem for Multi-dimensional Conservation Laws

This is the name given to the following type of problem:

$$\begin{aligned} &\frac{\partial u}{\partial t}(t,x) + \sum_{j=1}^{d} \frac{\partial}{\partial x_j}\big(f_j(u(t,x))\big) = 0, \quad t \geq 0, \quad x \in \mathbf{R}^d, \\ &u|_{t=0} = u_0, \end{aligned} \tag{1.5.1}$$

where $u : [0, +\infty) \times \mathbf{R}^d \to \mathbf{R}$, the functions $f_j : \mathbf{R} \to \mathbf{R}$ are C^2 functions, and the initial data u_0 is **real-valued**.

This is an equation similar to that defined in (1.3.1), since we can write

$$\frac{\partial}{\partial x_j}\big(f_j(u)\big) = f_j'(u)\frac{\partial u}{\partial x_j},$$

except that the coefficients in front of the derivatives in x of u do not depend on the variable (t,x). Thanks to this simplification, it will be possible to specify the maximal existence time of the solution, which we could not generally do for Eq. (1.3.1). We will set in what follows

$$f(u) = (f_1(u), f_2(u), \dots, f_d(u)),$$
$$F(u) = (F_1(u), F_2(u), \dots, F_d(u)), \text{ where } F_i(u) = f_i'(u).$$

Moreover, given u_0, we introduce, for $\xi \in \mathbf{R}^d$, the matrix

$$A(\xi) = \left(F_i'(u_0(\xi))\frac{\partial u_0}{\partial x_j}(\xi)\right)_{1\le i,j\le d}, \tag{1.5.2}$$

and we denote by Tr $A(\xi)$ its trace, that is the sum of its diagonal elements.

The aim of this paragraph is to prove the following result.

Theorem 1.5.1 *Let $u_0 \in C^1(\mathbf{R}^d) \cap L^\infty(\mathbf{R}^d)$ be real-valued. Set*

$$T^* =: \begin{cases} +\infty & \text{if } \inf_{\xi\in\mathbf{R}^d} Tr\, A(\xi) \ge 0, \\ \frac{-1}{\inf_{\xi\in\mathbf{R}^d} Tr\, A(\xi)} & \text{if } \inf_{\xi\in\mathbf{R}^d} Tr\, A(\xi) < 0. \end{cases} \tag{1.5.3}$$

Then the problem (1.5.1) *admits a unique solution $u(t,x)$ of class C^1 in the set $\{(t,x) : t \in [0,T^*), x \in \mathbf{R}^d\}$. Moreover, $u \in L^\infty([0,T^*) \times \mathbf{R}^d)$.*

Proof We begin with a necessary condition that will provide us with the uniqueness of the solution as well as the method of resolution.

Suppose that u is a C^1 solution of (1.5.1) on $[0,T) \times \mathbf{R}^d$ for some $T > 0$.

We define the characteristic curve, $(t, X(t))$ starting from a point $\xi \in \mathbf{R}^d$ by

$$\dot{X}(t) = F(u(t, X(t)), \quad X(0) = \xi. \tag{1.5.4}$$

By the Cauchy-Lipschitz Theorem, this curve exists for $t \in [0, T_1)$, for some $T_1 > 0$ and we have

$$\begin{aligned} \frac{d}{dt}\big[u(t,X(t))\big] &= \Big(\frac{\partial u}{\partial t} + \sum_{j=1}^d \dot{X}_j \frac{\partial u}{\partial x_j}\Big)(t, X(t)), \\ &= \Big(\frac{\partial u}{\partial t} + \sum_{j=1}^d f_j'(u) \frac{\partial u}{\partial x_j}\Big)(t, X(t)) = 0. \end{aligned}$$

We deduce that $u(t, X(t)) = u(0, X(0)) = u_0(\xi)$. It then follows from (1.5.4) that

$$\dot{X}(t) = F(u_0(\xi)), \quad X(0) = \xi,$$

so that, denoting by $X(t,\xi)$ the solution of (1.5.4),

$$X(t,\xi) = \xi + tF(u_0(\xi)).$$

Consequently,

$$u(t, \xi + tF(u_0(\xi))) = u_0(\xi). \tag{1.5.5}$$

This shows that to find the value of the function u at a point (t, x), it suffices to solve the equation $\xi + tF(u_0(\xi)) = x$, in other words to invert the mapping $\xi \mapsto \xi + tF(u_0(\xi))$.

Let us now proceed to the resolution of (1.5.1). Recall that, by hypothesis, we have $u_0 \in C^1(\mathbf{R}^d) \cap L^\infty(\mathbf{R}^d)$. □

Let us consider, for fixed t, the mapping from $\mathbf{R}^d$ to $\mathbf{R}^d$, $\xi \mapsto \xi + tF(u_0(\xi))$. To invert it, we will use the following theorem, due to J. Hadamard, a proof of which, in the case where G is of class C^2, is given in the appendix.

Theorem 1.5.2 *Let $G : \mathbf{R}^d \to \mathbf{R}^d$ be a C^1 function. The following two conditions are equivalent.*

(i) G is a C^1-diffeomorphism from $\mathbf{R}^d$ onto $\mathbf{R}^d$.
(ii) G is proper and, for every point $\xi \in \mathbf{R}^d$, $G'(\xi) : \mathbf{R}^d \to \mathbf{R}^d$ is invertible.

Recall that a mapping is said to be proper if the preimage of a compact set is compact, which is equivalent here (see Lemma 1.7.6 in the appendix) to the fact that

$$\lim_{|\xi| \to +\infty} |G(\xi)| = +\infty. \tag{1.5.6}$$

Let us set, for fixed $t \in [0, T]$,

$$G(\xi) = \xi + tF(u_0(\xi)). \tag{1.5.7}$$

Since $|u_0(\xi)| \le \|u_0\|_{L^\infty}$ we have $|F(u_0(\xi))| \le \sup_{|Y| \le \|u_0\|_{L^\infty}} |F(Y)| = M$. Consequently,

$$|G(\xi)| \ge |\xi| - tM \to +\infty, \quad \text{as} \quad |\xi| \to +\infty.$$

Next, the matrix of the differential of $G : \mathbf{R}^d \to \mathbf{R}^d$ is the matrix $\left(\frac{\partial G_i}{\partial \xi_j}(\xi)\right)_{1 \le i,j \le d}$. It follows from (1.5.7) that this matrix is equal to $Id + tA(\xi)$ where Id is the identity matrix and

$$A(\xi) = \left(F_i'(u_0(\xi)) \frac{\partial u_0}{\partial x_j}(\xi)\right)_{1 \le i,j \le d}. \tag{1.5.8}$$

Lemma 1.5.3 *We have*

$$\det(G'(\xi)) = \det(Id + tA(\xi)) = 1 + tTrA(\xi) = 1 + t \sum_{i=1}^{d} F_i'(u_0(\xi)) \frac{\partial u_0}{\partial x_i}(\xi),$$

where Tr denotes the trace of the matrix.

Proof Let us clarify some notations. Let $(e_1, \ldots, e_d)$ be the canonical basis of $\mathbf{R}^d$. We will set $a_j = F_i'(u_0(\xi)), b_j = \frac{\partial u_0}{\partial x_j}(\xi)$. Then $A = (a_i b_j)_{1 \le i,j \le d}$. Finally, let $\ell_i(t)$ denote the i-th row of the matrix $\mathrm{Id} + tA$, i.e., $\ell_i(t) = (\delta_{ij} + t a_i b_j)_{1 \le j \le d}$. We then have

$$\ell_i(0) = e_i, \quad \ell_i'(t) = (a_i b_j)_{1 \le j \le d} = a_i (b_j)_{1 \le j \le d}, \quad \ell_i''(t) \equiv 0,$$

and,

$$\Delta(t) = \det(\mathrm{Id} + tA) = \det(\ell_1(t), \ell_2(t), \ldots, \ell_d(t)).$$

We have $\Delta(0) = 1$ and

$$\Delta'(t) = \sum_{k=1}^{d} \det(\ell_1(t), \ldots, \ell_k'(t), \ldots, \ell_d(t)).$$

Since $\ell_i(0) = e_i$, and $\ell_k'(0) = a_k(\sum_{i \ne k} b_i e_i + b_k e_k)$, we have

$$\Delta'(0) = \sum_{k=1}^{d} \det(e_1, \ldots, \ell_k'(0), \ldots, e_d) = \sum_{k=1}^{d} a_k \sum_{i \ne k} b_i \det(e_1, \ldots, e_i, \ldots, e_d)$$
$$+ \sum_{k=1}^{d} a_k b_k \det(e_1, \ldots, e_k, \ldots, e_d) = \sum_{k=1}^{d} a_k b_k = \mathrm{Tr} A.$$

Finally, since $\ell_i''(t) = 0$ for all $1 \le i \le d$ we have

$$\Delta''(t) = \sum_{k=1}^{d} \sum_{i \ne k} \det(\ell_1(t), \ldots, \ell_i'(t), \ldots \ell_k'(t), \ldots, \ell_d(t)).$$

But $\ell_i'(t) = a_i (b_j)_{1 \le j \le d} = a_i b$ and $\ell_k'(t) = a_k (b_j)_{1 \le j \le d} = a_k b$, so that, for all t,

$$\Delta''(t) = \sum_{k=1}^{d} \sum_{i \ne k} a_i a_k \det(\ell_1(t), \ldots, b, \ldots b, \ldots, \ell_d(t)) = 0,$$

since all the determinants have two equal rows.

We deduce that $\Delta(t) = \Delta(0) + t\Delta'(0) = 1 + t\mathrm{Tr} A$. □

Let us set

$$T^* =: \begin{cases} +\infty, & \text{if } \inf_{\xi \in \mathbf{R}^d} \mathrm{Tr}\, A(\xi) \ge 0, \\ \dfrac{-1}{\inf_{\xi \in \mathbf{R}^d} \mathrm{Tr}\, A(\xi)}, & \text{if } \inf_{\xi \in \mathbf{R}^d} \mathrm{Tr}\, A(\xi) < 0. \end{cases} \tag{1.5.9}$$

We can deduce from Lemma 1.5.3 that

$$\text{for } t \in [0, T^*) \text{ we have } \det(G'(\xi)) > 0, \forall \xi \in \mathbf{R}^d.$$

Indeed, if $\inf_{\xi \in \mathbf{R}^d} \operatorname{Tr} A(\xi) \geq 0$ then $\operatorname{Tr} A(\xi) \geq 0$ for all $\xi \in \mathbf{R}^d$. Therefore we will have

$$\det(G'(\xi)) = 1 + t \operatorname{Tr} A(\xi) \geq 1, \quad \text{for } t \geq 0 \text{ and all } \xi \in \mathbf{R}^d.$$

If $\inf_{\xi \in \mathbf{R}^d} \operatorname{Tr} A(\xi) < 0$ we can write

$$\det(G'(\xi)) = 1 + t \operatorname{Tr} A(\xi) \geq 1 + t \inf_{\xi \in \mathbf{R}^d} \operatorname{Tr} A(\xi) > 0, \quad \text{if } t \in [0, T^*).$$

It follows then, from (1.5.6) and Theorem 1.5.2, that for every fixed $t \in [0, T^*)$ the map $\xi \mapsto \xi + tF(u_0(\xi))$ is a diffeomorphism from $\mathbf{R}^d$ onto itself.

For $t \in [0, T^*)$ we thus have

$$\xi + tF(u_0(\xi)) = x \iff \xi = \kappa(t, x), \tag{1.5.10}$$

where κ is a C^1 function from $[0, T^*) \times \mathbf{R}^d$ with values in $\mathbf{R}^d$. Taking into account (1.5.5) we then set

$$u(t, x) = u_0(\kappa(t, x)). \tag{1.5.11}$$

Then $u \in C^1([0, T^*) \times \mathbf{R}^d)$ and $u \in L^\infty([0, T^*) \times \mathbf{R}^d)$, since we have the estimate, $\|u\|_{L^\infty([0,T^*)\times\mathbf{R}^d)} \leq \|u_0\|_{L^\infty(\mathbf{R}^d)}$.

We will show that this is the unique solution to problem (1.5.1). Uniqueness follows immediately from the necessary condition (1.5.5). It is therefore sufficient to show that u as defined above is indeed a solution to problem (1.5.1). First, from (1.5.10) we have $\kappa(0, x) = x$ so that

$$u(0, x) = u_0(x). \tag{1.5.12}$$

Next, using (1.5.11) we have

$$\frac{\partial u}{\partial t}(t, x) = \sum_{k=1}^{d} \frac{\partial u_0}{\partial y_k}(\kappa(t, x)) \frac{\partial \kappa_k}{\partial t}(t, x), \tag{1.5.13}$$

$$\frac{\partial u}{\partial x_j}(t, x) = \sum_{k=1}^{d} \frac{\partial u_0}{\partial y_k}(\kappa(t, x)) \frac{\partial \kappa_k}{\partial x_j}(t, x). \tag{1.5.14}$$

On the other hand, from (1.5.10) we have $\kappa_k(t, \xi + tF(u_0(\xi))) = \xi_k$. Differentiating this equality with respect to t gives

$$\frac{\partial \kappa_k}{\partial t}(\xi + tF(u_0(\xi)) + \sum_{j=1}^{d} \frac{\partial \kappa_k}{\partial x_j}(\xi + tF(u_0(\xi))F_j(u_0(\xi)) = 0.$$

Using (1.5.10), (1.5.11), (1.5.13) and finally (1.5.14) we obtain

$$\begin{aligned}\frac{\partial u}{\partial t}(t,x) &= -\sum_{k=1}^{d}\sum_{j=1}^{d} \frac{\partial u_0}{\partial y_k}(\kappa(t,x))\frac{\partial \kappa_k}{\partial x_j}(t,x)F_j(u(t,x)),\\ &= -\sum_{j=1}^{d} \frac{\partial u}{\partial x_j}(\kappa(t,x))F_j(u(t,x)) = -\sum_{j=1}^{d} \frac{\partial u}{\partial x_j}(t,x)f_j'(u(t,x))\\ &= -\sum_{j=1}^{d} \frac{\partial}{\partial x_j}\big(f_j(u(t,x)\big),\end{aligned}$$

which, together with (1.5.12), proves that u is a solution to problem (1.5.1). □

A completely similar discussion can be made for negative times t.

Remark 1.5.4 (i) The existence time T^* is, in fact, the largest time for which there exists a C^1 solution. Indeed, we will show that the modulus of the gradient of the solution u tends to $+\infty$ as we approach this time. For this, we start from relation (1.5.5) that is

$$u(t, \xi + tF(u_0(\xi))) = u_0(\xi).$$

We can write this equality in another form. Indeed, if we set $x = \xi + tF(u_0(\xi))$ we have

$$\xi = x - tF(u_0(\xi)) = x - tF(u(t,x)),$$

so that we also have

$$u(t,x) = u_0(x - tF(u(t,x))).$$

Let us differentiate this equality with respect to x_k. We obtain

$$\frac{\partial u}{\partial x_k}(t,x) = \sum_{j=1}^{d} \frac{\partial u_0}{\partial x_j}(\xi)\big[\delta_{jk} - tF_j'(u(t,x))\frac{\partial u}{\partial x_k}(t,x)\big],$$

which is equivalent to

$$\big[1 + t\sum_{j=1}^{d} \frac{\partial u_0}{\partial x_j}(\xi)F_j'(u(t,x))\big]\frac{\partial u}{\partial x_k}(t,x) = \frac{\partial u_0}{\partial x_k}(\xi).$$

Since $u(t,x) = u_0(\xi)$ we can write

$$\left[1 + t\sum_{j=1}^{d} \frac{\partial u_0}{\partial x_j}(\xi) F_j'(u_0(\xi))\right] \frac{\partial u}{\partial x_k}(t,x) = \frac{\partial u_0}{\partial x_k}(\xi).$$

Recall that $A(\xi) = \left(F_i'(u_0(\xi)) \frac{\partial u_0}{\partial x_j}(\xi)\right)_{1\leq i,j\leq d}$. Setting $\nabla u = (\frac{\partial u}{\partial x_k})_{1\leq k\leq d}$ the above equality can be written

$$\nabla u(t,x) = \frac{1}{1 + t\mathrm{Tr}A(\xi)} \nabla u_0(\xi).$$

Let ξ_0 be such that $\mathrm{Tr}A(\xi_0) < 0$ then $\nabla u_0(\xi_0) \not\equiv 0$ (otherwise $A(\xi_0) = 0$) and the above formula shows that $|\nabla u(t,x)|$ tends to $+\infty$ as t tends to $-\frac{1}{\mathrm{Tr}A(\xi_0)}$.

The maximal existence time of the solution, as a C^1 function, is thus the smallest of these times, that is,

$$T^* = \inf_{\xi_0 \in \mathbf{R}^d} \left(-\frac{1}{\mathrm{Tr}A(\xi_0)}\right) = -\sup_{\xi_0\in\mathbf{R}^d} \frac{1}{\mathrm{Tr}A(\xi_0)} = \frac{-1}{\inf_{\xi_0\in\mathbf{R}^d} \mathrm{Tr}A(\xi_0)}.$$

Note that at time T^* the solution u does not blow up (since it is constant along each characteristic line) but its derivative does blow up.

(ii) At the beginning of this chapter, we emphasized that the data, in particular the coefficients of Eq. (1.3.1), had to be real. We will show, with a very simple example, that when these coefficients are no longer real, the framework of C^1 functions is no longer relevant. Indeed, consider the linear Cauchy problem in $\mathbf{R}^2$

$$\left(\frac{\partial u}{\partial t} - i\frac{\partial u}{\partial x}\right)(t,x) = 0, \quad u|_{t=0} = u_0, \tag{1.5.15}$$

where u_0 is a C^1 function in a neighborhood of the origin.

Suppose that this problem has a C^1 solution in the set

$$Q = \{(t,x) \in \mathbf{R}^2 : |t| < a, |x| < b\} \text{ for some } a, b > 0.$$

We will show that then the function u_0 must be real-analytic.

Indeed, since u is C^1 and $(\partial_x + i\partial_t)u = 0$ in Q, the function u satisfies the Cauchy-Riemann conditions. Therefore, it is a holomorphic function of the variable $z = x + it$, so, there exists φ, holomorphic in $\{z = x + it \in \mathbf{C} : |t| < a, |x| < b\}$, such that $u(t,x) = \varphi(x + it)$. It follows that $u_0(x) = u(0,x) = \varphi(x)$ must be real-analytic. Consequently, if u_0 is C^1 but not real-analytic (for example $u_0(x) = e^{-\frac{1}{x}}$ for $x > 0$ and $u_0(x) = 0$ for $x \leq 0$), the problem (1.5.15) has no solution.

It can be shown that the analytic framework is adequate for equations of the type (1.3.1) when all the data are analytic in the real sense. This is a particular case of a famous result known as the Cauchy-Kovalevski Theorem, which will be the subject of the next chapter.

1.6 A Classic Example

The equation we are going to study, which is a particular case of the equations considered previously, is known as the "Burgers equation". It was introduced in 1948 by the physicist Johannes Martinus Burgers, to model problems in gas dynamics. Here is the problem:

$$\begin{aligned} &\left(\frac{\partial u}{\partial t} + u\frac{\partial u}{\partial x}\right)(t, x) = 0, \quad t \in \mathbf{R}, \quad x \in \mathbf{R} \\ &u|_{t=0} = u_0. \end{aligned} \tag{1.6.1}$$

With the notations of the previous paragraph, we have

$$d = 1, \quad f(u) = \frac{1}{2}u^2, \quad F(u) = u, \quad F'(u) = 1, \quad A(\xi) = u_0',$$

so that an application of Theorem 1.5.1 provides the following result.

Theorem 1.6.1 *Let* $u_0 \in C^1(\mathbf{R}) \cap L^\infty(\mathbf{R})$. *Then problem* (1.6.1) *admits a unique solution on* $[0, T^*[\times\mathbf{R}$ *where* $T^* = +\infty$ *if* $\inf_{\mathbf{R}} u_0' \geq 0$ *and* $T^* \geq -\frac{1}{\inf_{\mathbf{R}} u_0'}$ *if* $\inf_{\mathbf{R}} u_0' < 0$.

Remark 1.6.2 1. The result obtained can be interpreted geometrically. Indeed, as above, the solution u is such that

$$u(s, y + su_0(y)) = u_0(y).$$

This shows that the function u is constant and equal to $u_0(y)$ on the line D_y given by the equation $x = y + su_0(y)$. If $\inf_{\mathbf{R}} u_0' \geq 0$ then u_0 is increasing so that for $y > y'$ the slope of the line D_y is greater than that of the line $D_{y'}$. These two lines will therefore never meet for positive s. On the contrary, if $\inf_{\mathbf{R}} u_0' < 0$ there exist points y and y' such that $y > y'$ but $u_0(y') > u_0(y)$ and in this case the two lines will meet at a certain point m_0. At this point, the solution u should have two different values $u_0(y)$ and $u_0(y')$, which is not possible.

This equation will be studied in more detail in Chap. 5, using more advanced tools.

1.6.1 Another Example

Here is another example of a nonlinear equation that can be treated by the method of characteristics. It is a variant of the Burgers equation studied at the beginning of the section.

In what follows, we denote by $C_b^1(\mathbf{R})$ the space of C^1 functions on $\mathbf{R}$ such that u and u' are in $L^\infty(\mathbf{R})$.

The problem to be solved is the following:

$$\left(\frac{\partial u}{\partial t} + u\frac{\partial u}{\partial x}\right)(t,x) = u^2(t,x), \quad t > 0, x \in \mathbf{R}, \quad u|_{t=0} = u_0. \tag{1.6.2}$$

We will show the following result.

Proposition 1.6.3 *Let $u_0 \in C_b^1(\mathbf{R})$ with $u_0(x) < 0$, $u_0'(x) \geq 0$ for all $x \in \mathbf{R}$. Then the problem* (1.6.2) *admits a unique solution u defined on* $(0,+\infty) \times \mathbf{R}$ *such that u and $\partial_x u$ belong to* $C^0 \cap L^\infty((0,+\infty) \times \mathbf{R})$.

Example 1.6.4 In Proposition 1.6.3 one can take, for instance, the following initial data : $u_0(x) = -\frac{1}{1+e^x}$, $u_0(x) = \frac{-1}{\pi+\arctan x}$.

Proof of the Proposition As indicated, we will use the method of characteristics.

The Necessary Condition

In this paragraph we show that if u is the solution of problem (1.6.2) it has a well-determined form.

Suppose therefore that this problem has a solution u such that u and $\partial_x u$ belong to the space $C^0 \cap L^\infty((0,+\infty) \times \mathbf{R})$. Consider the differential problem:

$$\dot{x}(s) = u(s, x(s)), \quad x(0) = x_0 \in \mathbf{R}. \tag{1.6.3}$$

According to the Cauchy-Lipschitz Theorem and the maximal interval theorem, (see Theorem 1.7.2 in the Appendix) this problem admits a unique solution defined on $(0,+\infty) \times \mathbf{R}$. Then, according to (1.6.3) and (1.6.2),

$$\frac{d}{ds}\big[u(x, x(s))\big] = (\partial_s u + \dot{x}(s)\partial_x u)(s, x(s)) = (\partial_s u + u\partial_x u)(s, x(s)) = u^2(s, x(s)).$$

Therefore, setting $F(s) = u(s, x(s))$ we see that F is a solution of the problem

$$\dot{F}(s) = F(s)^2, \quad F(0) = u_0(x_0).$$

According to the Cauchy-Lipschitz Theorem, this problem admits a unique solution on $(0, T^*)$. On this interval F cannot vanish, otherwise by uniqueness it would be identically zero, which is not possible since $F(0) < 0$. We can then write

$$\frac{\dot{F}(s)}{F(s)^2} = 1 \Rightarrow \frac{d}{ds}\left[-\frac{1}{F(s}\right] = 1 \Rightarrow \frac{1}{F(0)} - \frac{1}{F(s)} = s \Rightarrow F(s) = \frac{u_0(x_0)}{1 - su_0(x_0)}.$$

As $u_0 < 0$, the denominator of the expression giving $F(s)$ does not vanish for $s > 0$, so that $T^* = +\infty$ and $F(s)$ exists for all $s \in (0, +\infty)$. We thus have

$$u(s, x(s)) = \frac{u_0(x_0)}{1 - su_0(x_0)}, \quad \forall s \in (0, +\infty). \tag{1.6.4}$$

Using then (1.6.3) it follows

$$\dot{x}(s) = \frac{u_0(x_0)}{1 - su_0(x_0)}, \quad \forall s \in (0, +\infty), \quad x(0) = x_0, \tag{1.6.5}$$

which implies that

$$x(s) = x_0 - \ln(1 - su_0(x_0)), \quad s \in (0, +\infty). \tag{1.6.6}$$

Let us now denote by $x(s, x_0)$ the solution of (1.6.3) given by (1.6.6); we will show that for any fixed $s > 0$, the map $x_0 \mapsto x(s, x_0)$ is a bijection from $\mathbf{R}$ onto itself. To do this, we show that this map is strictly increasing from $\mathbf{R}$ to $\mathbf{R}$. Indeed, we have

$$\frac{\partial}{\partial x_0}[x(s, x_0)] = 1 + \frac{su_0'(x_0)}{1 - su_0(x_0)} > 0,$$

by hypothesis.

On the other hand, since $1 - su_0(x_0) > 1$, we have $\ln(1 - su_0(x_0)) > 0$ so that $x(s, x_0) < x_0$. This shows that $\lim_{x_0 \to -\infty} x(s, x_0) = -\infty$. Next, since u_0 is increasing and bounded above by 0, it admits a limit ℓ as $x \to +\infty$. The formula (1.6.6) then shows that $\lim_{x_0 \to +\infty} x(s, x_0) = +\infty$. We can therefore write

$$x(s, x_0) = y, \quad y \in \mathbf{R} \Longleftrightarrow x_0 = \kappa(s, y). \tag{1.6.7}$$

The formula (1.6.4) then implies that, necessarily

$$u(s, y) = \frac{u_0(\kappa(s, y))}{1 - su_0(\kappa(s, y))} \quad \forall (s, y) \in (0, +\infty) \times \mathbf{R}. \tag{1.6.8}$$

The Sufficient Condition

The next step is to show that the function given by formula (1.6.8) is well-defined and that it is a solution to problem (1.6.2). To define u we need to construct the function κ and give its properties.

We start from u_0 satisfying the hypotheses of the statement. We set

$$x(s, x_0) = x_0 - \ln(1 - su_0(x_0)), \quad (s, x_0) \in (0, +\infty) \times \mathbf{R}.$$

Then,

$$\frac{\partial x}{\partial x_0}(s, x_0) = \frac{1 - su_0(x_0) + su_0'(x_0)}{1 - su_0(x_0)}.$$

As previously, we show that the function $x_0 \mapsto x(s, x_0)$ is a C^1-diffeomorphism from $\mathbf{R}$ onto itself. By its form, the function x is also C^1 with respect to s. We then have the existence of a function $\kappa \in C^1((0, +\infty) \times \mathbf{R})$ satisfying the equivalence (1.6.7).

We then define the function u by (1.6.8). It belongs to $C^1((0, +\infty) \times \mathbf{R})$ and to $L^\infty((0, +\infty) \times \mathbf{R})$ because $u_0 \in L^\infty(\mathbf{R})$ and $1 - su_0(\kappa) \geq 1$. We will show later that $\partial_y u$ belongs to $L^\infty((0, +\infty) \times \mathbf{R})$. It remains to show that this is a solution to problem (1.6.2).

First, we have $u(0, y) = u_0(\kappa(0, y))$. But according to (1.6.7), $\kappa(0, y) = x_0$ is equivalent to $x(0, x_0) = y = x_0$. Thus $\kappa(0, y) = y$ hence $u(0, y) = u_0(y)$. We then show that u satisfies Eq. (1.6.2).

We will denote in what follows $f(\kappa)$ the function $(s, y) \mapsto f(\kappa(s, y))$.

Since $x(s, \kappa(s, y)) = y$ we have

$$\partial_y \kappa = \frac{1}{\frac{\partial x}{\partial x_0}(s, \kappa)} = \frac{1 - su_0(\kappa)}{1 + su_0'(\kappa) - su_0(\kappa)} \tag{1.6.9}$$

and, also using (1.6.5),

$$\partial_s \kappa = -\frac{\dot{x}(s, \kappa)}{\frac{\partial x}{\partial x_0}(s, \kappa)} = \frac{-u_0(\kappa)}{1 + su_0'(\kappa) - su_0(\kappa)}. \tag{1.6.10}$$

We deduce that

$$\begin{aligned} &(i) \quad \partial_y \kappa = s\partial_s \kappa + F, \quad F = \frac{1}{1 + su_0'(\kappa) - su_0(\kappa)}, \\ &(ii) \quad \partial_s \kappa = -u_0(\kappa)F, \\ &(iii) \quad 0 < \partial_y \kappa \leq 1. \end{aligned} \tag{1.6.11}$$

Assertion (iii) follows from (1.6.9) and from the fact that $u_0 < 0$ and $u_0' \geq 0$.

Next,

$$\begin{aligned}\partial_y u(s,y) &= \frac{u_0'(\kappa)\partial_y\kappa}{(1-su_0(\kappa))^2},\\ \partial_s u(s,y) &= \frac{u_0'(\kappa)\partial_s\kappa + u_0(\kappa)^2}{(1-su_0(\kappa))^2},\end{aligned} \tag{1.6.12}$$

The hypothesis made on u_0 and (1.6.11) (iii) show that $\partial_y u \in L^\infty((0,+\infty)\times\mathbf{R})$. Set

$$A = (1-su_0(\kappa))^2(\partial_s u + u\partial_y u)(s,y). \tag{1.6.13}$$

It follows from (1.6.12) and (1.6.8) that

$$\begin{aligned}A &= u_0'(\kappa)\partial_s\kappa + u_0(\kappa)^2 \; + \frac{u_0(\kappa)u_0'(\kappa)}{1-su_0(\kappa)}\partial_y\kappa,\\ &= \frac{u_0'(\kappa)\partial_s\kappa - u_0(\kappa)u_0'(\kappa)s\partial_s\kappa + u_0^2(\kappa)(1-su_0(\kappa)) + u_0(\kappa)u_0'(\kappa)\partial_y\kappa}{1-su_0(\kappa)}.\end{aligned}$$

Using (1.6.11) we deduce that

$$\begin{aligned}A = \frac{1}{1-su_0(\kappa)}\big[&u_0'(\kappa)\partial_s\kappa - u_0(\kappa)u_0'(\kappa)s\partial_s\kappa + u_0^2(\kappa)(1-su_0(\kappa))\\ &+ u_0(\kappa)u_0'(\kappa)s\partial_s\kappa + u_0(\kappa)u_0'(\kappa)F\big],\end{aligned}$$

so

$$A = \frac{u_0'(\kappa)\partial_s\kappa + u_0(\kappa)u_0'(\kappa)F}{1-su_0(\kappa)} + u_0^2(\kappa).$$

It follows from (1.6.11) that $u_0'(\kappa)\partial_s\kappa + u_0(\kappa)u_0'(\kappa)F = 0$. We deduce from (1.6.13), (1.6.8) that

$$\partial_s u + u\,\partial_y u = u^2.$$

The uniqueness of this solution u follows from the first part.

1.7 Appendix

1.7.1 Gronwall's Lemma

Lemma 1.7.1 *Let $T>0$. Let φ and k be two continuous functions from* $[0,T]$ *into* $\mathbf{R}^+$ *and $g:[0,T]\to\mathbf{R}^+$ of class C^1 such that $g'\geq 0$ on* $[0,T]$*. Suppose that*

$$\varphi(t) \leq g(t) + \int_0^t k(s)\varphi(s)\,ds, \quad \forall t\in(0,T).$$

Then,

$$\varphi(t) \leq g(t) \exp\Big(\int_0^t k(s)\, ds \Big) \quad \forall t \in (0, T).$$

Proof Let $F(t) = g(t) + \int_0^t k(s)\varphi(s)\, ds$. Then $F \in C^1(0, T)$, $F(0) = g(0)$ and $\varphi(t) \leq F(t)$. Now, on $(0, T)$, we can write

$$F'(t) = g'(t) + k(t)\varphi(t) \leq g'(t) + k(t)F(t),$$

and thus

$$\begin{aligned} \frac{d}{dt}\Big[\exp\Big(-\int_0^t k(s)\, ds \Big) F(t) \Big] &= \exp\Big(-\int_0^t k(s)\, ds \Big)\big(F'(t) - k(t)F(t)\big), \\ &\leq \exp\Big(-\int_0^t k(s)\, ds \Big) g'(t) \leq g'(t). \end{aligned}$$

Consequently,

$$\exp\Big(-\int_0^t k(s)\, ds \Big) F(t) - g(0) \leq g(t) - g(0),$$

from which we deduce the lemma since $\varphi(t) \leq F(t)$. □

1.7.2 *The Maximal Interval Theorem*

Let $-\infty \leq a < b \leq +\infty$ and let $f : (a, b) \times \mathbf{R}^d \to \mathbf{R}^d$ be a continuous function. Consider the differential system

$$\dot{x}(t) = f(t, x(t)), \quad x(0) = x_0 \in \mathbf{R}^d, \tag{1.7.1}$$

where $\dot{x}$ denotes the derivative with respect to the variable t.

Then we have the following result.

Theorem 1.7.2 *Let* (x, J), *where* $J = (T_*, T^*)$, *be a maximal solution of the system* (1.7.1). *Then,*

$$\begin{cases} T^* = b & \textit{or,} \\ T^* < b & \textit{and } \lim_{t \to T^*} |x(t)| = +\infty. \end{cases}$$

Similarily,

$$\begin{cases} T_* = a & \textit{or,} \\ T_* > a & \textit{and } \lim_{t \to T_*} |x(t)| = +\infty. \end{cases}$$

A proof of this result can be found in the book [1].

1.7.3 A Hadamard Theorem

Theorem 1.7.3 *Let $f : \mathbf{R}^d \to \mathbf{R}^d$ be a C^1 function. The following two conditions are equivalent.*

(i) f is a diffeomorphism from $\mathbf{R}^d$ onto $\mathbf{R}^d$.

(ii) f is proper and at every point $x \in \mathbf{R}^d$ its differential $df(x)$ is invertible.

Recall that a map is said to be proper if the preimage of a compact set is compact. A characterization of proper maps from $\mathbf{R}^d$ to itself can be found in Lemma 1.7.6 below.

The proof of Theorem 1.7.3 is not simple; however, if f is of class C^2 there exists an accessible proof, which uses the theory of differential equations and which we will detail.

Proof $(i) \Rightarrow ii)$ Since f^{-1} is continuous, it maps a compact set to a compact set. Thus f is proper. Next, since $f^{-1} \circ f = f \circ f^{-1} = Id$ we have

$$d(f^{-1})(f(x)) \circ df(x) = Id \quad \text{and} \quad df(f^{-1}(y)) \circ d(f^{-1})(y) = Id.$$

Therefore $df(x)$ is invertible at every point.

$(ii) \Rightarrow i)$ We assume $f \in C^2(\mathbf{R}^d)$. By setting $\widetilde{f}(x) = f(x) - f(0)$ we reduce to the case where $f(0) = 0$ with the same assumptions on f. Moreover, note that by hypothesis, for all $x \in \mathbf{R}^d$, the linear map $df(x) : \mathbf{R}^d \to \mathbf{R}^d$ is invertible. □

Lemma 1.7.4 *Suppose there exists a function $g : \mathbf{R}^d \to \mathbf{R}^d$ such that*

$$\begin{cases} (a) & g \text{ is continuous}, \\ (b) & f(g(y)) = y, \quad \forall y \in \mathbf{R}^d, \\ (c) & g \text{ is surjective}, \end{cases}$$

then f is a C^1 diffeomorphism from $\mathbf{R}^d$ onto $\mathbf{R}^d$.

Proof 1. f is injective. Indeed, let x, x' be two points in $\mathbf{R}^d$ such that $f(x) = f(x')$. According to (c) there exist $y, y' \in \mathbf{R}^d$ such that $g(y) = x$ and $g(y') = x'$. Then by (b) we have

$$f(x) = f(g(y)) = y = f(x') = f(g(y')) = y'.$$

Thus $y = y'$ which implies $g(y) = g(y')$ that is, $x = x'$.

2. f is surjective. This follows obviously from (b).

We deduce that f is bijective. Then, since the Jacobian of f is nowhere zero, the local inversion theorem shows that f is a local C^1-diffeomorphism; thus f and f^{-1} are of class C^1 on $\mathbf{R}^d$. (Being C^1 is a local property.) □

We have to construct g satisfying the hypotheses of the lemma. For that, we look for a regular function $x : \mathbf{R}_t \times \mathbf{R}^d_y \to \mathbf{R}^d$ such that $f(x(t, y)) = ty$ and then set $g(y) = x(1, y)$. By differentiating the previous equality with respect to t we see that a necessary condition is that $df(x(t, y))\dot{x}(t, y) = y$ and thus that $\dot{x}(t, y) = [df(x(t, y))]^{-1}(y)$.

Let $y \in \mathbf{R}^d$. Consider the differential system

$$\begin{cases} \dot{x}(t) = [df(x(t))]^{-1}(y), \\ x(0) = 0. \end{cases} \tag{1.7.2}$$

We first have the following lemma.

Lemma 1.7.5 *Let $f : \mathbf{R}^d \to \mathbf{R}^d$ be of class C^2 such that, for all $x \in \mathbf{R}^d$, the linear map $df(x)$ is invertible from $\mathbf{R}^d$ to $\mathbf{R}^d$. Then for any fixed $y \in \mathbf{R}^d$, the map $x \mapsto F(x) = [df(x)]^{-1}(y)$ from $\mathbf{R}^d$ to $\mathbf{R}^d$ is of class C^1 and for $x, z \in \mathbf{R}^d$ we have*

$$dF(x)(z) = -[df(x)]^{-1}\big\{d^2 f(x)(z, [df(x)]^{-1}(y))\big\}.$$

The proof of this lemma will be given at the end of this paragraph.

By the above lemma and the Cauchy-Lipschitz Theorem, the system (1.7.2) admits a maximal solution, denoted $x(t, y)$, defined on an interval $I = [0, T^*[$. On I we have

$$\frac{d}{dt}[f(x(t, y))] = df(x(t, y)) \circ \dot{x}(t, y) = y.$$

Since $x(0, y) = 0$ and $f(0) = 0$ we deduce that $f(x(t, y)) = ty$. If T^* is finite then we will have $|f(x(t, y))| \leq T^*|y|$ and thus, for $t \in I$, $x(t, y) \in f^{-1}(\overline{B}(0, T^*|y|))$ which is by hypothesis a fixed compact set; this contradicts the maximal interval Theorem (see Theorem 1.7.2 in the Appendix).Therefore $T^* = +\infty$. In particular, $x(t, y)$ exists on $[0, 1]$ and $x(t, y) \in f^{-1}(\overline{B}(0, |y|))$ for all $t \in [0, 1]$.

We set $g(y) = x(1, y)$. It remains to show that g satisfies conditions (a), (b), (c) of Lemma 1.7.4.

1. As (see above) we have $f(x(t, y)) = ty$, it follows that $f(g(y)) = y$, which proves (b).

2. g is continuous. Let $y_0 \in \mathbf{R}^d$. For $|y - y_0| \leq 1$ we have $|y| \leq |y_0| + 1$.

Set $K_0 = f^{-1}(\overline{B}(0, 1 + |y_0|))$. This is a compact subset of $\mathbf{R}^d$. Let B_0 be a closed ball in $\mathbf{R}^d$ centered at zero such that $K_0 \subset B_0$. As above, for all $t \in [0, 1]$ and all $\lambda \in [0, 1]$ we have $\lambda x(t, y_0) + (1 - \lambda)x(t, y) \in B_0$. We write

$$\dot{x}(t, y_0) - \dot{x}(t, y) = [df(x(t, y_0))]^{-1}(y_0) - [df(x(t, y))]^{-1}(y),$$

whence

$$x(t, y_0) - x(t, y) = \int_0^t [df(x(s, y_0))]^{-1}(y_0 - y)\,ds$$
$$+ \int_0^t \{[df(x(s, y_0))]^{-1} - [df(x(s, y))]^{-1}\}(y)\,ds = A + B.$$

For $t \in [0, 1]$,

$$|A| \le \sup_{z \in B_0} \|[df(z)]^{-1}\| |y - y_0| = M(y_0)|y - y_0|.$$

Next, with the notations of Lemma 1.7.5, the quantity inside the integral B is written as

$$F(x(s, y_0)) - F(x(s, y)) = \int_0^1 dF(\lambda x(s, y_0) + (1 - \lambda)(x(s, y)))(x(s, y_0) - x(s, y))\,d\lambda.$$

Thus,

$$|B| \le \int_0^t \sup_{X \in B_0} \|dF(X)\| |x(s, y_0) - x(s, y)|\,ds.$$

We deduce from Lemma 1.7.5 that there exists $C > 0$ depending only on y_0 such that

$$|B| \le C(y_0) \int_0^t |x(s, y_0) - x(s, y)|\,ds.$$

Using Gronwall's Lemma (see Lemma 1.7.1) we deduce that for $t \in [0, 1]$,

$$|x(t, y_0) - x(t, y)| \le M(y_0)|y - y_0|e^{C(y_0)t}$$

and thus $|g(y) - g(y_0)| \le M(y_0)|y - y_0|e^{C(y_0)}$, which proves the continuity of g, that is, property (a) of the lemma.

3. g is surjective. To show this, we prove that $g(\mathbf{R}^d)$ is both open and closed in $\mathbf{R}^d$.

Let us show that it is closed. Let $(x_k)_{k \in \mathbf{N}}$ be a sequence in $g(\mathbf{R}^d)$ converging to x in $\mathbf{R}^d$. We have $x_k = g(y_k)$ and thus $f(x_k) = f(g(y_k)) = y_k$ by (b). Since f is continuous, $y_k = f(x_k) \to f(x)$; since g is continuous, $x_k = g(y_k) \to g(f(x))$. But $x_k \to x$ so $x = g(f(x)) \in g(\mathbf{R}^d)$.

Let us show that the set $g(\mathbf{R}^d)$ is open. Let $x_0 \in g(\mathbf{R}^d)$, $x_0 = g(y_0)$. Then we have $f(x_0) = f(g(y_0)) = y_0$. There exist neighborhoods U_{x_0}, V_{y_0} such that the map $f : U_{x_0} \to V_{y_0}$ is a diffeomorphism. Since g is continuous at y_0 and $g(y_0) = x_0$ there exists an open neighborhood $W_{y_0} \subset V_{y_0}$ of y_0 such that $g(W_{y_0}) \subset U_{x_0}$. Let $y \in W_{y_0}$. We have $f^{-1}(y) \in U_{x_0}$, $g(y) \in U_{x_0}$, $f(f^{-1}(y)) = y$, $f(g(y)) = y$. Since f is injective on U_{x_0} we deduce that $f^{-1}(y) = g(y) \in g(\mathbf{R}^d)$ and thus $f^{-1}(W_{y_0}) \subset g(\mathbf{R}^d)$.

Since f is a diffeomorphism from U_{x_0} onto V_{y_0}, $f^{-1}(W_{y_0})$ is an open neighborhood of x_0. Thus, we have found an open neighborhood of x_0 contained in $g(\mathbf{R}^d)$, which proves that $g(\mathbf{R}^d)$ is open.

Proof of Lemma 1.7.5 Fix $x_0 \in \mathbf{R}^d$. Since $f \in C^2$ for $h \in \mathbf{R}^d$ we have the equality in $\mathcal{L}(\mathbf{R}^d, \mathbf{R}^d)$,

$$df(x_0 + h)y = df(x_0)y + d^2 f(x_0)(h, y) + o(|h|), \quad y \in \mathbf{R}^d.$$

We deduce that

$$\begin{aligned} df(x_0 + h) &= df(x_0)\Big(Id + [df(x_0)]^{-1} d^2 f(x_0)(h, \cdot) + [df(x_0)]^{-1}(o(|h|)\Big), \\ &= df(x_0)(Id + T_h), \end{aligned}$$

where $T_h = [df(x_0)]^{-1}\big[d^2 f(x_0)(h, \cdot) + (o(|h|)\big]$. Then $\|T_h\|_{\mathcal{L}(\mathbf{R}^d, \mathbf{R}^d)} \le C_0|h|$.

For $|h|$ small enough, the operator $Id + T_h$ is invertible and we have

$$[df(x_0 + h)]^{-1} = (Id + T_h)^{-1}[df(x_0)]^{-1} = (Id - T_h + \mathcal{O}(|h|^2))[df(x_0)]^{-1},$$

so that we have

$$F(x_0 + h) - F(x_0) = -[df(x_0)]^{-1}\{d^2 f(x_0)(h, [df(x_0)]^{-1} \cdot)\} + o(|h|).$$

This proves that the function $x \mapsto F(x)$ is differentiable at x_0 and that we have the formula stated in the theorem. The C^1 character follows from the equality giving the differential of F. □

1.7.4 Characterization of Proper Maps from $\mathbf{R}^d$ to $\mathbf{R}^d$

Lemma 1.7.6 *Let $G : \mathbf{R}^d \to \mathbf{R}^d$ be a continuous map. The following two conditions are equivalent.*

- (i) *The inverse image by G of every compact is compact.*
- (ii) $\lim_{|\xi| \to +\infty} |G(\xi)| = +\infty.$

Note that (ii) is equivalent to

$(ii)'$ for every sequence $(\xi_n)_{n \in \mathbf{N}}$ such that $\lim_{n \to +\infty} |\xi_n| = +\infty$ we have

$(\star)$ $\forall A > 0 \ \ \exists N \in \mathbf{N} : \forall n \ge N, \quad |G(\xi_n)| \ge A.$

Proof Recall that in $\mathbf{R}^d$ a set K is compact if and only if it is sequentially compact, that is: from any sequence in K one can extract a convergent subsequence.

$(i) \Longrightarrow (ii)'$. Suppose $(ii)'$ is false. There exists $(\xi_n)_{n\in\mathbf{N}} \subset \mathbf{R}^d$ such that $\lim_{n\to+\infty} |\xi_n| = +\infty$ and such that the sequence $(|G(\xi_n)|)_{n\in\mathbf{N}}$ does not tend to $+\infty$. Writing the contrapositive of $(\star)$, we see that there exists a subsequence $(\xi_{\sigma(n)})_{n\in\mathbf{N}}$ and $A_0 > 0$ such that $|G(\xi_{\sigma(n)})| \leq A_0$, $\forall n \in \mathbf{N}$. But the set $\{\eta \in \mathbf{R}^d : |\eta| \leq A_0\}$ is a compact subset of $\mathbf{R}^d$. We deduce from (i) that the sequence $(\xi_{\sigma(n)})_{n\in\mathbf{N}}$ belongs to a compact subset of $\mathbf{R}^d$. It is therefore a bounded sequence, which is absurd because every subsequence of the sequence $(|\xi_n|)_{n\in\mathbf{N}}$ must tend to $+\infty$.

$(ii)' \Longrightarrow (i)$. Let K be a compact subset of $\mathbf{R}^d$ and $(\xi_n)_{n\in\mathbf{N}}$ be a sequence in $G^{-1}(K)$. Since the sequence $(G(\xi_n))_{n\in\mathbf{N}}$ belongs to the compact set K, there exists a subsequence $(G(\xi_{\sigma(n)}))_{n\in\mathbf{N}}$ which converges to $\eta \in \mathbf{R}^d$. The sequence $(\xi_{\sigma(n)})_{n\in\mathbf{N}}$ is a bounded sequence in $\mathbf{R}^d$. Indeed, if this were not the case, there would exist a subsequence $(\xi_{\sigma_1(n)})_{n\in\mathbf{N}}$ such that $|\xi_{\sigma_1(n)}| \to +\infty$ and according to $(ii)'$, $\lim_{|\xi|\to+\infty} |G(\xi_{\sigma_1(n)})| = +\infty$, which contradicts the convergence to $|\eta|$. In $\mathbf{R}^d$ a bounded sequence admits a convergent subsequence. Therefore there exists a subsequence of the sequence $(\xi_n)_{n\in\mathbf{N}}$ which is convergent. The limit obviously belongs to $G^{-1}(K)$ since this set is closed as the preimage of a closed set by a continuous map. Thus $G^{-1}(K)$ is compact. □

Chapter 2
Cauchy-Kovalevski Theorem

2.1 Prerequisites

Holomorphic functions in $\mathbf{C}^d$ (see Appendix and reference [2]); Cauchy formula. Real-analytic functions.

2.2 The Context

In the previous chapter we described a method for solving certain first-order quasi-linear partial differential equations when all the data (coefficients, initial data) were real-valued. In the present chapter we are interested in the case where this data may be non-real. We must then work in the framework of holomorphic functions.

The results presented here are fundamental in analysis. They concern the existence of solutions to partial differential equations when all the data are holomorphic or real-analytic.

The first two sections of this chapter are devoted to the statements of the main results. These will be consequences of an abstract theorem described in the third section. Finally, all the statements are proved in the last section.

C. Zuily, *Selected Topics in Partial Differential Equations*, Universitext,
https://doi.org/10.1007/978-3-032-24082-8_2

2.3 Cauchy-Kovalevski Theorem (Holomorphic Version)

2.3.1 The Problem

In this section we aim to solve the following problem:

$$\begin{aligned} &\frac{\partial u}{\partial t}(t,z)+\sum_{j=1}^{d} a_j(t,z,u(t,z))\frac{\partial u}{\partial z_j}(t,z)=b(t,z,u(t,z)),\\ &u|_{t=0}=\varphi, \end{aligned} \tag{2.3.1}$$

for $(t, z) \in I \times \omega$ where $I = (0, \delta)$ is an interval in $\mathbf{R}$ and ω is an open subset of $\mathbf{C}^d$.

Before stating the results, we first need to specify the framework as well as the assumptions on the a_j and on b.

2.3.2 Notations and Assumptions

If Ω is an open subset of $\mathbf{C}^N$ we denote by $X(\Omega)$ the space of **bounded holomorphic functions** on Ω, equipped with the norm $\|u\| = \sup_\Omega |u|$. This is a Banach space.

For $\delta > 0$, $\rho > 0$, $R > 0$ we set, in what follows,

$$I_\delta = \{t \in \mathbf{R} : 0 < t < \delta\}, \quad D_\rho = \{z \in \mathbf{C}^d : |z| < \rho\}, \quad \Delta_R = \{Z \in \mathbf{C} : |Z| < R\}.$$

We will assume that

$$b, a_j \in C^0(\overline{I_\delta}, X(D_\rho \times \Delta_R)), \quad j = 1\dots, d, \tag{2.3.2}$$

and we set

$$\sup_{j=1,\dots,d}\ \sup_{(t,z,Z)\in I_\delta \times D_\rho \times \Delta_R} (|a_j(t,z,Z)| + |b(t,z,Z)|) = M. \tag{2.3.3}$$

For $\rho > 0$ we set

$$X_\rho = X(D_\rho), \quad \|f\|_\rho = \sup_{D_\rho} |f|, \quad B_\rho(0,R) = \{f \in X_\rho : \|f\|_\rho < R\}. \tag{2.3.4}$$

2.3.3 Statement of the Results

Theorem 2.3.1 *There exist $\delta_0 > 0$, $\rho_0 > 0$, such that for all $R > 0$, all $\delta \in (0, \delta_0)$ all $\rho \in (0, \rho_0)$ and all a_j and b satisfying hypothesis* (2.3.2), *there exists a unique function $u \in C^1(\overline{I_\delta}, B_\rho(0, R))$ solution in $I_\delta \times D_\rho$ to problem* (2.3.1), *with $\varphi \equiv 0$.*

Corollary 2.3.2 *There exist $\delta_0 > 0$, $\rho_0 > 0$, such that for all $R > 0$, all $\delta \in (0, \delta_0)$ all $\rho \in (0, \rho_0)$, all a_j and b satisfying hypothesis* (2.3.2), *for any $r \in (0, R)$ and any $\varphi \in X_\rho$ such that $\|\varphi\|_\rho \leq r$, there exists a unique function $v \in C^1(\overline{I_\delta}, X_\rho)$ solution in $I_\delta \times D_\rho$ to problem* (2.3.1).

2.4 Cauchy-Kovalevski Theorem (Real-Analytic Version)

In what follows, we will state the real version of the Cauchy-Kovalevski theorem.

2.4.1 Notations and Hypotheses

We shall work in the space of real-analytic functions, a notion we now recall.

Let $\mathcal{O} \subset \mathbf{R}^d$ be open. A function $f : \mathcal{O} \to \mathbf{C}$ is said to be **real-analytic** in $\mathcal{O}$ if there exists an open set Ω in $\mathbf{C}^d$ such that $\mathcal{O} \subset \{z \in \Omega : \operatorname{Im} z = 0\}$ and a function $\widetilde{f} : \Omega \to \mathbf{C}$ holomorphic in Ω such that $\widetilde{f}|_{\mathcal{O}} = f$.

In other words, these are the restrictions to the real of holomorphic functions defined in a neighborhood of the real.

They are also the functions which, in the neighborhood of any point $x_0 \in \mathcal{O}$, can be expanded in a power series.

For $\delta > 0$, $\rho > 0$, $R > 0$ recall that we set

$$I_\delta = \{t \in \mathbf{R} : |t| < \delta\}, \quad D_\rho = \{z \in \mathbf{C}^d : |z| < \rho\}, \quad \Delta_R = \{Z \in \mathbf{C} : |Z| < R\}.$$

We consider functions a_j, $j = 1 \ldots d$, b and φ such that

$$a_j(t, x, Z) = \widetilde{a}_j(t, x, Z), \quad b(t, x, Z) = \widetilde{b}(t, x, Z), \quad \varphi(x) = \widetilde{\varphi}(x) \quad x \in D_\rho \cap \mathbf{R}^d. \tag{2.4.1}$$

where

$$\widetilde{a}_j, \ \widetilde{b} \in C^0(\overline{I_\delta}, X(D_\rho \times \Delta_R)), \quad \widetilde{\varphi} \in X_\rho = X(D_\rho). \tag{2.4.2}$$

2.4.2 Statement of the Result

Theorem 2.4.1 *There exist $\delta_0 > 0$, $\rho_0 > 0$, such that for all $R > 0$, all $\delta \in (0, \delta_0)$ all $\rho \in (0, \rho_0)$, all b, a_j, $j = 1, \ldots, d$ satisfying the hypotheses* (2.4.1), (2.4.2), *any φ real-analytic in $D_\rho \cap \mathbf{R}^d$ such that $\|\widetilde{\varphi}\|_\rho \le r < R$, there exists a unique solution u of class C^1 on I_δ and real-analytic in x on $D_\rho \cap \mathbf{R}^d$ to the problem*

$$\begin{aligned}
&\frac{\partial u}{\partial t}(t,x) + \sum_{j=1}^{d} a_j(t,x,u(t,x))\frac{\partial u}{\partial x_j}(t,x) = b(t,x,u(t,x)),\\
&u|_{t=0} = \varphi
\end{aligned} \tag{2.4.3}$$

in $I_\delta \times D_\rho \cap \mathbf{R}^d$.

The previous theorems will be a consequence of the following abstract result.

2.5 An Abstract Theorem

Let $\mathcal{X} = (X_\rho)_{0<\rho<\rho_0}$ be a family of Banach spaces whose norms are denoted $\|\cdot\|_\rho$. We will assume

$$\rho' \le \rho \implies X_\rho \subset X_{\rho'} \quad \text{and} \quad \|\cdot\|_{\rho'} \le \|\cdot\|_\rho.$$

For $R > 0$, $\delta > 0$ we denote

$$B_\rho(0,R) = \{f \in X_\rho : \|f\|_\rho < R\}, \quad I_\delta = \{t \in \mathbf{R} : 0 < t < \delta\}.$$

We are interested in the problem

$$\frac{du}{dt}(t) = F(t, u(t)) \quad t > 0, \qquad u(0) = 0. \tag{2.5.1}$$

The function F above will be assumed to satisfy the following conditions. There exist $R > 0$, $\delta > 0$, $\rho_0 \in (0, 1]$ such that

1. for every pair (ρ, ρ') such that $0 < \rho' < \rho < \rho_0$,

$$\text{the map} \quad (t, f) \mapsto F(t, f) \text{ is continuous from } I_\delta \times B_\rho(0, R) \text{ to } X_{\rho'}, \tag{2.5.2}$$

2. there exists $C > 0$ such that for all (ρ, ρ') such that $0 < \rho' < \rho < \rho_0$, we have

$$\|F(t,f) - F(t,g)\|_{\rho'} \le \frac{C}{\rho - \rho'} \|f - g\|_\rho, \quad \forall f, g \in B_\rho(0,R), \quad \forall t \in I_\delta, \tag{2.5.3}$$

3. there exists $K > 0$ such that for all $0 < \rho < \rho_0$, the map $t \mapsto F(t, 0)$ is continuous from I_δ into X_ρ and satisfies

$$\|F(t,0)\|_\rho \leq \frac{K}{\rho_0 - \rho}. \tag{2.5.4}$$

We then have the following result.

Theorem 2.5.1 *Under assumptions 1., 2., 3. on F there exist* $a > 0$ *and a unique function* $u(t)$ *which, for all* $\rho \in (0, \rho_0)$ *is continuously differentiable from* $I_{a(\rho_0 - \rho)}$ *with values in* $B_\rho(0, R)$ *and satisfies* (2.5.1).

2.6 The Proofs

2.6.1 Proof of Theorem 2.3.1

The strategy of the proof is to use Theorem 2.5.1. The spaces X_ρ being defined by (2.3.4), it is necessary to define the function F and show that it satisfies conditions (2.5.2), (2.5.3) and (2.5.4).

The following lemma will be essential.

Lemma 2.6.1 *There exists* $C > 0$ *such that for all* $0 < \rho' < \rho$, *all* $f \in X_\rho$ *and all* $1 \leq j \leq d$ *we have*

$$\left\| \frac{\partial f}{\partial z_j} \right\|_{\rho'} \leq \frac{C}{\rho - \rho'} \|f\|_\rho.$$

Proof Indeed, let $z \in D_{\rho'}$ and $\gamma_j = \{\zeta_j \in \mathbf{C} : \zeta_j = z_j + r_0 e^{i\theta_j}, r_0 = \frac{1}{\sqrt{d}}(\rho - \rho')\}$. Then for $\zeta = (\zeta_1, \dots, \zeta_d), \zeta_j \in \gamma_j, j = 1, \dots, d$, we have $|\zeta - z| < \rho - \rho'$. The Cauchy formula allows us to write

$$f(z) = \frac{1}{(2i\pi)^d} \int_{\gamma_1} \cdots \int_{\gamma_d} \frac{f(\zeta)}{\prod_{j=1}^d (\zeta_j - z_j)} d\zeta_1 \wedge \ldots \wedge d\zeta_d.$$

We deduce that

$$\frac{\partial f}{\partial z_k}(z) = \frac{1}{(2i\pi)^d} \int_{\gamma_1} \cdots \int_{\gamma_d} \frac{f(\zeta)}{(\zeta_k - z_k) \prod_{j=1}^d (\zeta_j - z_j)} d\zeta_1 \wedge \ldots \wedge d\zeta_d.$$

Next, since $|z| < \rho'$ and $|\zeta - z| < \rho - \rho'$ we have $|\zeta| < \rho$. Moreover we have $d\zeta_j = i r_0 e^{i\theta_j} d\theta_j$. Thus,

$$\left\| \frac{\partial f}{\partial z_k} \right\|_{\rho'} \leq C(d) \Big(\int_0^{2\pi} \dots \int_0^{2\pi} \frac{r_0^d}{r_0^{d+1}} \, d\theta_1 \dots d\theta_d \Big) \|f\|_\rho \leq \frac{C'(d)}{r_0} \|f\|_\rho$$

which proves the result since $r_0 = \frac{1}{\sqrt{d}}(\rho - \rho')$. □

Definition and Properties of the Function F

We define the function $F : (t, f) \in I_\delta \times B_\rho(0, R) \mapsto F(t, f)$ by

$$F(t, f)(z) = \sum_{j=1}^{d} a_j(t, z, f(z)) \frac{\partial f}{\partial z_j}(z) + b(t, z, f(z)).$$

The goal of the rest of this paragraph is to show that this function satisfies properties (2.5.2), (2.5.3) and (2.5.4).

Verification of (2.5.2)

We first show that for fixed $t \in I_\delta$ and $f \in B_\rho(0, R)$, the function $z \mapsto F(t, f)(z)$ belongs to $X_{\rho'}$. This function is holomorphic in D_ρ (thus in $D_{\rho'}$) as a composition, product, and sum of holomorphic functions. Next, from (2.3.3) and Lemma 2.6.1, we have

$$\sup_{|z|<\rho'} |F(t, f)(z)| \leq M \sum_{j=1}^{d} \sup_{|z|<\rho'} \left| \frac{\partial f}{\partial z_j}(z) \right| + M \leq \frac{C(d)MR}{\rho - \rho'} + M.$$

It is therefore indeed a bounded holomorphic function in $D_{\rho'}$.

Let us now show continuity.

Let $(t, f) \in I_\delta \times B_\rho(0, R)$ be fixed and $(t_n, f_n) \in I_\delta \times B_\rho(0, R)$ a sequence that converges in $\mathbf{R} \times X_\rho$ to (t, f). We write, for $|z| < \rho'$,

$$\begin{aligned} F(t_n, f_n)(z) - F(t, f)(z) &= A_n(z) + B_n(z) + C_n(z), \\ A_n(z) &= \sum_{j=1}^{d} a_j(t, z, f(z)) \frac{\partial (f_n - f)}{\partial z_j}(z), \\ B_n(z) &= \sum_{j=1}^{d} \big(a_j(t_n, z, f_n(z)) - a_j(t, z, f(z))\big) \frac{\partial f_n}{\partial z_j}(z), \\ C_n(z) &= b(t_n, z, f_n(z)) - b(t, z, f(z)). \end{aligned} \tag{2.6.1}$$

Using (2.3.3) and Lemma 2.6.1 we can write

$$\|A_n\|_{\rho'} \le \frac{C(d)}{\rho-\rho'} M \|f_n - f\|_\rho. \tag{2.6.2}$$

Next, using Lemma 2.6.1 and the fact that $f_n \in B_\rho(0, R)$, we have, for $|z| \le \rho'$,

$$\begin{aligned}|B_n(z)| &\le \sum_{j=1}^d |a_j(t_n, z, f_n(z)) - a_j(t, z, f(z))| \left|\frac{\partial f_n}{\partial z_j}(z)\right|, \\ &\le \frac{C}{\rho-\rho'} \sum_{j=1}^d |a_j(t_n, z, f_n(z)) - a_j(t, z, f(z))| \|f_n\|_\rho\end{aligned}$$

so,

$$|B_n(z)| \le \frac{CR}{\rho-\rho'} \sum_{j=1}^d |a_j(t_n, z, f_n(z)) - a_j(t, z, f(z))|. \tag{2.6.3}$$

Set $G_n(z, Z) = c(t_n, z, Z) - c(t, z, Z)$ where $c = a_j$, $1 \le j \le d$ or $c = b$.
We have, for $|Z| < R$, $|G_n(z, Z)| \le \sup_{|Z|<R} |G_n(z, Z)|$. Since $f_n \in B_\rho(0, R)$ we have $|G_n(z, f_n(z))| \le \sup_{|Z|<R} |G_n(z, Z)|$ thus,

$$|c(t_n, z, f_n(z)) - c(t, z, f_n(z))| \le \sup_{|z|<\rho', |Z|<R} |c(t_n, z, Z) - c(t, z, Z)|.$$

Consequently, with the notations of (2.3.2),

$$\sup_{|z|<\rho'} |c(t_n, z, f_n(z)) - c(t, z, f_n(z))| \le \|c(t_n, \cdot, \cdot) - c(t, \cdot, \cdot)\|_{X(D_\rho \times \Delta_R)}. \tag{2.6.4}$$

Thanks to hypothesis (2.3.2) the right-hand side tends to zero as n tends to $+\infty$.
Next, since $f \in B_\rho(0, R)$ there exists $\varepsilon > 0$ such that $\|f\|_\rho < R - 2\varepsilon$. Then for n large enough we have $\|f_n\|_\rho < R - \varepsilon$. It follows that

$$|c(t, z, f_n(z) - c(t, z, f(z))| \le |f_n(z) - f(z)| \sup_{|Z|<R-\varepsilon} \left|\frac{\partial c}{\partial Z}(t, z, Z)\right|.$$

Using Lemma 2.6.1 with $\rho' = R - 2\varepsilon$, $\rho = R - \varepsilon$ it follows

$$|c(t, z, f_n(z) - c(t, z, f(z))| \le \frac{C}{\varepsilon} |f_n(z) - f(z)| \sup_{|Z|<R} |c(t, z, Z)|.$$

According to (2.3.3) we can finally write

$$\sup_{|z|<\rho'} |c(t, z, f_n(z) - c(t, z, f(z))| < M \|f_n - f\|_\rho. \tag{2.6.5}$$

Using (2.6.4), (2.6.5), (2.6.3) and (2.6.1) we deduce

$$\lim_{n\to+\infty}(\|B_n\|_{\rho'}+\|C_n\|_{\rho'})=0 \tag{2.6.6}$$

which, added to (2.6.2) and (2.6.1), proves the continuity of F, that is property (2.5.2).

Verification of (2.5.3)

Let $f, g \in B_\rho(0, R)$. We have

$$\begin{aligned}
&F(t,f)(z)-F(t,g)(z)=A_1+A_2+A_3,\\
&A_1(z)=\sum_{j=1}^{d}a_j(t,z,f(z))\frac{\partial(f-g)}{\partial z_j}(z),\\
&A_2(z)=\sum_{j=1}^{d}\big(a_j(t,z,f(z))-a_j(t,z,g(z))\big)\frac{\partial g}{\partial z_j}(z),\\
&A_3(z)=b(t_n,z,f(z))-b(t,z,g(z)).
\end{aligned} \tag{2.6.7}$$

The term A_1 has the same form as the term A_n defined in (2.6.1) by replacing f_n with g. It follows from (2.6.2) that

$$\|A_1\|_{\rho'}\leq\frac{C(d)}{\rho-\rho'}M\|f-g\|_{\rho}. \tag{2.6.8}$$

Next, for $|z|<\rho'$, using Lemma 2.6.1 and the fact that $g \in B_\rho(0, R)$, we have

$$\begin{aligned}
|A_2(z)|&\leq\sum_{j=1}^{d}|a_j(t,z,f(z))-a_j(t,z,g(z))|\left\|\frac{\partial g}{\partial z_j}\right\|_{\rho'},\\
&\leq\frac{CR}{\rho-\rho'}\sum_{j=1}^{d}|a_j(t,z,f(z))-a_j(t,z,g(z))|\|g\|_{\rho}.
\end{aligned}$$

It follows that

$$|A_2(z)|\leq\frac{CR}{\rho-\rho'}\sum_{j=1}^{d}|a_j(t,z,f(z))-a_j(t,z,g(z))|.$$

Using the estimate (2.6.5), we deduce

$$\|A_2\|_{\rho'}\leq\frac{CR}{\rho-\rho'}\|f-g\|_{\rho}. \tag{2.6.9}$$

Again by (2.6.5) we obtain

$$\|A_3\|_{\rho'} \leq C\|f - g\|_\rho \leq \frac{C}{\rho - \rho'}\|f - g\|_\rho, \tag{2.6.10}$$

since we can take $\rho - \rho' \leq 1$.

It follows from (2.6.8), (2.6.9), (2.6.10) that

$$\|F(t, f) - F(t, g)\|_{\rho'} \leq \frac{C}{\rho - \rho'}\|f - g\|_\rho$$

which shows that condition (2.5.3) is satisfied.

Verification of (2.5.4)

We have $F(t, 0) = b(t, z, 0)$. The continuity of the map $t \mapsto F(t, 0)$ from I_δ into X_ρ follows from hypothesis (2.3.2). Next, by the estimate (2.3.3) we have

$$\sup_{|t|<\delta} \|F(t, 0)\|_\rho \leq M \leq \frac{M}{\rho_0 - \rho},$$

since we can take $\rho_0 \leq 1$.

End of the Proof of Theorem 2.3.1

The function F satisfying conditions (2.5.2), (2.5.3), and (2.5.4), we can apply Theorem 2.5.1 and deduce the existence of a function $u(t) : z \mapsto u(t, z)$ of class C^1 on the interval I_δ with values in $B_\rho(0, R)$, solution to problem (2.5.1), which, given the definition of the function F, solves problem (2.3.1) with $\varphi \equiv 0$.

2.6.2 *Proof of Corollary 2.3.2*

We use Theorem 2.3.1. Define the functions $A_j(t, z, Z)$ and $B(t, z, Z)$ by

$$A_j(t, z, Z) = a_j(t, z, Z + \varphi(z)),$$

$$B(t, z, Z) = b(t, z, Z + \varphi(z)) - \sum_{j=1}^{d} a_j(t, z, Z + \varphi(z))\frac{\partial \varphi}{\partial z_j}(z),$$

Let $R_0 = R - r$. For $|Z| < R_0$ we have $|Z| + \sup_{D_\rho} |\varphi(z)| < R$. The functions A_j, B satisfy hypothesis (2.3.2) with R replaced by R_0.

We can therefore, according to Theorem 2.3.1, solve the problem

$$\frac{\partial u}{\partial t}(t,z)+\sum_{j=1}^{d}A_j(t,z,u(t,z))\frac{\partial u}{\partial z_j}(t,z)=B(t,z,u(t,z)),\quad u|_{t=0}=0,$$

in $I_\delta \times D_\rho$ and the function u belongs to $B_\rho(0,R_0)$. The function $v=u+\varphi$ is then the unique solution of (2.3.1).

2.6.3 Proof of Theorem 2.4.1

We use Corollary 2.3.2. Then the problem

$$\begin{aligned}&\frac{\partial v}{\partial t}(t,z)+\sum_{j=1}^{d}\widetilde{a}_j(t,z,v(t,z))\frac{\partial v}{\partial z_j}(t,z)=\widetilde{b}(t,z,v(t,z)),\\&v|_{t=0}=\widetilde{\varphi},\end{aligned}\tag{2.6.11}$$

in $I_\delta \times D_\rho$ admits a unique solution $v\in C^1(I_\delta, X_\rho)$.

Since $v(t,\cdot)$ is holomorphic in D_ρ we have $\frac{\partial v}{\partial \overline{z}_j}=0$ in $I_\delta\times D_\rho$ so that

$$\frac{\partial v}{\partial x_j}(t,z)=\frac{\partial v}{\partial z_j}(t,z)+\frac{\partial v}{\partial \overline{z}_j}(t,z)=\frac{\partial v}{\partial z_j}(t,z).$$

The equation in (2.6.11) then reads

$$\frac{\partial v}{\partial t}(t,z)+\sum_{j=1}^{d}\widetilde{a}_j(t,z,v(t,z))\frac{\partial v}{\partial x_j}(t,z)=\widetilde{b}(t,z,v(t,z))$$

for all $t\in I_\delta$ and $z=x+iy\in D_\rho$. Taking $y=0$ in the equation above and setting $u(t,x)=v(t,x)$ we see that u is C^1 in t, real-analytic in x (since it is the trace on $D_\rho\cap\mathbf{R}^d$ of a holomorphic function) and is a solution to problem (2.4.3).

2.6.4 Proof of Theorem 2.5.1

Recall that $I_\delta=\{t:0<t<\delta\}$. Let

$$\mathcal{B}=\{u\in C^0(I_{a(\rho_0-\rho)},X_\rho),\forall\rho\in(0,\rho_0)\},\tag{2.6.12}$$

which we equip with the norm

$$N_a(u) = \sup_{0<\rho<\rho_0} \sup_{0<t<a(\rho_0-\rho)} \|u(t)\|_\rho \Big(\frac{a(\rho_0 - \rho)}{t} - 1\Big), \tag{2.6.13}$$

where $a > 0$ is a small constant to be specified later. This is a Banach space.

Remark 2.6.2 Notice that for fixed u the function $a \mapsto N_a(u)$ is decreasing. Indeed, if $0 < a < b$ we have $\{0 < t < a(\rho_0 - \rho)\} \subset \{0 < t < b(\rho_0 - \rho)\}$ and the inequality $\frac{a(\rho_0-\rho)}{t} - 1 < \frac{b(\rho_0-\rho)}{t} - 1$.

We propose to first solve the equation

$$u(t) = \int_0^t F(s, u(s))\, ds, \tag{2.6.14}$$

where, for $a > 0$ small enough, $N_a(u)$ is finite.

We introduce the strictly decreasing sequence $(a_k)_{k\in\mathbf{N}}$, defined by

$$a_0 > 0, \quad a_{k+1} = a_k\Big(1 - \frac{1}{(k+2)^2}\Big), \quad k \in \mathbf{N}, \tag{2.6.15}$$

and we set

$$a = a_0 \prod_{k=0}^{+\infty} \Big(1 - \frac{1}{(k+2)^2}\Big) > 0. \tag{2.6.16}$$

where a_0 will be chosen later small enough, in particular so that $a_0\rho_0 < \delta$ where $\delta > 0$ is the constant appearing in the hypotheses (2.5.2), (2.5.3), and (2.5.4).

The solution of (2.6.14) will be obtained as the limit of the sequence $(u_k(t))_{k\in\mathbf{N}}$ defined recursively for $t \in (0, a_k(\rho_0 - \rho))$ by

$$u_0(t) \equiv 0, \quad u_{k+1}(t) = \int_0^t F(s, u_k(s))\, ds, \tag{2.6.17}$$

and we set

$$v_k(t) = u_{k+1}(t) - u_k(t) \quad \text{and} \quad \lambda_k = N_{a_k}(v_k). \tag{2.6.18}$$

We can estimate λ_0 as follows. By definition,

$$\lambda_0 = \sup_{0<\rho<\rho_0} \sup_{0<t<a_0(\rho_0-\rho)} \Big(\frac{a_0(\rho_0 - \rho)}{t} - 1\Big) \|v_0(t)\|_\rho.$$

By (2.6.18) and (2.6.17) we have $v_0(t) = \int_0^t F(s, 0)\, ds$, so that, using the hypothesis (2.5.4) on F, we have for $0 < t < a_0(\rho_0 - \rho)$,

$$\|v_0(t)\|_\rho \leq \int_0^t \|F(s, 0)\|_\rho \, ds \leq \frac{Kt}{\rho_0 - \rho}.$$

We deduce that

$$\lambda_0 \leq K a_0. \tag{2.6.19}$$

The following proposition describes properties of the sequence $(u_k)_{k \in \mathbf{N}}$ that are essential for what follows.

Proposition 2.6.3

(1) $u_k, v_k \in C^0(I_{a_k(\rho_0 - \rho)}, X_\rho), \quad \forall \rho \in (0, \rho_0), \forall k \in \mathbf{N}.$

(2) *If* a_0 *is small enough we have* $\lambda_k \leq \dfrac{2^4 a_0 K}{(k+2)^4} \quad \forall k \in \mathbf{N}$, *where* $K > 0$ *is the constant appearing in the hypothesis* (2.5.4).

(3) *We have* $\|u_k(t)\|_\rho < \dfrac{1}{2} R, \quad \forall t \in I_{a_k(\rho_0 - \rho)}, \forall k \in \mathbf{N}.$

(4) *The sequence* $(u_k)_{k \in \mathbf{N}}$ *converges in* $\mathcal{B}$ *to a function* u *which solves* (2.6.14).

Proof Let us first note that, according to (2.6.17) and (2.6.19), assertions (1), (2), (3) are true for $k = 0$. We proceed by induction, assuming that $u_0, \ldots, u_k$ satisfy points (1), (2), (3) and showing that they are still true at rank $k + 1$.

Point (1). The function u_{k+1} satisfies (1), thanks to the hypothesis (2.5.2) and the fact that $I_{a_{k+1}(\rho_0 - \rho)}$ is contained in $I_{a_k(\rho_0 - \rho)}$.

Point (2). It follows from (2.6.17) and (2.6.18) that for $0 < t < a_{k+1}(\rho_0 - \rho)$ we have

$$v_{k+1}(t) = \int_0^t \big[F(s, u_{k+1}(s)) - F(s, u_k(s))\big]\, ds.$$

The following lemma will be useful. □

Lemma 2.6.4 *For* $0 < s < a_{k+1}(\rho_0 - \rho)$ *set*

$$\rho(s) = \frac{1}{2}\Big(\rho_0 - \frac{s}{a_{k+1}} + \rho\Big).$$

Then,

$$\begin{aligned}
&(i) \quad \rho < \rho(s),\\
&(ii) \quad \rho_0 - \rho(s) = \frac{1}{2}(\rho_0 - \rho) + \frac{1}{2}\frac{s}{a_{k+1}} > 0,\\
&(iii) \quad s < a_k(\rho_0 - \rho(s)),\\
&(iv) \quad \rho(s) - \rho = \frac{1}{2}(\rho_0 - \rho) - \frac{1}{2}\frac{s}{a_{k+1}},
\end{aligned}$$

Proof (i) By hypothesis we have $-s > -a_{k+1}(\rho_0 - \rho)$, from which we deduce that $\rho(s) > \frac{1}{2}(\rho_0 - (\rho_0 - \rho) + \rho) = \rho$.

(ii) We have $\rho_0 - \rho(s) = \frac{1}{2}(\rho_0 - \rho) + \frac{1}{2}\frac{s}{a_{k+1}} > 0$.

(iii) We have $a_k(\rho_0 - \rho(s)) = \frac{1}{2}a_k(\rho_0 - \rho) + \frac{1}{2}s\frac{a_k}{a_{k+1}} > \frac{1}{2}a_k(\rho_0 - \rho) + \frac{1}{2}s$, since $a_k > a_{k+1}$. Next we have $a_k(\rho_0 - \rho) > a_{k+1}(\rho_0 - \rho) > s$ by hypothesis. It follows that $a_k(\rho_0 - \rho(s)) > \frac{1}{2}s + \frac{1}{2}s = s$.

(iv) is easy to check. □

We will use the hypothesis (2.5.3) with the pair $(\rho', \rho) = (\rho, \rho(s))$, which is possible according to (i) and (ii) of Lemma 2.6.4. We write

$$\|v_{k+1}(t)\|_\rho \le \int_0^t \|F(s, u_{k+1}(s)) - F(s, u_k(s))\|_\rho \, ds \le C \int_0^t \frac{\|v_k(s)\|_{\rho(s)}}{\rho(s) - \rho} ds,$$

because, according to (2.6.18), we have $v_k(s) = u_{k+1}(s) - u_k(s)$. By definition of $\lambda_k = N_{a_k}(v_k)$ (see (2.6.13)) and since $\rho < \rho(s) < \rho_0$ and $0 < s < a_k(\rho_0 - \rho(s))$ we have

$$\lambda_k \ge \|v_k(s)\|_{\rho(s)}\Big(\frac{a_k(\rho_0 - \rho(s))}{s} - 1\Big) \ge \|v_k(s)\|_{\rho(s)}\Big(\frac{a_{k+1}(\rho_0 - \rho(s))}{s} - 1\Big),$$

whence

$$\|v_k(s)\|_{\rho(s)} \le \frac{s\lambda_k}{a_{k+1}(\rho_0 - \rho(s)) - s}.$$

We deduce that

$$\|v_{k+1}(t)\|_\rho \le C\lambda_k \int_0^t \frac{s}{(a_{k+1}(\rho_0 - \rho(s)) - s)(\rho(s) - \rho)} ds.$$

Using the points (ii) and (iv) of Lemma 2.6.4 we obtain

$$\begin{aligned}
\big(a_{k+1}(\rho_0 - \rho(s)) - s\big)\big(\rho(s) - \rho\big) &= \frac{1}{4a_{k+1}}\big(a_{k+1}(\rho_0 - \rho) - s\big)\big(a_{k+1}(\rho_0 - \rho) - s\big),\\
&= \frac{1}{4a_{k+1}}\big(a_{k+1}(\rho_0 - \rho) - s\big)^2.
\end{aligned}$$

We deduce

$$\begin{aligned}
\|v_{k+1}(t)\|_\rho &\leq 4C\lambda_k a_{k+1} t \int_0^t \frac{ds}{\big(a_{k+1}(\rho_0-\rho)-s\big)^2},\\
&\leq 4C\lambda_k a_{k+1} t \Big[\frac{1}{a_{k+1}(\rho_0-\rho)-s}\Big]_0^t,\\
&\leq 4C\lambda_k a_{k+1} t \Big(\frac{1}{a_{k+1}(\rho_0-\rho)-t}-\frac{1}{a_{k+1}(\rho_0-\rho)}\Big),\\
&\leq 4C\lambda_k a_{k+1} t \,\frac{t}{(a_{k+1}(\rho_0-\rho)-t)a_{k+1}(\rho_0-\rho)},\\
&\leq \frac{4C\lambda_k a_{k+1} t}{a_{k+1}(\rho_0-\rho)}\,\frac{t}{a_{k+1}(\rho_0-\rho)-t} \leq \frac{4C\lambda_k a_{k+1} t}{a_{k+1}(\rho_0-\rho)}\,\frac{1}{\frac{a_{k+1}(\rho_0-\rho)}{t}-1}.
\end{aligned}$$

It follows that

$$\Big(\frac{a_{k+1}(\rho_0-\rho)}{t}-1\Big)\|v_{k+1}(t)\|_\rho \leq \frac{4C\lambda_k a_{k+1} t}{a_{k+1}(\rho_0-\rho)} \leq 4C\lambda_k a_{k+1},$$

if $0 < t < a_{k+1}(\rho_0-\rho)$.

Consequently, as by definition

$$\lambda_{k+1} = \sup_{0<\rho<\rho_0}\ \sup_{0<t<a_{k+1}(\rho_0-\rho)} \Big(\frac{a_{k+1}(\rho_0-\rho)}{t}-1\Big)\|v_{k+1}(t)\|_\rho,$$

we have

$$\lambda_{k+1} \leq 4Ca_{k+1}\lambda_k \leq 4Ca_0\lambda_k. \tag{2.6.20}$$

Recall that we must show assertion (2) of Proposition 2.6.3, that is to say: if a_0 is small enough we have

$$\lambda_k \leq \frac{2^4 a_0 K}{(k+2)^4} \quad \forall k \in \mathbf{N}. \tag{2.6.21}$$

We proceed by induction. The estimate (2.6.21) being true for $k = 0$, suppose it is true at order k. Using (2.6.20) we can write

$$\lambda_{k+1} \leq 4Ca_0 \frac{2^4 a_0 K}{(k+2)^4} \leq \frac{2^4 a_0 K}{(k+3)^4} 4Ca_0 \frac{(k+3)^4}{(k+2)^4}.$$

Since $k+3 \leq 2(k+2)$ we have $Ca_0 \frac{(k+3)^4}{(k+2)^4} \leq 4Ca_0 2^4$. It suffices to take a_0 small enough so that $4Ca_0 2^4 \leq 1$ and then we have $\lambda_{k+1} \leq \frac{2^4 a_0 K}{(k+3)^4}$, which completes the induction and proves Point (2).

Point (3). By the definition of the norm N_a, we have, for $0 < t < a_k(\rho_0 - \rho)$,

$$\|v_k(t)\|_\rho \leq \frac{1}{\frac{a_k(\rho_0-\rho)}{t} - 1} N_{a_k}(v_k).$$

If $0 < t < a_{k+1}(\rho_0 - \rho) < a_k(\rho_0 - \rho)$, then $\frac{a_k(\rho_0-\rho)}{|t|} - 1 > \frac{a_k}{a_{k+1}} - 1 > 0$ so that

$$\|v_k(t)\|_\rho \leq \frac{\lambda_k}{\frac{a_k}{a_{k+1}} - 1}, \quad \text{if } 0 < t < a_{k+1}(\rho_0 - \rho). \tag{2.6.22}$$

We deduce that

$$\|u_{k+1}(t)\|_\rho \leq \frac{\lambda_k}{\frac{a_k}{a_{k+1}} - 1} + \|u_k(t)\|_\rho, \quad \text{if } 0 < t < a_{k+1}(\rho_0 - \rho).$$

By induction and since $u_0 = 0$ we have

$$\|u_{k+1}(t)\|_\rho \leq \sum_{j=0}^{k} \frac{\lambda_j}{\frac{a_j}{a_{j+1}} - 1}, \quad \text{if } 0 < t < a_{k+1}(\rho_0 - \rho). \tag{2.6.23}$$

Note that if $\alpha > \beta > 0$ we have $\frac{\alpha}{\beta} + \frac{\beta}{\alpha} \geq 2$. Indeed, this inequality is equivalent to $\alpha^2 + \beta^2 \geq 2\alpha\beta$. Taking $\alpha = a_j$ and $\beta = a_{j+1}$ we obtain

$$\frac{a_j}{a_{j+1}} - 1 \geq 1 - \frac{a_{j+1}}{a_j},$$

so that

$$\|u_{k+1}(t)\|_\rho \leq \sum_{j=0}^{k} \frac{\lambda_j}{1 - \frac{a_{j+1}}{a_j}}.$$

Recall (see (2.6.15)) that $\frac{a_{j+1}}{a_j} = 1 - \frac{1}{(j+2)^2}$.
We deduce that

$$\|u_{k+1}(t)\|_\rho \leq \sum_{j=0}^{k} (j+2)^2 \lambda_j.$$

Using Point (2) we obtain

$$\|u_{k+1}(t)\|_\rho \leq 2^4 a_0 K \sum_{j=0}^{k} \frac{1}{(k+2)^2} \leq 2^4 a_0 K \sum_{j=0}^{+\infty} \frac{1}{(k+2)^2}.$$

If we choose a_0 small enough, we finally obtain

$$\|u_{k+1}(t)\|_\rho < \frac{1}{2}R, \quad \text{for all } t \in (0, a_{k+1}(\rho_0 - \rho)), \tag{2.6.24}$$

which completes the proof of Point (3).

Point (4). By construction of the sequence (a_k), its limit a defined by (2.6.16) satisfies $a \leq a_k$ for all k. The map $a \mapsto N_a(\cdot)$ being decreasing according to Remark 2.6.2, we have

$$N_a(u_{k+1}(t) - u_k(t)) \leq N_{a_k}(u_{k+1}(t) - u_k(t)) = N_{a_k}(v_k(t)) \leq \lambda_k$$

for $0 < t < a(\rho_0 - \rho)$. As, according to the estimate in Point (2), the series $\sum \lambda_k$ is convergent, we deduce that the sequence (u_k) converges in the space $\mathcal{B}$ defined in (2.6.12) to a function u.

This completes the proof of Proposition 2.6.3.

End of the Proof of Theorem 2.5.1

We use Point (4) of Proposition 2.6.3 as well as hypothesis (2.6.1), which allows us to pass to the limit in both sides of the equality $u_{k+1}(t) = \int_0^t F(s, u_k(s))\, ds$. We deduce that the function u is the unique solution of the equation

$$u(t) = \int_0^t F(s, u(s))\, ds. \tag{2.6.25}$$

According to (2.6.24) we have $\|u(t)\| < \frac{1}{2}R$ for $0 < t < a(\rho_0 - \rho)$.

Next, by hypothesis (2.5.2) of Theorem 2.5.1, the right-hand side of (2.6.25) is a C^1 function on $\{0 < t < a(\rho_0 - \rho)\}$ with values in $X_{\rho'}$ for all $\rho' < \rho_0$. The function $t \mapsto u(t)$ is therefore continuously differentiable from $I_{a(\rho_0 - \rho)}$ with values in $B_\rho(0, R)$, and by differentiating both sides of (2.6.25) we see that u is a solution of problem (2.5.1). This completes the proof of Theorem 2.5.1. □

2.7 Appendix

2.7.1 Holomorphic Functions in Open Subsets of $\mathbf{C}^d$

In this paragraph, we recall some elementary definitions and results (without proof) about holomorphic functions in several variables, necessary for the understanding of what precedes, without, however, developing a complete theory of this subject.

The reader, looking for more information, may consult the book [2].

Let Ω be an open subset of $\mathbf{C}^d$. We will denote by $z = (z_1, \dots, z_d)$ the variable in Ω. We set $z_j = x_j + iy_j, \quad (x_j, y_j) \in \mathbf{R} \times \mathbf{R}$ and

$$\mathcal{O} = \{(x_1, \dots, x_d, y_1, \dots, y_d) \in \mathbf{R}^d \times \mathbf{R}^d : z = (z_1, \dots z_d) \in \Omega\}.$$

If $f : \Omega \to \mathbf{C}$ is a function, we denote by $g : \mathcal{O} \to \mathbf{C}$ the function defined by

$$g(x_1, \dots, x_d, y_1, \dots, y_d) = f(z_1, \dots, z_d). \tag{2.7.1}$$

We then set

$$\overline{\partial}_j = \frac{1}{2}\Big(\frac{\partial}{\partial x_j} + i\frac{\partial}{\partial y_j}\Big), \quad \partial_j = \frac{1}{2}\Big(\frac{\partial}{\partial x_j} - i\frac{\partial}{\partial y_j}\Big). \tag{2.7.2}$$

Definition 2.7.1 A function $f : \Omega \to \mathbf{C}$ is said to be holomorphic in Ω if the function g defined in (2.7.1) has continuous partial derivatives with respect to x_j, y_j for $j = 1, \dots, d$ and

$$\overline{\partial}_j g(x_1, \dots, x_d, y_1, \dots, y_d) = 0, \quad \forall (x_1, \dots, x_d, y_1, \dots, y_d) \in \mathcal{O}, \quad \forall j = 1, \dots, d.$$

We denote by $\mathcal{H}(\Omega)$ the space of holomorphic functions in Ω.

The Cauchy Formula in a Polydisc

In $\mathbf{C}^d$ an (open) polydisc is an open subset of $\mathbf{C}^d$ of the form

$$D = \{z = (z_1, \dots z_d) \in \mathbf{C}^d : |z_j - z_j^0| < r_j, \ 1 \leq j \leq d\},$$

where $z_j^0 \in \mathbf{C}$ and $r_j > 0$. The closed set

$$\partial D = \{z = (z_1, \dots z_d) \in \mathbf{C}^d : |z_j - z_j^0| = r_j, \ 1 \leq j \leq d\}$$

is called the distinguished boundary of D and the point $(z_1^0, \dots, z_d^0)$ the center of the polydisc.

Proposition 2.7.2 *Let D be a polydisc in $\mathbf{C}^d$, and $f : \overline{D} \to \mathbf{C}$ a continuous function that is holomorphic with respect to each variable z_j, the others being fixed. For all $z \in D$ we have*

$$f(z) = \Big(\frac{1}{2i\pi}\Big)^d \int_{\partial D} \frac{f(\zeta_1, \dots \zeta_d)}{\prod_{j=1}^d (\zeta_j - z_j)} d\zeta_1 \dots d\zeta_d.$$

Consequences

If $\alpha = (\alpha_1, \ldots, \alpha_d) \in \mathbf{N}^d$, $z = (z_1, \cdots, z_d) \in \mathbf{C}^d$ we denote

$$|\alpha| = \alpha_1 + \cdots + \alpha_d, \quad \alpha! = \alpha_1! \cdots \alpha_d!, \quad z^\alpha = z_1^{\alpha_1} \cdots z_d^{\alpha_d}.$$

Proposition 2.7.3 *A function $f : \Omega \to \mathbf{C}$ belongs to $\mathcal{H}(\Omega)$ if and only if for every $z_0 \in \Omega$ there exists $r > 0$ such that for all z with $|z - z_0| < r$,*

$$f(z) = \sum_{\alpha \in \mathbf{N}^d} a_\alpha (z - z_0)^\alpha, \quad \textit{where } a_\alpha \in \mathbf{C}.$$

Proposition 2.7.4 *If $f \in \mathcal{H}(\Omega)$ the function g defined in* (2.7.1) *belongs to $C^\infty(\mathcal{O})$.*

Proposition 2.7.5 *Let $(f_k)_{k \in \mathbf{N}}$ be a sequence in $\mathcal{H}(\Omega)$ that converges uniformly on every compact subset of Ω to a function f. Then $f \in \mathcal{H}(\Omega)$.*

Proposition 2.7.6 (Cauchy inequalities) *Let f be a holomorphic function in the polydisc D where $D = \{z \in \mathbf{C}^d : |z_j| < r_j,\ 1 \le j \le d\}$. If $|f| \le M$ in D then for every multi-index $\alpha \in \mathbf{N}^d$ we have*

$$|\partial^\alpha f(0)| \le M\alpha! r^{-\alpha}.$$

Theorem 2.7.7 (Hartogs' theorem) *Let Ω be an open subset of $\mathbf{C}^d$ and f a function from Ω to $\mathbf{C}$. Suppose that f is holomorphic with respect to each variable z_j, the others being fixed. Then $f \in \mathcal{H}(\Omega)$.*

2.8 Comments

1. Augustin Louis Cauchy (1789–1857) is a French mathematician whose work is so pervasive that he needs no introduction. He is buried in the Sceaux cemetery (near Paris). Sophie Kovalevski (Sophia Kovalevskaya, her name in Russian) (1850–1891) is a Russian mathematician who was the first woman to obtain a doctorate in mathematics from a German university. She is the author of numerous works on partial differential equations.
2. The proof method for the Cauchy-Kovalevski Theorem presented here, which is based on an abstract result, is not the original method used by the historical authors of this result. That method was rather based on majorant series techniques. The method developed in this chapter is quite recent. It was established by a famous American mathematician, Louis Nirenberg in 1972 and then improved in 1977 by a Japanese mathematician, Takaaki Nishida, to whom the abstract theorem in

this chapter is due. It should be noted that this method made it possible to assume only continuity with respect to the variable t for the coefficients a_j and b, whereas previous works required holomorphy with respect to all variables.

3. The results of this chapter are not limited to first-order equations. Analogous results are also valid for systems of first-order equations as well as for nonlinear equations of any order. However, in the latter case, the proof method essentially consists in reducing to first-order systems.
4. In this chapter, we have taken as the initial surface the hypersurface $\{t = 0\}$ while the coefficient of $\frac{\partial}{\partial t}$ was taken equal to 1. We could have taken a more general initial surface $S = \{z \in \mathbf{C}^d : \varphi(z) = 0\}$ where the function φ is holomorphic and satisfies $d\varphi(z) \neq 0$ on S.

Chapter 3
The Method of Stationary Phase

3.1 Prerequisites

Fourier transform of $e^{i\langle Bx,x\rangle}$ in $\mathcal{S}'(\mathbf{R}^d)$. Diagonalization of symmetric matrices. Convergent integrals on $\mathbf{R}$. Taylor's formula with integral remainder in $\mathbf{R}$. Lebesgue's dominated convergence theorem.

3.2 The Context

Let us consider, for example, the differential equation on $\mathbf{R}$ (called the Airy equation)

$$u''(t) - tu(t) = 0. \tag{3.2.1}$$

We will show in Sect. 3.3.1, using integration by parts, that the integral $\int_{-\infty}^{+\infty} e^{ity+i\frac{y^3}{3}}\,dy$ is convergent and defines an element of $C^\infty(\mathbf{R})$ which is a solution of the Airy equation. The question that then arises is to describe the behavior of this solution as t tends to $\pm\infty$. The same question also arises for solutions of other differential equations (Bessel equations, for example).

3.2.1 *The Problem*

In this chapter we will describe a method that allows us to study the behavior as $\lambda \to +\infty$ of integrals of the form

$$I(\lambda) = \int_{\mathbf{R}^d} e^{i\lambda\varphi(x)}\, a(x)\, dx, \tag{3.2.2}$$

C. Zuily, *Selected Topics in Partial Differential Equations*, Universitext,
https://doi.org/10.1007/978-3-032-24082-8_3

where φ is a C^∞ function on $\mathbf{R}^d$ with real values, called the "phase", and $a \in C_0^\infty(\mathbf{R}^d, \mathbf{C})$ called the "symbol".

Setting $y = |t|^{\frac{1}{2}} x$, the solution of the Airy equation can be written as

$$|t|^{\frac{1}{2}} \int_{-\infty}^{+\infty} e^{i|t|^{\frac{3}{2}}(\pm x + \frac{x^3}{3})}\, dx.$$

This integral is essentially of the form (3.2.2) with $\lambda = |t|^{\frac{3}{2}}$, $\varphi(x) = \pm x + \frac{x^3}{3}$ and we will see how the general results apply to this case and allow us to describe the behavior of the solutions of (3.2.1) as t tends to $\pm\infty$.

To conclude, a point of terminology. The word "stationary" refers to the fact that the framework in which this method applies is that where the function φ has so-called "critical" or "stationary" points, that is, points where its differential vanishes. In the case where it does not have any, the study of the asymptotic behavior of the integral (3.2.2) will, as we shall see, be simpler.

3.2.2 *The Outline*

In the following section, we will present a classical result where the phase φ is assumed, first not to have any critical point then to have non-degenerate critical points. In the latter case, we will have a complete expansion of $I(\lambda)$, as λ tends to $+\infty$, in powers of λ with a remainder. This will allow us, as an example, to study the solutions of equation (3.2.1).

However, when the phase φ depends on additional parameters, which happens in certain cases, the previous result does not allow us to satisfactorily control the dependence, with respect to the additional parameters, of the constants appearing in the remainder. It is to this problem that the last section is devoted.

3.3 The Classical Framework

The asymptotics of $I(\lambda)$ will depend on the behavior of φ on the support of a.

1st case. If $\varphi'(x) \neq 0$ for all $x \in \operatorname{supp} a$, then, for all $N \in \mathbf{N}$, there exists $C_N > 0$ such that

$$|I(\lambda)| \leq C_N\, \lambda^{-N}, \text{ for all } \lambda \geq 1. \tag{3.3.1}$$

Indeed, consider, for $x \in \operatorname{supp} a$, the differential operator

$$L = \frac{1}{i} \sum_{j=1}^{d} \frac{1}{\|\varphi'(x)\|^2} \frac{\partial \varphi}{\partial x_j} \frac{\partial}{\partial x_j}, \tag{3.3.2}$$

where $\|\varphi'(x)\|^2 = \sum_{j=1}^d \left(\frac{\partial\varphi}{\partial x_j}\right)^2$. We then have $L\, e^{i\lambda\varphi} = \lambda\, e^{i\lambda\varphi}$ and thus,

$$\frac{1}{\lambda^N} L^N(e^{i\lambda\varphi}) = e^{i\lambda\varphi}\,, \text{ for all } N \in \mathbf{N}\,, \tag{3.3.3}$$

where $L^N = L \circ \cdots \circ L$ (N times).

Set ${}^tL\, u = -\frac{1}{i}\sum_{j=1}^d \frac{\partial}{\partial x_j}\left(\frac{1}{\|\varphi'(x)\|^2}\frac{\partial\varphi}{\partial x_j}\, u\right)$ the transpose of L. Since the function a belongs to $C_0^\infty(\mathbf{R}^d)$, successive integrations by parts show that, for $N \in \mathbf{N}$,

$$I(\lambda) = \int_{\mathbf{R}^d} \frac{1}{\lambda^N} L^N\,(e^{i\lambda\varphi})\, a(x)\, dx = \frac{1}{\lambda^N}\int_{\mathbf{R}^d} e^{i\lambda\varphi}({}^tL)^N\, a(x)\, dx\,,$$

hence,

$$|I(\lambda)| \le \frac{1}{\lambda^N}\int_{\mathbf{R}^d} |({}^tL)^N\, a(x)|\, dx = C_N\, \lambda^{-N}.$$

2nd case. There exists $x_0 \in \overset{\circ}{\widehat{\operatorname{supp}}\, a}$ (the interior of $\operatorname{supp} a$) such that $\varphi'(x_0) = 0$. Such a point is called a stationary (or critical) point of φ, which gives its name to the method. In this case, the discussion must be refined.

The next interesting case is when the second derivative at the critical point x_0, that is,

$$\varphi''(x_0) = \left(\frac{\partial^2\varphi}{\partial x_i \partial x_j}(x_0)\right)_{1\le i,j\le d}$$

is non-degenerate. Recall that a real symmetric matrix M is said to be "non-degenerate" if all its eigenvalues (which are real) are nonzero. The signature of such a matrix is then defined by

$$\operatorname{sgn} M = \text{number of positive eigenvalues - number of negative eigenvalues.}$$

We will make the following assumption in what follows:

$$\begin{aligned}&\text{there exists a unique } x_0 \in \overset{\circ}{\widehat{\operatorname{supp}}\, a} \text{ such that,}\\ &\qquad\varphi'(x_0) = 0 \text{ and } \varphi''(x_0) \text{ is non-degenerate}\,.\end{aligned} \tag{3.3.4}$$

Remark 3.3.1 Note that non-degenerate critical points are isolated. This follows from Morse's Lemma (see Lemma 3.3.3 below).

Under assumption (3.3.4), the behavior of $I(\lambda)$ is given by the following theorem.

Theorem 3.3.2 *For every $N \in \mathbf{N}$, there exist $a_0, \dots, a_N \in \mathbf{C}$ (depending on a and φ), a function R_N, and a constant $C_N > 0$ such that, for all $\lambda \geq 1$, we have*

$$I(\lambda) = e^{i\lambda\varphi(x_0)} \sum_{k=0}^{N} a_k \, \lambda^{-\frac{d}{2}-k} + R_N(\lambda), \quad |R_N(\lambda)| \leq C_N \, \lambda^{-\frac{d}{2}-N-1}. \tag{3.3.5}$$

Moreover, $a_0 = \frac{(2\pi)^{d/2}}{\sqrt{|\det \varphi''(x_0)|}} \, e^{i\frac{\pi}{4} sgn\, \varphi''(x_0)} \, a_0(x_0)$, *where det denotes the determinant and sgn the signature.*

Proof In what follows we shall denote by $\mathcal{F}$ the Fourier transform and by $\mathcal{F}^{-1}$ its inverse, which is given on $\mathcal{S}(\mathbf{R}^d)$ by $\mathcal{F}^{-1}\psi(x) = (2\pi)^{-d} \int_{\mathbf{R}^d} e^{i\xi\cdot x} \psi(\xi) \, d\xi$.

We begin by treating the special case where $x_0 = 0$ and

$$\varphi(x) = \frac{1}{2} \, (Bx, x) \, , \tag{3.3.6}$$

where B is a symmetric matrix in $GL(d, \mathbf{R})$. The general case will be reduced to this one. We then have $I(\lambda) = \int e^{i\frac{\lambda}{2}(Bx,x)} \, a(x) \, dx = \langle e^{i\frac{\lambda}{2}(Bx,x)}, a \rangle$, where we have considered $e^{i\frac{\lambda}{2}(Bx,x)}$ as an element of the space $\mathcal{S}'(\mathbf{R}^d)$. Since $\mathcal{F}^{-1}\mathcal{F} = \mathrm{Id}$, we can write $I(\lambda) = \langle \mathcal{F} \, e^{i\frac{\lambda}{2}(Bx,x)}, \mathcal{F}^{-1} \, a \rangle$. Now we have the formula (see for instance [3])

$$\mathcal{F} \, e^{i\frac{\lambda}{2}(Bx,x)} = \frac{(2\pi)^{d/2}\lambda^{-d/2}}{\sqrt{|\det B|}} \, e^{i\frac{\pi}{4}\mathrm{sgn}\, B} e^{-\frac{i}{2\lambda}(B^{-1}\xi,\xi)}. \tag{3.3.7}$$

It follows that

$$I(\lambda) = \frac{(2\pi)^{d/2}\lambda^{-d/2}}{\sqrt{|\det B|}} \, e^{i\frac{\pi}{4}\mathrm{sgn}\, B} \langle e^{-\frac{i}{2\lambda}(B^{-1}\xi,\xi)}, \mathcal{F}^{-1} \, a \rangle. \tag{3.3.8}$$

Now, the Taylor formula with integral remainder shows that, for $\theta \in \mathbf{R}$, we have

$$e^{i\theta} = \sum_{k=0}^{N} \frac{(i\theta)^k}{k!} + \frac{\theta^{N+1}}{N!} \int_0^1 (1-t)^N \, i^{N+1} \, e^{it\theta} \, dt.$$

Taking $\theta = -\frac{1}{2\lambda}(B^{-1}\xi, \xi) \in \mathbf{R}$, we deduce that

$$e^{-\frac{i}{2\lambda}(B^{-1}\xi,\xi)} = \sum_{k=0}^{N} \frac{1}{k!}\Big(- \frac{i}{2\lambda} \, (B^{-1}\xi, \xi)\Big)^k + \lambda^{-(N+1)} \, F_N(\lambda, \xi) \, , \tag{3.3.9}$$

with, $|F_N(\lambda, \xi)| \leq C_N \, |(B^{-1}\xi, \xi)|^{N+1}$.

In particular, the term corresponding to $k = 0$ is equal to 1.

Using (3.3.8) and (3.3.9), we can write

$$I(\lambda) = \frac{(2\pi)^{d/2}\lambda^{-d/2}}{\sqrt{|\det B|}}\, e^{i\frac{\pi}{4}\operatorname{sgn} B}\Big[\sum_{k=0}^{N}\frac{\lambda^{-k}}{k!}\int\Big(-\frac{i}{2}(B^{-1}\xi,\xi)\Big)^k \mathcal{F}^{-1}a(\xi)\,d\xi$$
$$+\lambda^{-(N+1)}\int F_N(\lambda,\xi)\,\mathcal{F}^{-1}a(\xi)\,d\xi\Big].$$

We thus obtain formula (3.3.5) with

$$a_k = \frac{(2\pi)^{d/2}}{\sqrt{|\det B|}}\, e^{i\frac{\pi}{4}\operatorname{sgn} B}\,\frac{(-i)^k}{2^k\,k!}\int (B^{-1}\xi,\xi)^k\,\mathcal{F}^{-1}a(\xi)\,d\xi,$$

$$R_N(\lambda) = \frac{(2\pi)^{d/2}}{\sqrt{|\det B|}}\,\lambda^{-\frac{d}{2}-N-1}\, e^{i\frac{\pi}{4}\operatorname{sgn} B}\int F_N(\lambda,\xi)\,\mathcal{F}^{-1}a(\xi)\,d\xi.$$

When $k = 0$, the quantity $\int \mathcal{F}^{-1}a(\xi)\,d\xi = [\mathcal{F}(\mathcal{F}^{-1}a)](0) = a(0)$ appears in a_0. To treat the general case, we use the following result.

Lemma 3.3.3 (Morse) *Let $\varphi \in C^2(\mathbf{R}^d)$ and $x_0 \in \mathbf{R}^d$. Assume that*

$$\varphi'(x_0) = 0, \quad \textit{and} \quad \det\Big(\frac{\partial^2\varphi}{\partial x_j\partial x_k}(x_0)\Big) \neq 0.$$

Then there exists a neighborhood V of x_0 and a diffeomorphism χ from V onto a neighborhood U of zero in $\mathbf{R}^d$, such that for all $y \in U$,

$$\varphi(\chi^{-1}(y)) = \varphi(x_0) + \frac{1}{2}(y_1^2 + \cdots + y_r^2 - y_{r+1}^2 - \cdots - y_d^2), \tag{3.3.10}$$

where sgn $\varphi''(x_0) = 2r - d$.

Let us admit this lemma for a moment and finish the proof of Theorem 3.3.2. Let $\theta \in C_0^\infty(V)$, $\theta = 1$ in a neighborhood of x_0, and write

$$I(\lambda) = \int e^{i\lambda\varphi(x)}\,\theta(x)\,a(x)\,dx + \int e^{i\lambda\varphi(x)}\,(1-\theta(x))\,a(x)\,dx = I_1(\lambda) + I_2(\lambda).$$

According to the assumption, on the support of $(1-\theta)\,a$, φ has no stationary point, i.e. $\varphi'(x) \neq 0$. We deduce from the first case that $|I_2(\lambda)| \le C_N\,\lambda^{-N}$, for all $N \in \mathbf{N}$ and $\lambda \ge 1$. This term will be included in the remainder.

In the integral I_1, set $c(x) = \theta(x)\,a(x)$ then $x = \chi^{-1}(y)$. Using Lemma 3.3.3, we can write

$$I_1(\lambda) = e^{i\lambda\varphi(x_0)}\int e^{i\frac{\lambda}{2}(By,y)}\,c(\chi^{-1}(y))\,|\det J(\chi^{-1})(y)|\,dy\,,$$

where B is the diagonal matrix (b_{ii}) with

$$b_{ii} = 1, \ \text{ if } \ 1 \le i \le r, \ \text{ and } \ b_{ii} = -1 \text{ if } \ r+1 \le i \le d.$$

Using (3.3.8) we deduce (3.3.5). Let us compute a_0. From the equality $|\det B| = 1$ we deduce that we have $a_0 = (2\pi)^{-\frac{d}{2}}\, e^{i\frac{\pi}{4}\operatorname{sgn} B}\, c(\chi^{-1}(0))|\det J(\chi^{-1})(0)|$. Next we have

$$c(\chi^{-1}(0)) = c(x_0) = \theta(x_0)\, a(x_0) = a(x_0),$$

and $\operatorname{sgn} B = \operatorname{sgn} \varphi''(x_0)$. Finally, by differentiating twice the equality (3.3.10) which can be written as $\varphi(\chi^{-1}(y)) = \varphi(x_0) + \frac{1}{2}\,(By, y)$, we obtain

$$\varphi''(\chi^{-1}(y))(\chi^{-1})'(y)(\chi^{-1})'(y) + \varphi'(\chi^{-1}(y))(\chi^{-1})''(y) = B.$$

If $y = 0$ we have $\chi^{-1}(0) = x_0$ and $\varphi'(x_0) = 0$, hence

$$\varphi''(x_0)(\chi^{-1})'(0)(\chi^{-1})'(0) = B,$$

thus $|\det \varphi''(x_0)||\det J(\chi^{-1})(0)|^2 = 1$, which gives the desired value of a_0. □

Proof of Lemma 3.3.3 We make some reductions. Set $\psi(x) = \varphi(x + x_0) - \varphi(x_0)$. We obtain $\psi(0) = d\psi(0) = 0$ and $\det \psi''(0) \neq 0$. Next, since $\psi''(0) = \varphi''(x_0)$ is real, symmetric and non-degenerate, there exists a matrix A such that ${}^t A\psi''(0)\, A = B = (b_{ii})$, with $b_{ii} = 1$ if $1 \le i \le r$, $b_{ii} = -1$ if $r+1 \le i \le d$. Set $\psi_1(x) = \psi(A\,x)$. Then $\psi_1(0) = 0$, $\psi_1'(0) = \psi'(0)A = 0$ and $\psi''(0) = {}^t A\psi''(0)\, A = B$ and it suffices to prove the lemma for ψ_1. According to Taylor's formula with integral remainder we can write

$$\psi_1(x) = \frac{1}{2}\,(x,\, Q(x)\, x)\,, \ \text{ where } \ Q(0) = \Big(\frac{\partial^2 \psi_1}{\partial x_j \partial x_k}\,(0)\Big) = B. \tag{3.3.11}$$

Suppose that we find, for x near zero, a matrix $A(x)$ in $M_d(\mathbf{R})$, with C^∞ coefficients such that $A(0) = \mathrm{Id}$ and, with B defined above,

$${}^t\!A(x)\, B\, A(x) = Q(x). \tag{3.3.12}$$

Set then $\chi(x) = A(x)\, x$. We deduce from (3.3.11), (3.3.12) that

$$\begin{aligned}\psi_1(x) &= \frac{1}{2}\,(x, {}^t A(x)\, B\, A(x)\, x) = \frac{1}{2}\,(A(x)\, x,\ B\, A(x)\, x)\\ &= \frac{1}{2}\,(\chi(x),\ B\,\chi(x)).\end{aligned}$$

We obtain, if χ is invertible near zero, $\psi_1(\chi^{-1}(y)) = \frac{1}{2}\,(y, By)$ which is exactly (3.3.10). It therefore suffices to solve (3.3.12).

Let us consider the map $F : M_d(\mathbf{R}) \to \mathrm{Sym}_d(\mathbf{R})$ (where the second space is that of real symmetric matrices), $F(A) = {}^tA\, B\, A$. Let us compute its differential at the point $A = \mathrm{Id}$. We have $F(\mathrm{Id} + \lambda A) - F(\mathrm{Id}) = \lambda dF(\mathrm{Id})\, A + O(\lambda^2)$. Now,

$$F(I + \lambda A) = (I + \lambda {}^tA)\, B(I + \lambda A) = B + \lambda(B\, A + {}^tAB) + O(\lambda^2),$$

so $dF(\mathrm{Id})\, A = B\, A + {}^tA\, B$. The map $dF(\mathrm{Id}) : M_d(\mathbf{R}) \to \mathrm{Sym}_d(\mathbf{R})$ is surjective. Indeed, let $C \in \mathrm{Sym}_d(\mathbf{R})$. Set $A = \frac{1}{2}\, B^{-1}\, C$, which makes sense since B is invertible. Then $B\, A = \frac{1}{2}\, C$, ${}^t(B\, A) = {}^tA\, B = \frac{1}{2}{}^tC = \frac{1}{2}\, C$, hence $B\, A + {}^tA\, B = C$. We deduce that F admits a right inverse G defined near $M_0 \in \mathrm{Sym}_d(\mathbf{R})$ such that $G(M_0) = \mathrm{Id}$. Since $F(G(M_0)) = M_0$ and $F(G(M_0)) = F(\mathrm{Id}) = B$, we have $M_0 = B = Q(0)$. Set then for x near zero, $A(x) = G(Q(x))$. We have

$$F(A(x)) = {}^tA(x)\, B\, A(x) = F(G(Q(x))) = Q(x),$$

and $A(0) = G(Q(0)) = G(B) = \mathrm{Id}$. This solves (3.3.12) and completes the proof of Lemma 3.3.3. □

3.3.1 Asymptotic Study of the Solutions of the Airy Equation

Let us set

$$u(t) = \int_{-\infty}^{+\infty} e^{i\varphi_t(y)}\, dy, \quad \varphi_t(y) = ty + \frac{y^3}{3}, \quad t \in \mathbf{R}. \tag{3.3.13}$$

Proposition 3.3.4 *For all $t \in \mathbf{R}$ the integral on the right-hand side of* (3.3.13) *is convergent and defines a C^∞ function on $\mathbf{R}$, which is a solution of the Airy equation, $u''(t) - tu(t) = 0$.*

Proof Let us fix $t_0 \in \mathbf{R}$. It suffices to show that for $a = (1 + |t_0|)^{\frac{1}{2}}$ the integrals $\int_a^{+\infty} e^{i\varphi_{t_0}(y)}\, dy$ and $\int_{-\infty}^{-a} e^{i\varphi_{t_0}(y)}\, dy$ are convergent. Since the reasoning is the same for both, it is enough to consider the first integral. Let us set, for $N > 0$, $I_N = \int_a^N e^{i\varphi_{t_0}(y)}\, dy$. We have

$$\frac{1}{i}\left(e^{i\varphi_{t_0} y)}\right)' = (t_0 + y^2)e^{i\varphi_{t_0} y)} \text{ and } \quad t_0 + y^2 \geq t_0 + 1 + |t_0| \geq 1 \text{ for } y \geq a.$$

We can therefore write

$$I_N = \frac{1}{i} \int_a^N \frac{1}{t_0 + y^2} \left(e^{i\varphi_{t_0}(y)}\right)' dy.$$

By integrating by parts we obtain

$$iI_N = \frac{e^{i\varphi_{t_0}(N)}}{t_0 + N^2} - \frac{e^{i\varphi_{t_0}(a)}}{t_0 + a^2} + \int_a^N \frac{2y}{(t_0 + y^2)^2} e^{i\varphi_{t_0}(y)}\, dy. \tag{3.3.14}$$

Since

$$\left|\frac{y}{(t_0 + y^2)^2} e^{i\varphi_{t_0}(y)}\right| \leq \frac{|y|}{(t_0 + y^2)^2} \in L^1(a, +\infty),$$

the right-hand side of (3.3.14) admits a limit as N tends to $+\infty$ by the dominated convergence theorem.

Next, it suffices to show that $u \in C^\infty(I)$ where $I = (-\ell^2, \ell^2)$ for all $\ell \geq 1$ and that the equation is satisfied on I. In the following, we fix $\ell \geq 1$. Let us set, for $\varepsilon > 0$,

$$u_\varepsilon(t) = \int_{\mathbf{R}} e^{-\varepsilon y^2} e^{i\varphi_t(y)}\, dy. \tag{3.3.15}$$

Lemma 3.3.5 *The function u_ε belongs to $C^\infty(\mathbf{R})$ and satisfies the following equation on* $\mathbf{R}$

$$u''_\varepsilon(t) - 2\varepsilon u'_\varepsilon(t) - tu_\varepsilon(t) = 0.$$

Proof The fact that u_ε belongs to $C^\infty(\mathbf{R})$ is a consequence of the theorem on differentiation under the integral sign for Lebesgue integrals, since for all $k \in \mathbf{N}$

$$\left|\left(\frac{d}{dt}\right)^k \left[e^{-\varepsilon y^2} e^{i\varphi_t(y)}\right]\right| = |y|^k e^{-\varepsilon y^2} \in L^1(\mathbf{R}).$$

Next,

$$\begin{aligned} u''_\varepsilon(t) - 2\varepsilon u'_\varepsilon(t) - tu_\varepsilon(t) &= -\int_{\mathbf{R}} (y^2 + 2i\varepsilon y + t) e^{-\varepsilon y^2} e^{i\varphi_t(y)}\, dy, \\ &= i \int_{\mathbf{R}} \frac{d}{dy}\left[e^{-\varepsilon y^2} e^{i\varphi_t(y)}\right] dy = 0. \end{aligned}$$

□

Lemma 3.3.6 *1. The sequence (u_ε) converges uniformly on I to the function u defined in* (3.3.13).

2. There exists $M > 0$ such that $\sup_{t \in I} |u'_\varepsilon(t)| \leq M$ for all $\varepsilon > 0$.

Proof Let $\theta \in C^\infty(\mathbf{R})$ such that $\theta(y) = 1$ if $|y| < 3\ell$, $\theta(y) = 0$ if $|y| \geq 4\ell$ and $0 \leq \theta \leq 1$. We write

$$\begin{aligned} u_\varepsilon(t) - u(t) &= \int_{\mathbf{R}} (e^{-\varepsilon y^2} - 1) e^{i\varphi_t(y)} \theta(y)\, dy + \int_{\mathbf{R}} (e^{-\varepsilon y^2} - 1) e^{i\varphi_t(y)} (1 - \theta(y))\, dy, \\ &=: A_\varepsilon(t) + B_\varepsilon(t). \end{aligned} \tag{3.3.16}$$

We have

$$\sup_{t\in I} |A_\varepsilon(t)| \le \int_{\mathbf{R}} |e^{-\varepsilon y^2} - 1|\theta(y)\, dy, \tag{3.3.17}$$

and the right-hand side converges to zero by the dominated convergence theorem since the integrand tends to zero for fixed y and $|e^{-\varepsilon y^2} - 1|\theta(y) \le 2\theta(y) \in L^1(\mathbf{R})$.

Using the fact that $e^{i\varphi_t(y)} = \frac{-i}{t+y^2}\left(e^{i\varphi_t(y)}\right)'$ and integrating by parts, we can write

$$B_\varepsilon(t) = i \int_{\mathbf{R}} \frac{d}{dy}\Big[\frac{1}{t+y^2}(e^{-\varepsilon y^2} - 1)(1-\theta(y))\Big] e^{i\varphi_t(y)}\, dy.$$

We then have

$$\begin{aligned}
i B_\varepsilon(t) &= B_\varepsilon^1(t) + B_\varepsilon^2(t) + B_\varepsilon^3(t),\\
B_\varepsilon^1(t) &= \int_{\mathbf{R}} \frac{2y}{(t+y^2)^2}(e^{-\varepsilon y^2} - 1)(1-\theta(y)) e^{i\varphi_t(y)}\, dy,\\
B_\varepsilon^2(t) &= \int_{\mathbf{R}} \frac{2\varepsilon y}{t+y^2} e^{-\varepsilon y^2}(1-\theta(y)) e^{i\varphi_t(y)}\, dy,\\
B_\varepsilon^3(t) &= \int_{\mathbf{R}} \frac{1}{t+y^2}(e^{-\varepsilon y^2} - 1)\theta'(y) e^{i\varphi_t(y)}\, dy.
\end{aligned}$$

Note that on the supports of $1-\theta$ and θ', we have

$$t + y^2 \ge \frac{1}{2}(y^2 + \ell^2) \quad \text{for } t \in I.$$

We can then write

$$\sup_{t\in I} |B_\varepsilon^1(t)| \le C \int_{\mathbf{R}} \frac{|y|}{(\ell^2+y^2)^2}(e^{-\varepsilon y^2} - 1)\, dy,$$

and the right-hand side tends to zero by the dominated convergence theorem.

Next, since $\sqrt{\varepsilon}|y|e^{-\varepsilon y^2} \le 1$ we can write

$$\sup_{t\in I} |B_\varepsilon^2(t)| \le 4\sqrt{\varepsilon} \int_{\mathbf{R}} \frac{1}{\ell^2+y^2}\, dy,$$

and the right-hand side tends to zero as $\varepsilon \to 0$.

Finally, on the support of θ' we have $3\ell \le |y| \le 4\ell$.

$$\sup_{t\in I} |B_\varepsilon^2(t)| \le 2 \int_{\mathbf{R}} \frac{1}{\ell^2+y^2}(e^{-\varepsilon y^2} - 1)|\theta'(y)|\, dy,$$

and the right-hand side tends to zero as ε goes to zero.

Therefore, $\sup_{t\in I}|B_\varepsilon(t)|$ tends to zero as ε goes to zero. Using (3.3.17) and (3.3.16) we deduce point 1. of Lemma 3.3.6.

Let us move to point 2. We can write

$$\begin{aligned} u'_\varepsilon(t) &= \int_{\mathbf{R}} \left(e^{-\varepsilon y^2}-1\right)e^{i\varphi_t(y)}y\theta(y)\,dy + \int_{\mathbf{R}} \left(e^{-\varepsilon y^2}-1\right)e^{i\varphi_t(y)}y(1-\theta(y))\,dy,\\ &=: J_\varepsilon(t)+K_\varepsilon(t). \end{aligned} \tag{3.3.18}$$

We use the following facts: $\operatorname{supp}\theta \subset \{y : |y|\le 4\ell\}$, $|e^{-\varepsilon y^2}-1|\le 2$ and $0\le\theta\le 1$. We obtain

$$\sup_{t\in I}|J_\varepsilon(t)| \le 8\ell \int_{|y|<4\ell} dy.$$

To handle the term K_ε we use the same method as for the previous term B_ε. We have

$$K_\varepsilon(t) = i\int_{\mathbf{R}} \frac{d}{dy}\Big[\frac{1}{t+y^2}(e^{-\varepsilon y^2}-1)y(1-\theta(y))\Big]e^{i\varphi_t(y)}\,dy.$$

We have

$$\begin{aligned} iK_\varepsilon(t) &= \sum_{j=1}^{4} K_\varepsilon^j(t),\\ K_\varepsilon^1(t) &= \int_{\mathbf{R}} \frac{2y^2}{(t+y^2)^2}(e^{-\varepsilon y^2}-1)(1-\theta(y))e^{i\varphi_t(y)}\,dy,\\ K_\varepsilon^2(t) &= \int_{\mathbf{R}} \frac{2\varepsilon y^2}{t+y^2}\, e^{-\varepsilon y^2}(1-\theta(y))e^{i\varphi_t(y)}\,dy,\\ K_\varepsilon^3(t) &= \int_{\mathbf{R}} \frac{y}{t+y^2}(e^{-\varepsilon y^2}-1)\theta'(y)e^{i\varphi_t(y)}\,dy,\\ K_\varepsilon^4(t) & \int_{\mathbf{R}} \frac{1}{t+y^2}(e^{-\varepsilon y^2}-1)(1-\theta(y))e^{i\varphi_t(y)}\,dy. \end{aligned}$$

On the supports of $1-\theta$ and θ', as well as for $t\in I$, we have $t+y^2\ge\frac{1}{2}(\ell^2+y^2)$. Thus,

$$\sup_{t\in I}|K_\varepsilon^1(t)| \le C\int_{\mathbf{R}} \frac{1}{\ell^2+y^2}\,dy.$$

To estimate $K_\varepsilon^2(t)$, we note that $\varepsilon y^2 e^{-\varepsilon y^2}\le 1$. Thus,

$$\sup_{t\in I}|K_\varepsilon^2(t)| \le C\int_{\mathbf{R}} \frac{1}{\ell^2+y^2}\,dy.$$

Next, since supp $\theta' \subset \{y : |y| \leq 4\ell\}$ we have

$$\sup_{t\in I} |K_\varepsilon^3(t)| \leq C \int_{|y|\leq 4\ell} \frac{|y|}{\ell^2 + y^2}\, dy.$$

Eventually,

$$\sup_{t\in I} |K_\varepsilon^4(t)| \leq C \int_{\mathbf{R}} \frac{1}{\ell^2 + y^2}\, dy.$$

We deduce from the above that $\sup_{t\in I} |K_\varepsilon(t)| \leq M$ where M is independent of ε. This, taking into account (3.3.18), proves point 2. of Lemma 3.3.6. □

End of the Proof of Proposition 3.3.4

It follows from Lemmas 3.3.5 and 3.3.6 that the sequence (u''_ε) converges uniformly on I to the function tu.

According to Lemma 3.3.6, the sequence $(u'_\varepsilon(0))$ is a bounded sequence in $\mathbf{C}$. Therefore, there exists a subsequence $(u'_{\varepsilon_k}(0))$ converging to $\alpha \in \mathbf{C}$. On the other hand we have

$$u'_{\varepsilon_k}(t) = u'_{\varepsilon_k}(0) + \int_0^t u''_{\varepsilon_k}(s)\, ds.$$

Next we have

$$\sup_{t\in I} \int_0^t |u''_{\varepsilon_k}(s) - su(s)|\, ds \leq \ell^2 \sup_{t\in I} |u''_{\varepsilon_k}(t) - tu(t)| \to 0.$$

We deduce that the sequence (u'_{ε_k}) converges uniformly on I to $\alpha + \int_0^t su(s)\, ds$. Since the sequence (u_{ε_k}) converges uniformly on I to u, this implies that we have $u \in C^1(I)$ and $u'(t) = \alpha + \int_0^t su(s)\, ds$. This shows that u' is C^1 and $u'' = tu$. Since $u \in C^2(I)$, we deduce from the equation that $u'' \in C^2(I)$, hence $u \in C^4(I)$, and so on. Therefore, $u \in C^\infty(I)$, which completes the proof of Proposition 3.3.4.

Let us now move on to the asymptotic study of the solution u described in (3.3.1). The main result is the following.

Theorem 3.3.7 *1. For $t \to +\infty$, $u(t)$ decays rapidly, that is,*

$$\forall N \in \mathbf{N} \quad \exists C_N > 0 : |u(t)| \leq C_N t^{-N}, \quad \forall t \geq 1. \tag{3.3.19}$$

2. For $t \to -\infty$, we have

$$u(t) = 2\sqrt{\pi}|t|^{-\frac{3}{4}} \cos\Big(\frac{2}{3}|t|^{\frac{3}{2}} - \frac{\pi}{4}\Big) + O(|t|^{-\frac{9}{4}}). \tag{3.3.20}$$

Proof 1. Set $y = \sqrt{t}\, x$ in the integral defining $u(t)$. We obtain

$$u(t) = \sqrt{t} \int_{-\infty}^{+\infty} e^{it^{\frac{3}{2}}(x+\frac{x^3}{3})}\, dx.$$

If we set $\varphi(x) = x + \frac{x^3}{3}$, we have $\varphi'(x) = 1 + x^2 \neq 0$. We then proceed as in the first case of Sect. 3.3, that is by integration by parts using the fact that

$$e^{it^{\frac{3}{2}}\varphi(x)} = \frac{1}{it^{\frac{3}{2}}(1+x^2)}\left(e^{it^{\frac{3}{2}}\varphi(x)}\right)'. \tag{3.3.21}$$

We write

$$u(t) = \sqrt{t} \int_{-\infty}^{+\infty} \frac{1}{it^{\frac{3}{2}}(1+x^2)}\left(e^{it^{\frac{3}{2}}\varphi(x)}\right)' dx,$$

and integrate by parts. We obtain

$$u(t) = \frac{2i}{t} \int_{-\infty}^{+\infty} \frac{x}{(1+x^2)^2} e^{it^{\frac{3}{2}}\varphi(x)}\, dx.$$

We repeat this process using (3.3.21). At each integration by parts, we gain a factor $\frac{1}{t^{\frac{3}{2}}}$. We finally obtain the estimate (3.3.19).

2. When t is negative, we set $y = \sqrt{-t}x$ in the integral defining $u(t)$. We obtain

$$u(t) = \sqrt{|t|} \int_{-\infty}^{+\infty} e^{i|t|^{\frac{3}{2}}(-x+\frac{x^3}{3})}\, dx.$$

Here, $\varphi(x) = -x + \frac{x^3}{3}$ so that $\varphi'(x) = x^2 - 1$. There are thus two critical points $x = 1$ and $x = -1$. They are non-degenerate since $\varphi''(\pm 1) = \pm 2$.

Let, for $j = 1, 2$, $\theta_j \in C_0^\infty(\mathbf{R})$, $0 \leq \theta_j \leq 1$ such that

$$\begin{aligned} \theta_1(x) &= 1, \text{ if } |x-1| \leq \frac{1}{4}, \quad \theta_1(x) = 0, \text{ if } |x-1| \geq \frac{1}{2}, \\ \theta_2(x) &= 1, \text{ if } |x+1| \leq \frac{1}{4}, \quad \theta_2(x) = 0, \text{ if } |x+1| \geq \frac{1}{2}. \end{aligned}$$

and write

$$\begin{aligned} \frac{1}{\sqrt{|t|}}u(t) &= \int_{-\infty}^{+\infty} e^{i|t|^{\frac{3}{2}}\varphi(x)}\theta_1(x)\, dx + \int_{-\infty}^{+\infty} e^{i|t|^{\frac{3}{2}}\varphi(x)}\theta_2(x)\, dx \\ &\quad + \int_{-\infty}^{+\infty} e^{i|t|^{\frac{3}{2}}\varphi(x)}(1 - \theta_1(x) - \theta_2(x))\, dx = F_1(t) + F_2(t) + F_3(t). \end{aligned} \tag{3.3.22}$$

On the support of θ_j for $j = 1, 2$ the phase φ has a single non-degenerate critical point. On the support of $1 - \theta_1 - \theta_2$ the phase has no critical point. We can therefore apply the results of Sect. 3.3.

According to the first case of this section (and also by the proof of case 1. of Theorem 3.3.7) we have

$$\forall N \in \mathbf{N} \quad \exists C_N > 0 : |F_3(t)| \leq C_N |t|^{-N}, \quad \forall |t| \geq 1. \tag{3.3.23}$$

To handle F_j, $j = 1, 2$, we use formula (3.3.5) from Theorem 3.3.2 with

$$d = 1, \; N = 0, \; \lambda = |t|^{\frac{3}{2}}, \; x_0 = \pm 1, \; \varphi(x_0) = \mp \frac{2}{3}, \; \varphi''(x_0) = \pm 2, \; \theta_j(x_0) = 1.$$

We obtain

$$F_1(t) = \sqrt{\pi} e^{-i(\frac{2}{3}|t|^{\frac{2}{3}} - \frac{\pi}{4})} |t|^{-\frac{3}{4}} + O|t|^{-\frac{9}{4}}, \qquad F_2(t) = \sqrt{\pi} e^{i(\frac{2}{3}|t|^{\frac{2}{3}} - \frac{\pi}{4})} |t|^{-\frac{3}{4}} + O|t|^{-\frac{9}{4}}.$$

Using (3.3.22), (3.3.23) and the above equalities, we obtain (3.3.20), which completes the proof of Theorem 3.3.7.

Remark 3.3.8 Although it appears quite simple, the Airy equation possesses a surprising property. Before describing it, let us recall, for example, that first-order differential equations or second-order equations with constant coefficients can be solved explicitly using elementary operations: use of standard functions, classical operations such as integration, etc. In contrast, the Airy equation belongs to the class of equations that cannot be solved by these operations. It is said not to be solvable by quadratures. This is the analogue, for differential equations, of the non-solvability by radicals of certain algebraic equations of degree at least five. In fact, there exists a theory called "differential Galois theory," analogous to Galois theory for algebraic equations, which allows one to prove this result.

For a very brief overview of this theory, the reader may consult the article (in French) by J. Roques "Tell me... about differential Galois theory" in La gazette des mathématiciens, Volume 52, April 2017.

3.4 A Precise Stationary Phase Estimate

The purpose of this section is to present a precise stationary phase estimate well suited to phases depending on parameters. In particular, we make no assumptions about the localization or behavior of the critical points, and we do not assume any lower or upper bound on the gradient of the phase. Moreover, the dependence of the constants on the derivatives of the phase and the symbol is explicit.

3.4.1 Preliminaries

- We consider

$$I(\lambda) = \int_{\mathbf{R}^d} e^{i\lambda\Phi(\xi)} b(\xi)\, d\xi,$$

where Φ is a complex phase, $\Phi = \Phi_1 + i\Phi_2 \in C^\infty$ where the Φ_j are real-valued and $b \in C_0^\infty(\mathbf{R}^d)$.

- Let $K = \operatorname{supp} b$ and $\varepsilon_0 \in (0, 1]$ a small constant. We set

$$K_{\varepsilon_0} = \{\xi \in \mathbf{R}^d : \operatorname{dist}(\xi, K) \le \varepsilon_0\}, \tag{3.4.1}$$

where "dist" denotes the sup distance on $\mathbf{R}^d$.

- We assume that Φ is regular on a neighborhood of K_{ε_0} and we set

$$\mathcal{M}_k := \sum_{2\le|\alpha|\le k} \sup_{\xi\in K_{\varepsilon_0}} |D_\xi^\alpha \Phi(\xi)|, \quad \mathcal{N}_l := \sum_{|\alpha|\le l} \sup_{\xi\in K} |D_\xi^\alpha b(\xi)|$$

for $k \ge 2$ and for $l \ge 0$.

- Let $\operatorname{Hess}\Phi_1 = \left(\frac{\partial^2\Phi_1}{\partial\xi_i\partial\xi_j}\right)_{1\le i,j\le d}$ be the Hessian matrix of Φ_1. We set

$$a_0 = \inf_{\xi\in K_{\varepsilon_0}} |\det \operatorname{Hess}\Phi_1(\xi)|. \tag{3.4.2}$$

- We now fix some constants. First, since $\operatorname{Hess}\Phi_1$ is a real symmetric matrix, its eigenvalues $(\lambda_j)_{j=1,\dots,d}$ are real and satisfy

$$|\lambda_j| \le C_1\mathcal{M}_2,$$

where C_1 depends only on the dimension d. Since the determinant is the product of the eigenvalues we deduce from (3.4.2) that

$$|\lambda_j| \ge \frac{a_0}{(C_1\mathcal{M}_2)^{d-1}}. \tag{3.4.3}$$

This implies that

$$|\operatorname{Hess}\Phi_1(\xi)X| \ge \frac{a_0}{(C_1\mathcal{M}_2)^{d-1}}|X|, \quad \forall X \in \mathbf{R}^d. \tag{3.4.4}$$

- Now, if B is an open ball contained in the interior of K_{ε_0}, there exists $C_2 > 0$, depending only on the dimension d, such that for all $\xi, \eta \in B$,

$$\begin{aligned} \nabla\Phi_1(\xi) - \nabla\Phi_1(\eta) &= \operatorname{Hess}\Phi_1(\eta)(\xi-\eta) + R(\xi,\eta), \\ |R(\xi,\eta)| &\le C_2\mathcal{M}_3|\xi-\eta|^2. \end{aligned} \tag{3.4.5}$$

- We set, when $\mathcal{M}_2 > 0$,

$$\delta_{\varepsilon_0} = \frac{a_0}{4(C_1\mathcal{M}_2)^{d-1}C_2\mathcal{M}_3}, \quad \delta = \min(\delta_{\varepsilon_0}, \frac{\varepsilon_0}{4}). \tag{3.4.6}$$

3.4.2 The Main Result

Theorem 3.4.1 *Suppose*

$$\begin{aligned} &(i) \quad \mathcal{M}_{d+2} < +\infty, \quad \mathcal{N}_{d+1} < +\infty, \\ &(ii) \quad a_0 > 0, \\ &(iii) \quad \Phi_2 \geq 0 \text{ on } K. \end{aligned} \tag{3.4.7}$$

Then there exists $C > 0$ depending only on d such that, for all $\lambda \geq 1$,

$$|I(\lambda)| \leq \frac{C|K_{\varepsilon_0}|}{a_0\delta^d}\left(1 + \mathcal{M}_{d+2}^{\frac{d}{2}}\right)\mathcal{N}_{d+1}\lambda^{-\frac{d}{2}}, \tag{3.4.8}$$

where $|K_{\varepsilon_0}|$ is the Lebesgue measure of the set K_{ε_0}.

Remark 3.4.2 1. Note that no upper or lower bound on $\nabla_\xi \Phi$ is required. This is particularly important in the case where the phase Φ depends on parameters. For example, in some cases the phase Φ is of the form

$$\Phi(x, y, \xi) = (x - y) \cdot \xi + \phi(x, y, \xi)$$

where x, y are in $\mathbf{R}^d$. In this case, $\nabla_\xi \Phi = x - y + \nabla_\xi \phi$ and there is no natural upper or lower bound for this quantity.

2. Here is another example. Suppose that $\Phi(x, y, \xi) = (x - y) \cdot \xi + t\theta(x, y, \xi)$ where $t \in (0, T)$ and x, y are in $\mathbf{R}^d$. Suppose that Φ and b satisfy the conditions (i), (ii) and that $|\det \text{Hess}\, \theta| \geq c > 0$ where c depends only on d. Then, setting $X = \frac{x}{t}$, $Y = \frac{y}{t}$, we write $i\lambda\Phi = i\lambda t\big((X - Y) \cdot \xi + \theta(t, tX, tY, \xi)\big)$ and we can apply Theorem 3.4.1 with $a_0 = c$ and λ replaced by λt. We obtain an estimate of $I(\lambda)$ by $t^{-\frac{d}{2}}\lambda^{-\frac{d}{2}}$ as soon as $t \geq \lambda^{-1}$.

3. When a_0 is very small (that is, $\delta = \frac{a_0}{4(C_1\mathcal{M}_2)^{d-1}C_2\mathcal{M}_3}$) then $\frac{1}{\delta^d}$ is proportional to $\frac{1}{a_0^d}$. The term $\frac{1}{a_0^d}$ in (3.4.8) comes from the possible existence of a_0^{-d} critical points of the phase on the support of b.

In the case where Φ has only one non-degenerate critical point, this term could be replaced by $\frac{1}{a_0}$. In this direction, we have the following result.

Theorem 3.4.3 *Suppose that* Φ *and* b *satisfy the hypotheses* (3.4.7) *and that the mapping*

$$K_{\varepsilon_0} \to \mathbf{R}^d : \quad \xi \mapsto \nabla_\xi \Phi_1(\xi) \tag{3.4.9}$$

is injective.

There exists a constant $C > 0$ *depending only on* d *such that*

$$|I(\lambda)| \le \frac{C}{a_0}\big(1 + \mathcal{M}_{d+2}^{\frac{d}{2}}\big)\mathcal{N}_{d+1}\lambda^{-\frac{d}{2}}. \tag{3.4.10}$$

Here are two examples of application of Theorem 3.4.3.

Example 3.4.4 1. Suppose in addition to (3.4.7) that K_{ε_0} is convex and that

$$\langle \text{Hess}\, \Phi_1(\xi)X, X\rangle > 0, \quad \forall \xi \in K_{\varepsilon_0}, \quad \forall X \in \mathbf{R}^d.$$

Then (3.4.9) is satisfied.

Indeed, since the symmetric matrix Hess Φ_1 is non-negative, its eigenvalues are non-negative. It follows from (3.4.3) that

$$\langle \text{Hess}\, \Phi_1(\xi)X, X\rangle \ge \frac{a_0}{(C_1\mathcal{M}_2)^{d-1}}|X|^2, \quad \forall \xi \in K_{\varepsilon_0}, \quad \forall X \in \mathbf{R}^d. \tag{3.4.11}$$

For $\xi, \eta \in K_{\varepsilon_0}$ we write

$$\begin{aligned}
\nabla_\xi \Phi_1(\xi) - \nabla_\xi \Phi_1(\eta) &= \int_0^1 \frac{d}{ds}\big[\nabla_\xi \Phi_1(s\xi + (1-s)\eta)\big]\, ds,\\
&= \int_0^1 \text{Hess}\, \Phi_1(s\xi + (1-s)\eta)\cdot(\xi - \eta)\, ds.
\end{aligned}$$

It follows from (3.4.11) that

$$\langle \nabla_\xi \Phi_1(\xi) - \nabla_\xi \Phi_1(\eta), \xi - \eta\rangle \ge \frac{a_0}{(C_d\mathcal{M}_2)^{d-1}}|\xi - \eta|^2,$$

from which we deduce that

$$|\nabla_\xi \Phi_1(\xi) - \nabla_\xi \Phi_1(\eta)| \ge \frac{a_0}{(C_d\mathcal{M}_2)^{d-1}}|\xi - \eta|,$$

which completes the proof.

2. Let A be a $d \times d$ real, symmetric, non-singular matrix, Ψ a regular phase which satisfies $\mathcal{M}_{d+2}(\Psi) < +\infty$. Set $\Phi(\xi) = \frac{1}{2}\langle A\xi, \xi\rangle + \varepsilon\Psi(\xi)$. Then, if ε is small enough, the hypotheses of Theorem 3.4.3 are satisfied.

Remark 3.4.5 Note that the estimates (3.4.8), (3.4.10) do not seem optimal with respect to the power of a_0 (for small a_0) since, according to the classical method, one might expect $\frac{1}{\sqrt{a_0}}$ on the right-hand side. However, such an estimate is probably false as shown by the following argument. Indeed, suppose that we have an estimate of the type,

$$\left| \int e^{i\lambda\Phi_1(\xi)} b(\xi)\, d\xi \right| \leq C \mathcal{M}_N(\Phi_1) \mathcal{N}_M(b) \frac{1}{\sqrt{a_0}} \lambda^{-\frac{d}{2}},$$

for certain $M, N \in \mathbf{N}$, where $a_0 = \inf_{K_{\varepsilon_0}} |\det \text{Hess } \Phi_1|$.

Let us apply this inequality with λ replaced by $t\lambda$ and Φ by $\frac{1}{t}\Phi$ where $t > 0$ is to be chosen. Then $\mathcal{M}_N(\Phi_1)$ is replaced by $\frac{1}{t}\mathcal{M}_N(\Phi_1)$, a_0 is replaced by $t^{-d}a_0$ so that we would have the inequality

$$\left| \int e^{i\lambda\Phi_1(\xi)} b(\xi)\, d\xi \right| \leq \frac{C}{t} \mathcal{M}_N(\Phi_1) \mathcal{N}_M(b)\, \frac{t^{\frac{d}{2}}}{\sqrt{a_0}} t^{-\frac{d}{2}} \lambda^{-\frac{d}{2}}.$$

Taking $t = \mathcal{M}_N(\Phi_1)$ we would have the inequality

$$\left| \int e^{i\lambda\Phi_1(\xi)} b(\xi)\, d\xi \right| \leq C \mathcal{N}_M(b) \frac{1}{\sqrt{a_0}} \lambda^{-\frac{d}{2}}.$$

The right-hand side of this estimate depends on Φ_1 only through a_0, with no dependence on higher-order derivatives of Φ_1. This is highly unlikely.

Let us return to the proof of the main results. In fact, it suffices to prove the following (apparently weaker) result.

Theorem 3.4.6 *1. Under the hypotheses of Theorem 3.4.1, there exists a non-decreasing function $\mathcal{F} : \mathbf{R}^+ \to \mathbf{R}^+$ such that for all $\lambda \geq 1$*

$$|I(\lambda)| \leq |K_{\varepsilon_0}| \mathcal{F}(\mathcal{M}_{d+2}) \mathcal{N}_{d+1} \frac{1}{a_0 \delta^d} \lambda^{-\frac{d}{2}}. \tag{3.4.12}$$

2. Under the hypotheses of Theorem 3.4.3, there exists a non-decreasing function $\mathcal{F} : \mathbf{R}^+ \to \mathbf{R}^+$ such that for all $\lambda \geq 1$

$$|I(\lambda)| \leq \mathcal{F}(\mathcal{M}_{d+2}) \mathcal{N}_{d+1} \frac{1}{a_0} \lambda^{-\frac{d}{2}}. \tag{3.4.13}$$

Proof of Theorems 3.4.1 *and* 3.4.3 *assuming Theorem* 3.4.6
Suppose that the estimate (3.4.12) is proved. Our goal is to deduce (3.4.8) with $C = \mathcal{F}(1)$.

Set $t = 1 + \mathcal{M}_{d+2}$ and suppose $\lambda \geq 1$. Since $t\lambda \geq 1$, we can apply (3.4.12) with (λ, Φ) replaced by $(t\lambda, \Phi(\xi)/t)$. Then a_0 is changed to $t^{-d}a_0$ and $\mathcal{M}_k(\frac{\Phi}{t}) = \frac{1}{t}\mathcal{M}_k(\Phi)$. Consequently, δ remains unchanged. It follows that we can write

$$\begin{aligned}\left|\int_{\mathbf{R}^d} e^{i\lambda\Phi(\xi)} b(\xi)\, d\xi\right| &= \left|\int_{\mathbf{R}^d} e^{it\lambda\frac{\Phi(\xi)}{t}} b(\xi)\, d\xi\right|,\\ &\leq |K_{\varepsilon_0}|\mathcal{F}\left(\mathcal{M}_{d+2}\left(\frac{\Phi}{t}\right)\right)\mathcal{N}_{d+1}\frac{1}{t^{-d}a_0\delta^d}(t\lambda)^{-\frac{d}{2}},\\ &\leq |K_{\varepsilon_0}|\mathcal{F}\left(\frac{\mathcal{M}_{d+2}(\Phi)}{t}\right)\mathcal{N}_{d+1}\frac{1}{a_0\delta^d}\lambda^{-\frac{d}{2}}t^{\frac{d}{2}},\end{aligned}$$

which gives the desired estimate. The case 2. is similar. □

It remains to prove Theorem 3.4.6.

3.4.3 Proof of Theorem 3.4.6

Proof of 1.

In what follows, C_d will denote a positive constant depending only on d, and $\mathcal{F}$ a non-decreasing function from $\mathbf{R}^+$ to $\mathbf{R}^+$, both of which may change from line to line.

Point 1. First, we may assume that

$$\lambda^{\frac{1}{2}}\delta \geq 1. \tag{3.4.14}$$

Indeed, suppose $\lambda^{\frac{1}{2}}\delta \leq 1$; then, by (3.4.7), (ii), we have $a_0 \leq C\mathcal{M}_2^d$, so that

$$|I(\lambda)| \leq \|b\|_{L^1(\mathbf{R}^d)} \leq \|b\|_{L^1(\mathbf{R}^d)}\frac{a_0}{a_0}\frac{1}{(\lambda^{\frac{1}{2}}\delta)^d} \leq C|K|\mathcal{N}_0\mathcal{M}_2^d\frac{1}{a_0\delta^d}\lambda^{-\frac{d}{2}}.$$

Point 2. We localize the problem in small balls. Let δ be defined in (3.4.6). We can write

$$\mathbf{R}^d = \cup_{j\in\mathbf{N}} B(\xi_j^*, \delta)$$

where the $B(\xi_j^*, \delta)$ are cubes of size δ such that there exists k_0, depending only on the dimension d, such that every point of $\mathbf{R}^d$ belongs to at most k_0 cubes. This implies in particular that for any finite subset $J \subset \mathbf{N}$ we have

$$\sum_{j\in J}|B(\xi_j^*, \delta)| \leq k_0\left|\bigcup_{j\in J} B(\xi_j^*, \delta)\right|, \tag{3.4.15}$$

where $|A|$ denotes the Lebesgue measure of the set A.

Let $(\chi_j)_{j\in\mathbf{N}}$ be a partition of unity associated to this covering. Note that χ_j can be taken of the form $\chi_0(\frac{\xi-\xi_j^*}{\delta})$ so that

$$|\partial_\xi^\alpha \chi_j(\xi)| \leq C_\alpha \delta^{-|\alpha|}. \tag{3.4.16}$$

Let

$$\Lambda = \{j \in \mathbf{N} : B(\xi_j^*, \delta) \cap K \neq \emptyset\}. \tag{3.4.17}$$

This is a finite set whose cardinality we denote by $|\Lambda|$. Then K is contained in $\cup_{j\in\Lambda} B(\xi_j^*, \delta)$.

Next, by definition, if $\xi \in B(\xi_j^*, \delta)$ where $j \in \Lambda$, we can find $\eta \in B(\xi_j^*, \delta) \cap K$. Then by (3.4.6) we have

$$\text{dist}(\xi, K) \leq |\xi - \eta| \leq 2\delta \leq \frac{\varepsilon_0}{2}.$$

It follows that $\cup_{j\in\Lambda} B(\xi_j^*, \delta)$ is contained in the interior of K_{ε_0} defined in (3.4.1). Using (3.4.15) we obtain $\sum_{j\in\Lambda} |B(\xi_j^*, \delta)| \leq k_0 |K_{\varepsilon_0}|$ which implies that

$$|\Lambda| \leq k_0 |K_{\varepsilon_0}| \, \delta^{-d}. \tag{3.4.18}$$

Lemma 3.4.7 *On each ball $B(\xi_j^*, \delta)$ the map $\xi \mapsto \nabla_\xi \Phi_1(\xi)$ is injective.*

Proof Indeed, as we have seen previously, these balls are contained in the interior of K_{ε_0} where the hypotheses (3.4.7) are satisfied. By (3.4.5) and (3.4.4), if ξ, η belong to $B(\xi_j^*, \delta)$ we can write

$$|\nabla_\xi \Phi_1(\xi) - \nabla_\xi \Phi_1(\eta)| \geq \Big(\frac{a_0}{(C_1 \mathcal{M}_2)^{d-1}} - 2\delta C_2 \mathcal{M}_3\Big)|\xi - \eta| \geq \frac{a_0}{2(C_1 \mathcal{M}_2)^{d-1}} |\xi - \eta|.$$

Setting $b_j = \chi_j b$ we can write

$$I(\lambda) = \sum_{j\in\Lambda} I_j(\lambda), \quad I_j(\lambda) = \int_{\mathbf{R}^d} e^{i\lambda\Phi(\xi)} b_j(\xi)\, d\xi. \tag{3.4.19}$$

Now we fix $j \in \Lambda$ and estimate the integral $I_j(\lambda)$.

Lemma 3.4.8 *Let $i \in \{1, \ldots, d\}$ and $A_i = \frac{\partial_i \overline{\Phi}}{|\nabla_\xi \Phi|^2}$. We can find $\mathcal{F} : \mathbf{R}^+ \to \mathbf{R}^+$ increasing such that*

$$|D_\xi^\alpha A_i(\xi)| \leq \mathcal{F}(\mathcal{M}_{1+|\alpha|}) \sum_{k=2}^{1+|\alpha|} \frac{1}{|\nabla_\xi \Phi(\xi)|^k}, \quad |\alpha| \geq 1. \tag{3.4.20}$$

Proof We proceed by induction on $|\alpha|$. A simple calculation shows that (3.4.20) is true for $|\alpha| = 1$. Suppose it is true for $|\alpha| \leq l$ and let $|\gamma| = l + 1 \geq 2$. Differentiating $|\gamma|$ times the equality $|\nabla\Phi|^2 A_i = \partial_i \overline{\Phi}$ we obtain

$$
\begin{aligned}
&|\nabla_\xi \Phi|^2 D_\xi^\gamma A_i = (1) - (2) - (3), \quad \text{where,} \\
&\qquad (1) = D_\xi^\gamma \partial_i \overline{\Phi}, \quad (2) = \sum_{|\beta|=1} \binom{\gamma}{\beta} (D_\xi^\beta |\nabla_\xi \Phi|^2) D^{\gamma-\beta} A_i, \\
&\qquad (3) = \sum_{2 \leq |\beta| \leq |\gamma|} \binom{\gamma}{\beta} (D_\xi^\beta |\nabla_\xi \Phi|^2) D^{\gamma-\beta} A_i.
\end{aligned}
\tag{3.4.21}
$$

We have $|(1)| \leq \mathcal{M}_{l+2}$. By induction we have

$$
|(2)| \leq C \mathcal{M}_2 |\nabla_\xi \Phi| \mathcal{F}(M_{l+1}) \sum_{k=2}^{l+1} \frac{1}{|\nabla_\xi \Phi|^k}.
$$

Let us estimate (3). If $|\beta| \geq 2$ we have $|\gamma| - |\beta| \leq l - 1$. Since we have the inequality, $|D_\xi^\beta |\nabla_\xi \Phi|^2| \leq C_2 \mathcal{M}_{|\beta|+1} |\nabla_\xi \Phi| + \mathcal{F}(\mathcal{M}_{|\beta|})$ the induction shows that

$$
|(3)| \leq C \mathcal{F}(\mathcal{M}_{l+2})(1 + |\nabla_\xi \Phi|) \sum_{k=2}^{l} \frac{1}{|\nabla \Phi|^k}.
$$

Dividing both sides of the first equation in (3.4.21) by $|\nabla_\xi \Phi|^2$ we finally obtain

$$
|D_\xi^\gamma A_i| \leq \mathcal{F}(\mathcal{M}_{l+2}) \sum_{k=2}^{l+2} \frac{1}{|\nabla_\xi \Phi|^k}, \quad |\gamma| = l + 1.
$$

This completes the proof of (3.4.20). □

Lemma 3.4.9 *Let $\mathcal{L} = A \cdot \nabla_\xi$ where $A_i = \frac{\partial_i \overline{\Phi}}{|\nabla_\xi \Phi|^2}$. For all $N \in \mathbf{N}$ we have*

$$
\begin{aligned}
&({}^t\mathcal{L})^N = \sum_{|\alpha| \leq N} c_{\alpha,N} \partial^\alpha, \quad \textit{with} \\
&|\partial_\xi^\beta c_{\alpha,N}| \leq \mathcal{F}(\mathcal{M}_{N-|\alpha|+|\beta|+1}) \sum_{k=N}^{2N-|\alpha|+|\beta|} \frac{1}{|\nabla_\xi \Phi_1|^k}.
\end{aligned}
\tag{3.4.22}
$$

(Here we set $\mathcal{M}_1 = 1$ which occurs when $\beta = 0$, $|\alpha| = N$.)

Proof We again proceed by induction on N. We will show the estimate in (3.4.22) with Φ instead of Φ_1 on the right-hand side, then use the estimate $|\nabla_\xi \Phi| \geq |\nabla_\xi \Phi_1|$. For $N = 1$ we have $c_{\alpha,N} = A_i$ if $|\alpha| = 1$ and $c_{\alpha,N} = \operatorname{div} A$ if $|\alpha| = 0$. Then (3.4.22) follows immediately from (3.4.20). Suppose that (3.4.22) is true at order N and prove it for $N + 1$. We write

$$
\begin{aligned}
({}^t\mathcal{L})^{N+1} &= {}^t\mathcal{L}({}^t\mathcal{L})^N = -(\nabla_\xi \cdot A)({}^t\mathcal{L})^N = -\sum_{|\alpha|\le N}\sum_{i=1}^{d} \partial_i (A_i c_{\alpha,N}\partial^\alpha),\\
&= -\sum_{|\alpha|\le N} (\operatorname{div} A) c_{\alpha,N}\partial^\alpha - \sum_{|\alpha|\le N} A\cdot\nabla_\xi c_{\alpha,N}\partial^\alpha - \sum_{|\alpha|\le N}\sum_{i=1}^{d} A_i c_{\alpha,N}\partial_i\partial^\alpha,\\
&= \sum_{|\gamma|\le N+1} c_{\gamma,N+1}\partial^\gamma,
\end{aligned}
$$

where

$$
\begin{aligned}
c_{0,N+1} &= -(\operatorname{div} A)c_{0,N} - A\cdot\nabla_\xi c_{0,N}, \quad \text{if } \ |\gamma| = 0,\\
c_{\gamma,N+1} &= -(\operatorname{div} A)c_{\gamma,N} - A\cdot\nabla_\xi c_{\gamma,N} - A_i c_{\alpha,N} \quad |\alpha| = |\gamma| - 1, \text{ if } 1 \le |\gamma| \le N,\\
c_{\gamma,N+1} &= -A_i c_{\alpha,N}, \quad \text{if } \partial^\gamma = \partial_i\partial^\alpha, |\alpha| = N, \text{ if } |\gamma| = N+1.
\end{aligned}
$$

We estimate each coefficient. First, $\partial^\beta c_{0,N+1}$ is a finite sum of terms of the form $(\partial^{\beta_1}\partial_i A_i)(\partial^{\beta_2}c_{0,N})$ and $(\partial^{\beta_1}A_i)(\partial^{\beta_2}\partial_i c_{0,N})$ with $\beta = \beta_1 + \beta_2$. Using (3.4.20) and the induction, the first term is bounded by

$$
\mathcal{F}(M_{|\beta_1|+2}) \sum_{k=2}^{|\beta_1|+2} |\nabla_\xi\Phi|^{-k} \mathcal{F}(M_{N+|\beta_2|+1}) \sum_{l=N}^{2N+|\beta_2|} |\nabla_\xi\Phi|^{-l}.
$$

For the second term, if $\beta_1 = 0$, $\beta_2 = \beta$ it is bounded by

$$
\frac{1}{|\nabla_\xi\Phi|}\mathcal{F}(M_{N+|\beta|+2}) \sum_{l=N}^{2N+|\beta|+1} |\nabla_\xi\Phi|^{-l} \le \mathcal{F}(M_{N+1+|\beta|+1}) \sum_{l=N+1}^{2(N+1)+|\beta|} |\nabla_\xi\Phi|^{-l}.
$$

If $\beta_1 \neq 0$ it is bounded by

$$
\mathcal{F}(M_{|\beta_1|+1}) \sum_{k=2}^{|\beta_1|+1} |\nabla_\xi\Phi|^{-k} \mathcal{F}(M_{N+|\beta_2|+1}) \sum_{l=N}^{2N+|\beta_2|+1} |\nabla_\xi\Phi|^{-l}.
$$

Since $N+2 \le k+l \le 2N+2+|\beta_1|+|\beta_2| = 2(N+1)+|\beta|$ we see that $\partial^\beta c_{0,N+1}$ satisfies the estimate in (3.4.22) with N replaced by $N+1$.

Let us examine the term $\partial^\beta c_{\gamma,N+1}$ with $|\gamma| = N+1$. This term is also a finite sum of terms of the form $(\partial^{\beta_1}A_i)(\partial^{\beta_2}c_{\alpha,N})$, $|\alpha| = |\gamma| - 1$. As before, if $\beta_1 = 0$, using (3.4.20) and the induction, it is bounded by

$$\frac{1}{|\nabla\Phi|}\mathcal{F}(\mathcal{M}_{N-|\gamma|+1+|\beta|+1})\sum_{l=N}^{2N-|\gamma|+1+|\beta|}|\nabla_\xi\Phi|^{-l}$$
$$\leq \mathcal{F}(\mathcal{M}_{N+1-|\gamma|+|\beta|+1})\sum_{l=N+1}^{2(N+1)-|\gamma|+|\beta|}|\nabla_\xi\Phi|^{-l}.$$

If $\beta_1 \neq 0$ it is bounded by

$$\mathcal{F}(\mathcal{M}_{|\beta_1|+1})\sum_{k=2}^{|\beta_1|+1}|\nabla_\xi\Phi|^{-k}\mathcal{F}(\mathcal{M}_{N-|\gamma|+1+|\beta_2|})\sum_{l=N}^{2N-|\gamma|+1+|\beta_2}|\nabla_\xi\Phi|^{-l}.$$

Since $N+2 \leq k+l \leq 2N+2-|\gamma|+|\beta|$, we see that $\partial^\beta c_{\gamma,N+1}$ also satisfies the estimate of (3.4.22). The estimates for the other terms are similar and left to the reader. □

Let $\psi \in C_0^\infty(\mathbf{R}^d)$ such that $\psi(x) = 1$ if $|x| \leq 1$, $\psi(x) = 0$ if $|x| \geq 2$. With the notation of (3.4.19), with j fixed, we write

$$\begin{aligned} I_j(\lambda) &= \int e^{i\lambda\Phi(\xi)}\psi\big(\lambda^{\frac{1}{2}}\nabla_\xi\Phi_1(\xi)\big)\chi_j(\xi)b(\xi)\,d\xi \\ &+ \int e^{i\lambda\Phi(\xi)}(1-\psi\big(\lambda^{\frac{1}{2}}\nabla_\xi\Phi_1(\xi)\big))\chi_j(\xi)b(\xi)\,d\xi =: K_j(\lambda)+L_j(\lambda). \end{aligned} \tag{3.4.23}$$

Let us estimate K_j. We will use (see Lemma 3.4.7) the fact that on the support of χ_j the map $\xi \mapsto \nabla_\xi\Phi_1(\xi)$ is injective. We write

$$|K_j(\lambda)| \leq \int |\psi\big(\lambda^{\frac{1}{2}}\nabla_\xi\Phi_1(\xi)\big)\chi_j(\xi)b(\xi)|\,d\xi.$$

Setting $\eta = \lambda^{\frac{1}{2}}\nabla_\xi\Phi_1(\xi)$, we have $d\eta = \lambda^{\frac{d}{2}}|\det\mathrm{Hess}\,\Phi_1(\xi)|\,d\xi$. Using (3.4.7) (ii) and its notations, we obtain

$$|K_j(\lambda)| \leq \frac{C_d}{a_0}\mathcal{N}_0\lambda^{-\frac{d}{2}}. \tag{3.4.24}$$

To estimate L_j we introduce the vector field $X = \frac{1}{i\lambda}\frac{\nabla_\xi\overline{\Phi}}{|\nabla_\xi\Phi|^2}\cdot\nabla_\xi$ which satisfies

$$Xe^{i\lambda\Phi} = e^{i\lambda\Phi}.$$

Now, with $N \geq 1$ to be chosen, we write

$$L_j(\lambda) = \int e^{i\lambda\Phi(\xi)}({}^tX)^N\Big\{(1-\psi\big(\lambda^{\frac{1}{2}}\nabla_\xi\Phi_1(\xi)\big))\chi_j(\xi)b(\xi)\Big\}\,d\xi.$$

Since $X = \frac{1}{i\lambda}\mathcal{L}$ we can use (3.4.22) and obtain

$$|L_j(\lambda)| \le C_N \sum_{\substack{\alpha=\alpha_1+\alpha_2+\alpha_3\\ |\alpha|\le N}} S_{\alpha,N}, \quad \text{where,}$$

$$S_{\alpha,N} = \lambda^{-N}\int |c_{\alpha,N}| \left|\partial_\xi^{\alpha_1}\left[1-\psi\left(\lambda^{\frac{1}{2}}\nabla_\xi\Phi_1(\xi)\right)\right]\right| \left|\partial_\xi^{\alpha_2}\chi_j(\xi)\right| \left|\partial_\xi^{\alpha_3}b(\xi)\right| d\xi.$$

Our goal is to prove that, by choosing N appropriately, we have

$$|L_j(\lambda)| \le \mathcal{F}(\mathcal{M}_{d+2})\mathcal{N}_{d+1}\frac{1}{a_0}\lambda^{-\frac{d}{2}}. \tag{3.4.25}$$

Step 1. $\alpha_1 = 0$. Here we integrate over the set $|\nabla_\xi\Phi_1(\xi)| \ge \lambda^{-\frac{1}{2}}$. We use (3.4.9), the bounds (3.4.16), (3.4.22), (3.4.7) (ii) and make the change of variables $\eta = \nabla_\xi\Phi_1(\xi)$; then, $d\eta = |\det \mathrm{Hess}\,\Phi_1(\xi)|\,d\xi$ and

$$|S_{\alpha,N}| \le \frac{\lambda^{-N}}{a_0}\delta^{-|\alpha_2|}\mathcal{N}_{|\alpha_3|}\mathcal{F}\left(\mathcal{M}_{N+1}\right)\sum_{k=N}^{2N-|\alpha|}\int_{|\eta|>\lambda^{-\frac{1}{2}}}\frac{d\eta}{|\eta|^k},$$

thus,

$$|S_{\alpha,N}| \le \frac{C_d\lambda^{-N}}{a_0}\delta^{-|\alpha_2|}\mathcal{N}_{|\alpha_3|}\mathcal{F}\left(\mathcal{M}_{N+1}\right)\sum_{k=N}^{2N-|\alpha|}\int_{\lambda^{-\frac{1}{2}}}^{+\infty} r^{d-1-k}\,dr.$$

We take $N = d+1$. Since $|\alpha_j| \le |\alpha| \le N$ we see that

$$\begin{aligned}|S_{\alpha,N}| &\le \frac{\lambda^{-N}}{a_0}\delta^{-|\alpha|}\mathcal{N}_{d+1}\mathcal{F}(\mathcal{M}_{d+2})\lambda^{N-\frac{1}{2}|\alpha|-\frac{d}{2}},\\ &\le \frac{1}{a_0}\mathcal{N}_{d+1}\mathcal{F}(\mathcal{M}_{d+2})\lambda^{-\frac{d}{2}}\frac{1}{(\lambda^{\frac{1}{2}}\delta)^{|\alpha|}}.\end{aligned}$$

Since, (see (3.4.14)) we are in the case where $\lambda^{\frac{1}{2}}\delta \ge 1$, we obtain

$$|S_{\alpha,N}| \le \frac{1}{a_0}\mathcal{N}_{d+1}\mathcal{F}(\mathcal{M}_{d+2})\lambda^{-\frac{d}{2}}.$$

Step 2. $\alpha_1 \ne 0$. Here, since we differentiate ψ, we integrate over the set

$$\lambda^{-\frac{1}{2}} \le |\nabla_\xi\Phi_1(\xi)| \le 2\lambda^{-\frac{1}{2}} \le 2\delta \le \frac{\varepsilon_0}{2} \le 1.$$

We need to estimate

$$S_{\alpha,N} - \lambda^{-N}\int |c_{\alpha,N}| \left|\partial_\xi^{\alpha_1}\left[\psi\left(\lambda^{\frac{1}{2}}\nabla_\xi\Phi_1(\xi)\right)\right]\right| \left|\partial_\xi^{\alpha_2}\chi_j(\xi)\right| \left|\partial_\xi^{\alpha_3}b(\xi)\right| d\xi.$$

By the Faa-di-Bruno formula (see the Appendix) we have

$$\partial_\xi^{\alpha_1}\left[\psi\left(\lambda^{\frac{1}{2}}\nabla\Phi_1(\xi)\right)\right] = \sum_{1\le|\beta|\le|\alpha_1|} a_{\alpha_1,\beta}\psi^{(\beta)}\left(\lambda^{\frac{1}{2}}\nabla_\xi\Phi_1(\xi)\right)\prod_{i=1}^{s}\left(\lambda^{\frac{1}{2}}\partial_\xi^{l_i}\nabla_\xi\Phi_1\right)^{k_i},$$

where $a_{\alpha_1,\beta}$ are absolute constants, $\sum_{i=1}^{s} k_i = \beta$, $\sum_{i=1}^{s} k_i|l_i| = \alpha_1$. Then,

$$\begin{aligned}\left|\partial_\xi^{\alpha_1}\left[\psi\left(\lambda^{\frac{1}{2}}\nabla_\xi\Phi_1(\xi)\right)\right]\right| &\le \mathcal{F}(\mathcal{M}_{|\alpha_1|+1})\sum_{1\le|\beta|\le|\alpha_1|}\lambda^{\frac{|\beta|}{2}}\left|\psi^{(\beta)}\left(\lambda^{\frac{1}{2}}\nabla_\xi\Phi_1(\xi)\right)\right|,\\ &\le \mathcal{F}(\mathcal{M}_{|\alpha_1|+1})\lambda^{\frac{|\alpha_1|}{2}}\sum_{1\le|\beta|\le|\alpha_1|}\left|\psi^{(\beta)}\left(\lambda^{\frac{1}{2}}\nabla_\xi\Phi_1(\xi)\right)\right|.\end{aligned}$$

Using (3.4.22) and the fact that $|\nabla\Phi_1(\xi)| \le 1$, we can write

$$|c_{\alpha,N}| \le \mathcal{F}(\mathcal{M}_{N+1})|\nabla_\xi\Phi_1(\xi)|^{|\alpha|-2N}.$$

Finally, we have $\left|\partial_\xi^{\alpha_2}\chi_j(\xi)\right| \le C_\alpha\delta^{-|\alpha_2|}$.

Performing the change of variables $\eta = \nabla_\xi\Phi_1(\xi)$, we obtain

$$\begin{aligned}|S_{\alpha,N}| &\le \frac{\lambda^{-N}\delta^{-|\alpha_2|}}{a_0}\mathcal{F}(\mathcal{M}_{N+1})\mathcal{N}_{|\alpha|}\lambda^{\frac{|\alpha_1|}{2}}\int_{\lambda^{-\frac{1}{2}}\le|\eta|\le 2\lambda^{-\frac{1}{2}}}|\eta|^{|\alpha|-2N}\,d\eta,\\ &\le \frac{1}{a_0}\mathcal{F}(\mathcal{M}_{N+1})\mathcal{N}_{|\alpha|}\lambda^{-\frac{d}{2}}(\lambda^{\frac{1}{2}}\delta)^{-|\alpha_2|}.\end{aligned}$$

Since $\lambda^{\frac{1}{2}}\delta \ge 1$, we get $|S| \le \frac{1}{a_0}\mathcal{F}(\mathcal{M}_{N+1})\mathcal{N}_N\lambda^{-\frac{d}{2}}$. Consequently, (3.4.25) is proved. Using (3.4.23), (3.4.24), (3.4.25) and since $N = d+1$, we obtain

$$|I_j| \le \frac{1}{a_0}\mathcal{F}(\mathcal{M}_{d+2})\mathcal{N}_{d+1}\lambda^{-\frac{d}{2}}, \quad 1 \le j \le J.$$

Now, by (3.4.18) we have $|\Lambda| \le C_d|K_{\varepsilon_0}|\delta^{-d}$ (where C_d depends only on d). Using (3.4.19), we finally obtain

$$|I(\lambda)| \le \frac{|K_{\varepsilon_0}|}{a_0\delta^d}\mathcal{F}(\mathcal{M}_{d+2})\mathcal{N}_{d+1}\lambda^{-\frac{d}{2}},$$

which completes the proof of the first case of the theorem.

Proof of 2.

In this case, it is not necessary to localize $I(\lambda)$ in small balls of size δ as in the first case.

Then, as previously, we write

$$I(\lambda) = \int e^{i\lambda\Phi(\xi)}\psi\big(\lambda^{\frac{1}{2}}|\nabla_\xi\Phi_1(\xi)|\big)b(\xi)\,d\xi \\ + \int e^{i\lambda\Phi(\xi)}\big(1-\psi\big(\lambda^{\frac{1}{2}}|\nabla_\xi\Phi_1(\xi)|\big)\big)b(\xi)\,d\xi =: K(\lambda)+L(\lambda),$$

and the final estimate follows from (3.4.24) and (3.4.25).

3.5 Appendix

Here is a version (not very precise regarding the coefficients but often sufficient) of the Faa-di-Bruno formula.

Lemma 3.5.1 *Let $g : \mathbf{R}^d_x \to \mathbf{R}^m_y$ and $F : \mathbf{R}^m_y \to \mathbf{C}$ be two C^∞ functions. Then for all $\alpha \in \mathbf{N}^d$ such that $|\alpha| \geq 1$ we have*

$$\partial_x^\alpha\big(F(g(x)\big) = \sum_{1\leq|\beta|\leq|\alpha|} a_{\alpha,\beta}(\partial_y^\beta F)\big(g(x)\big)\prod_{j=1}^{s}\big(\partial_x^{\ell_j}g(x)\big)^{\gamma_j},$$

where the $a_{\alpha,\beta}$ are real constants,

$$\gamma_j \in \mathbf{N}^m,\quad \ell_j \in \mathbf{N}^d,\quad |\ell_j| \geq 1\quad \Sigma_{j=1}^s\gamma_j = \beta,\quad \Sigma_{j=1}^s\gamma_j|\ell_j| = \alpha.$$

Chapter 4
Continuity of Nonlinear Maps on Sobolev and Lebesgue Spaces

4.1 Prerequisites

Fourier transform in $\mathcal{S}'(\mathbf{R}^d)$. Convolution. Young's theorem. Sobolev spaces. Spaces $L^p(\mathbf{R}^d)$.

4.2 The Context

Let E be a space of functions with values in $\mathbf{C}$ (for example a Sobolev space or an L^p space). There are two ways to conceive the action of a map on E. The first is by composition. Let $f : \mathbf{C} \to \mathbf{C}$ be a function. For $u \in E$ we define $f(u)$ by $f(u) = f \circ u$, that is, $f(u)(x) = f(u(x))$. Next, suppose that u is the unique solution of an evolution partial differential equation for which the initial data u_0 is known at time $t = 0$. We can write $u = \Phi(u_0)$ and thus consider the map $u_0 \mapsto \Phi(u_0)$. This map is called the "flow" of the problem, and the continuity of Φ thus raises the question of the continuity of the flow. It is mainly this second case that is of interest in this chapter.

4.2.1 *The Problem*

The aim of this chapter is to study the action and the **continuity** of functions on Sobolev spaces and then on L^p spaces.

C. Zuily, *Selected Topics in Partial Differential Equations*, Universitext,
https://doi.org/10.1007/978-3-032-24082-8_4

4.2.2 Outline

In what follows, we begin by developing an elementary version of what is called "Littlewood-Paley theory." This theory, using a decomposition into annuli of the space $\mathbf{R}^d$, allows us to characterize Sobolev spaces as well as many other functional spaces. This theory, widely used in harmonic analysis, proves to be very useful for our purposes and in many other contexts. We then present some classical results on the continuity of composed functions before addressing the main result of this chapter, which provides a simple criterion for studying the continuity of flows of partial differential equations. Note that at the time of writing, this criterion is new and does not appear in the classical literature. We conclude with an example of continuity of composed functions on Lebesgue spaces.

4.3 Preliminaries on Sobolev Spaces

Definition 4.3.1 Let $d \geq 1$ and $s \in \mathbf{R}$. The Sobolev space of order s is defined by

$$H^s(\mathbf{R}^d) = \left\{u \in \mathcal{S}'(\mathbf{R}^d) : \|u\|_{H^s}^2 := \int_{\mathbf{R}^d} \langle\xi\rangle^{2s} |\widehat{u}(\xi)|^2 \, d\xi \; < +\infty\right\},$$

where $\langle\xi\rangle = (1 + |\xi|^2)^{\frac{1}{2}}$.

Note that for $r > 0$ there exists $C_r > 0$ such that for all $\xi, \eta \in \mathbf{R}^d$,

$$\langle\xi\rangle^r \leq C_r(\langle\xi - \eta\rangle^r + \langle\eta\rangle^r). \tag{4.3.1}$$

4.3.1 A Characterization of Sobolev Spaces

For $j \in \mathbf{N}$ we set

$$C_j = \left\{\xi \in \mathbf{R}^d : 2^{j-1} < |\xi| < 2^{j+1}\right\}.$$

Note that

$$|j - k| \geq 2 \implies C_j \cap C_k = \emptyset. \tag{4.3.2}$$

Indeed, if $j - k \geq 2$ we have $\frac{1}{2}2^j \geq \frac{1}{2}2^{k+2} \geq 2\,2^k$. Similarly if $k - j \geq 2$.

We will also use in the following other annuli of the form,

$$\widetilde{C}_j(a, b) = \left\{\xi \in \mathbf{R}^d : a2^j \leq |\xi| \leq b\,2^j\right\}, \text{ where } 0 < a < b.$$

Let $\psi \in C_0^\infty(\mathbf{R}^d)$ such that

$$0 \le \psi \le 1, \quad \operatorname{supp} \psi \subset \left\{\xi \in \mathbf{R}^d : |\xi| \le 1\right\}, \quad \psi(\xi) = 1 \text{ if } |\xi| \le \frac{1}{2},$$

and set

$$\varphi(\xi) = \psi\left(\frac{\xi}{2}\right) - \psi(\xi). \tag{4.3.3}$$

Then,

$$\psi(\xi) + \sum_{j=0}^{N-1} \varphi(2^{-j}\xi) = \psi(2^{-N}\xi), \quad \forall \xi \in \mathbf{R}^d, \forall N \ge 1, \tag{4.3.4}$$

which implies that

$$\psi(\xi) + \sum_{j=0}^{+\infty} \varphi(2^{-j}\xi) = 1, \quad \forall \xi \in \mathbf{R}^d, \forall N \ge 1. \tag{4.3.5}$$

We then have for $u \in \mathcal{S}'(\mathbf{R}^d)$

$$\widehat{u} = \psi(\xi)\widehat{u} + \sum_{j=0}^{+\infty} \varphi(2^{-j}\xi)\widehat{u}. \tag{4.3.6}$$

Denoting by $\mathcal{F}^{-1}$ the inverse Fourier transform we set, for $u \in \mathcal{S}'(\mathbf{R}^d)$,

$$\Delta_{-1} u = \mathcal{F}^{-1}\left(\psi(\xi)\widehat{u}\right), \quad \Delta_j u = \mathcal{F}^{-1}\left(\varphi(2^{-j}\xi)\widehat{u}\right) \quad \text{for} \quad j \ge 0, \tag{4.3.7}$$

$$S_k(u) = \sum_{j=-1}^{k-1} \Delta_j u = \mathcal{F}^{-1}\left(\psi(2^{-k}\xi)\widehat{u}\right). \tag{4.3.8}$$

By convention we set $\Delta_j u = 0$ if $j \le -2$.

Here are some important points to note.

- We have $\operatorname{supp}(\psi(\xi)\widehat{u}) \subset \{\xi : |\xi| \le 1\}$ and $\operatorname{supp}(\varphi(2^{-j}\xi)\widehat{u}) \subset C_j$.
- For $u \in \mathcal{S}'(\mathbf{R}^d)$, $\Delta_j u$ and $S_k(u)$ are C^∞ functions on $\mathbf{R}^d$, as inverse Fourier transforms of distributions with compact support.
- For $u \in \mathcal{S}'(\mathbf{R}^d)$, (4.3.2) implies that

$$\Delta_j \Delta_k u = 0, \quad \text{if } |j - k| \ge 2. \tag{4.3.9}$$

- Let $\theta = \mathcal{F}^{-1}(\varphi) \in \mathcal{S}(\mathbf{R}^d)$ and $h = \mathcal{F}^{-1}(\psi) \in \mathcal{S}(\mathbf{R}^d)$ then

$$\begin{aligned} \Delta_j u &= \mathcal{F}^{-1}(\varphi(2^{-j}\cdot)) * u = \left[2^{jd}\theta(2^j\cdot)\right] * u, \\ S_k(u) &= \mathcal{F}^{-1}(\psi(2^{-k}\cdot)) * u = \left[2^{kd}h(2^k\cdot)\right] * u. \end{aligned} \tag{4.3.10}$$

It follows from (4.3.6) that for $u \in \mathcal{S}'(\mathbf{R}^d)$ we have

$$u = \sum_{j=-1}^{+\infty} \Delta_j u. \tag{4.3.11}$$

This is *the Littlewood-Paley decomposition* of u.
The main result of this paragraph is the following.

Theorem 4.3.2 *1. Let $s \in \mathbf{R}$. If $u \in H^s(\mathbf{R}^d)$. We have*

$$\|\Delta_j u\|_{L^2(\mathbf{R}^d)} \leq c_j 2^{-js}, \ j \geq -1, \quad \textit{where,} \quad \sum_{j=-1}^{+\infty} c_j^2 \leq C\, \|u\|^2_{H^s(\mathbf{R}^d)}. \tag{4.3.12}$$

2. For $j \in \mathbf{Z}$, $j \geq -1$ let $u_j \in L^2(\mathbf{R}^d)$ be such that supp $\widehat{u}_j \subset \widetilde{C}_j(a,b)$ and

$$\|u_j\|_{L^2(\mathbf{R}^d)} \leq c_j 2^{-js}, \quad \textit{where} \quad \sum_{j=-1}^{+\infty} c_j^2 < +\infty. \tag{4.3.13}$$

Let $u = \sum_{j=-1}^{+\infty} u_j$. Then,

$$u \in H^s(\mathbf{R}^d) \quad \textit{and} \quad \|u\|_{H^s(\mathbf{R}^d)} \leq C \left(\sum_{j=-1}^{+\infty} 2^{2js} \|u_j\|^2_{L^2(\mathbf{R}^d)} \right)^{\frac{1}{2}}. \tag{4.3.14}$$

3. If $s > 0$, in statement 2. one can replace the annulus $\widetilde{C}_j$ by the ball $B(0, C2^j)$.

Proof 1. Set $c_j = 2^{js}\|\Delta_j u\|_{L^2}$. For $j \geq 0$ we can write

$$c_j^2 = 2^{2js} \int_{\mathbf{R}^d} \langle\xi\rangle^{-2s} |\varphi(2^{-j}\xi)|^2 \langle\xi\rangle^{2s} |\widehat{u}(\xi)|^2 \, d\xi.$$

On the support of the function φ we have $\frac{1}{2}\,2^j \leq |\xi| \leq 2\,2^j$, in particular $|\xi| \geq \frac{1}{2}$ so $1 \leq 4|\xi|^2$. We deduce that

$$\frac{1}{2}\,2^j \leq |\xi| \leq \langle\xi\rangle \leq 3|\xi| \leq 6\,2^j.$$

Consequently, for all $s \in \mathbf{R}$, there exist two positive constants C_1, C_2 depending only on s such that

$$C_1 \leq \left(\frac{\langle \xi \rangle}{2^j}\right)^{-s} \leq C_2.$$

Since $\varphi \in C_0^\infty(\mathbf{R}^d)$ we finally obtain

$$c_j^2 \leq C_3 \int_{\mathbf{R}^d} \varphi(2^{-j}\xi)\langle \xi \rangle^{2s} |\widehat{u}(\xi)|^2 \, d\xi. \tag{4.3.15}$$

Next,

$$c_{-1}^2 = 2^{-2s} \int_{\mathbf{R}^d} \langle \xi \rangle^{-2s} |\psi(\xi)|^2 \langle \xi \rangle^{2s} |\widehat{u}(\xi)|^2 \, d\xi.$$

On the support of ψ we have $|\xi| \leq 1$, so $1 \leq \langle \xi \rangle \leq \sqrt{2}$. Therefore for all $s \in \mathbf{R}$ we have $\langle \xi \rangle^{2s} \leq C_4$. Consequently,

$$c_{-1}^2 \leq C_5 \int_{\mathbf{R}^d} \psi(\xi)\langle \xi \rangle^{2s} |\widehat{u}(\xi)|^2 \, d\xi. \tag{4.3.16}$$

Using (4.3.15) and (4.3.16) we obtain

$$\sum_{j=-1}^{+\infty} c_j^2 \leq C_6 \int_{\mathbf{R}^d} \Big(\psi(\xi) + \sum_{j=0}^{+\infty} \varphi(2^{-j}\xi)\Big)\langle \xi \rangle^{2s} |\widehat{u}(\xi)|^2 \, d\xi \leq C_6 \|u\|_{H^s}^2.$$

according to (4.3.5).

2. Let n_0 be an integer such that $n_0 + 1 > \ln \frac{b}{a}$. It is easy to see that $\widetilde{C}_j \cap \widetilde{C}_k = \emptyset$ if $|j-k| \geq n_0 + 1$. Then, since $\operatorname{supp} \widehat{u}_j \subset \widetilde{C}_j$ we have

$$\begin{aligned}
\|u\|_{H^s}^2 &= \sum_{\substack{j,k \geq -1 \\ |j-k| \leq n_0}} \int_{\mathbf{R}^d} \langle \xi \rangle^{2s} \widehat{u}_j(\xi) \overline{\widehat{u}_k(\xi)} \, d\xi, \\
&\leq \sum_{\substack{j,k \geq -1 \\ |j-k| \leq n_0}} \Big(\int_{\widetilde{C}_j} \langle \xi \rangle^{2s} |\widehat{u}_j(\xi)|^2 \, d\xi\Big)^{\frac{1}{2}} \Big(\int_{\widetilde{C}_k} \langle \xi \rangle^{2s} |\widehat{u}_k(\xi)|^2 \, d\xi\Big)^{\frac{1}{2}}.
\end{aligned}$$

Using the inequality $2ab \leq a^2 + b^2$ we obtain

$$\begin{aligned}
\|u\|_{H^s}^2 \leq \frac{1}{2} \sum_{j=-1}^{+\infty} \sum_{k=j-n_0}^{j+n_0} \Big(\int_{\widetilde{C}_j} \langle \xi \rangle^{2s} |\widehat{u}_j(\xi)|^2 \, d\xi\Big) \\
+ \frac{1}{2} \sum_{k=-1}^{+\infty} \sum_{j=k-n_0}^{k+n_0} \Big(\int_{\widetilde{C}_k} \langle \xi \rangle^{2s} |\widehat{u}_k(\xi)|^2 \, d\xi\Big).
\end{aligned}$$

For $s \in \mathbf{R}$, on $\widetilde{C}_j$ (resp. $\widetilde{C}_k$) we have as previously

$$0 < m \leq \frac{\langle \xi \rangle^s}{2^{js}} \leq M \quad (\text{resp. } 0 < m \leq \frac{\langle \xi \rangle^s}{2^{ks}} \leq M). \tag{4.3.17}$$

We deduce

$$\|u\|_{H^s}^2 \leq C_1 \sum_{j=-1}^{+\infty} \sum_{k=j-n_0}^{j+n_0} 2^{2js} \Big(\int_{\widetilde{C}_j} |\widehat{u}_j(\xi)|^2 \, d\xi \Big) \\ + C_1 \sum_{k=-1}^{+\infty} \sum_{j=k-n_0}^{k+n_0} 2^{2ks} \Big(\int_{\widetilde{C}_k} \widehat{u}_k(\xi)|^2 \, d\xi \Big).$$

Therefore,

$$\|u\|_{H^s}^2 \leq C_2(2n_0 + 1) \sum_{j=-1}^{+\infty} 2^{2js} \|u_j\|_{L^2}^2 \leq C_3 \sum_{j=-1}^{+\infty} c_j^2.$$

3. If $s > 0$ and $\operatorname{supp} \widehat{u}_j \subset B(0, C2^j)$ we replace everywhere $\widetilde{C}_j$ by $B(0, C2^j)$, and we use the fact that for $\xi \in B(0, C2^j)$ we have as in (4.3.17), $\frac{\langle \xi \rangle^s}{2^{js}} \leq C^s$ and we conclude as in the previous case. □

4.4 Boundedness and Continuity

Algebra Property

Proposition 4.4.1 *For $s > \frac{d}{2}$ the space $H^s(\mathbf{R}^d)$ is an algebra and there exists $C > 0$ such that*

$$\|uv\|_{H^s} \leq C \|u\|_{H^s} \|v\|_{H^s} \tag{4.4.1}$$

for all $u, v \in H^s(\mathbf{R}^d)$.

Proof The functions u and v belong in particular to $L^2(\mathbf{R}^d)$ (since $s > \frac{d}{2}$). Therefore, for almost every $\xi \in \mathbf{R}^d$ the function $\eta \mapsto \widehat{u}(\xi - \eta)$ belongs to $L^2(\mathbf{R}^d)$ and the function $\eta \mapsto \widehat{v}(\eta)$ belongs to $L^2(\mathbf{R}^d)$. Therefore, the function $\eta \mapsto \widehat{u}(\xi - \eta)\widehat{v}(\eta)$ belongs to $L^1(\mathbf{R}^d)$. We can thus write

$$(\widehat{uv})(\xi) = (\widehat{u} * \widehat{v})(\xi) = \int_{\mathbf{R}^d} \widehat{u}(\xi - \eta)\widehat{v}(\eta) \, d\eta.$$

Using (4.3.1) it follows

$$\langle\xi\rangle^s|(\widehat{uv})(\xi)| \le C_s\Big(\int_{\mathbf{R}^d}\langle\xi-\eta\rangle^s|\widehat{u}(\xi-\eta)||\widehat{v}(\eta)|\,d\eta + \int_{\mathbf{R}^d}|\widehat{u}(\xi-\eta)|\langle\eta\rangle^s|\widehat{v}(\eta)|\,d\eta\Big),$$
$$\le (1)+(2).$$

We will estimate, in the same way, the L^2 norms of the terms (1) and (2).

The term (1) can be written as $C_s(\langle\cdot\rangle^s|\widehat{u}|) * |\widehat{v}|$ where $*$ denotes convolution. We then use Young's Theorem which ensures that the mapping,

$$L^2(\mathbf{R}^d)\times L^1(\mathbf{R}^d)\to L^2(\mathbf{R}^d),\quad (f,g)\mapsto f*g$$

is continuous and that there exists $C>0$ such that

$$\|f*g\|_{L^2(\mathbf{R}^d)} \le \|f\|_{L^2(\mathbf{R}^d)}\|g\|_{L^1(\mathbf{R}^d)},$$

for all $(f,g)\in L^2(\mathbf{R}^d)\times L^1(\mathbf{R}^d)$.

We deduce that

$$\|(1)\|_{L^2(\mathbf{R}^d)} \le C\|u\|_{H^s}\|\widehat{v}\|_{L^1}.$$

Now, by the Cauchy-Schwarz inequality we can write

$$\int_{\mathbf{R}^d}|\widehat{v}(\eta)|\,d\eta \le \Big(\int_{\mathbf{R}^d}\langle\eta\rangle^{2s}|\widehat{v}(\eta)|^2\,d\eta\Big)^{\frac12}\Big(\int_{\mathbf{R}^d}\frac{1}{\langle\eta\rangle^{2s}}\,d\eta\Big)^{\frac12} \le C_s\|v\|_{H^s},$$

since $2s>d$. Therefore, the L^2 norm of term (1) is bounded by the right-hand side of (4.4.1). The same holds for term (2). □

Corollary 4.4.2 *Let $s>\frac{d}{2}$ and P be a polynomial of degree $m\ge 1$ on $\mathbf{C}$ such that $P(0)=0$. There exists a polynomial Q on $\mathbf{R}$, of degree $m-1$ such that*

$$\|P(u)\|_{H^s} \le Q(\|u\|_{H^s})\|u\|_{H^s}. \tag{4.4.2}$$

Proof We deduce from Proposition 4.4.1 and by induction on the integer $j\ge 1$ that $\|u^j\|_{H^s}\le C\|u\|_{H^s}^{j-1}\|u\|_{H^s}$. The corollary follows. □

Corollary 4.4.3 *Let $s>\frac{d}{2}$ and $B_s(0,R)=\{u\in H^s(\mathbf{R}^d):\|u\|_{H^s}\le R\}$, for $R>0$. Let P be a polynomial of degree $m\ge 1$ on $\mathbf{C}$ such that $P(0)=0$. Then there exists $C>0$ such that*

$$\|P(u)-P(v)\|_{H^s} \le C\Big(\sum_{\ell=0}^{m-1}R^\ell\Big)\|u-v\|_{H^s}\quad \forall u,v\in B_s(0,R).$$

Proof If $P(z) = \sum_{j=1}^{m} a_j z^j$, we have using Proposition 4.4.1,

$$\begin{aligned}\|P(u) - P(v)\|_{H^s} &\le C_1 \sum_{j=1}^{m} \|u^j - v^j\|_{H^s} \le C_2 \sum_{j=1}^{m}\sum_{k=0}^{j-1} \|(u^{j-k-1}v^k)(u-v)\|_{H^s},\\ &\le C_3 \sum_{j=1}^{m}\sum_{k=0}^{j-1} \|u\|_{H^s}^{j-k-1}\|v\|_{H^s}^{k}\|u-v\|_{H^s},\\ &\le C_4\Big(\sum_{j=1}^{m} R^{j-1}\|u-v\|_{H^s}\Big).\end{aligned}$$

□

The Case $0 < s < 1$

For $0 < s < 1$, the norm on $H^s(\mathbf{R}^d)$ given in Definition 4.3.1 is equivalent to the following norm

$$N(u) = \Big(\|u\|_{L^2(\mathbf{R}^d)}^2 + \iint_{\mathbf{R}^d\times\mathbf{R}^d} \frac{|u(x)-u(y)|^2}{|x-y|^{d+2s}}\, dx\, dy\Big)^{\frac{1}{2}}, \tag{4.4.3}$$

(see the proof in the appendix).

Let Lip(**C**, **C**) be the space of functions $F : \mathbf{C} \to \mathbf{C}$ such that $F \in L^\infty(\mathbf{C})$ and there exists $C > 0$ such that

$$|F(z) - F(z')| \le C|z - z'|, \quad \forall z, z' \in \mathbf{C}. \tag{4.4.4}$$

In what follows, we write $F(u) = F \circ u$.

Theorem 4.4.4 *Let $s \in]0, 1[$ and $F \in Lip(\mathbf{C}, \mathbf{C})$ such that $F(0) = 0$. Then, for any u in $H^s(\mathbf{R}^d)$, $F(u)$ belongs to $H^s(\mathbf{R}^d)$ and there exists $C > 0$ independent of u such that*

$$\|F(u)\|_{H^s} \le C\|u\|_{H^s}, \quad \forall u \in H^s(\mathbf{R}^d). \tag{4.4.5}$$

Proof We use the norm (4.4.3) on $H^s(\mathbf{R}^d)$. Since $F(0) = 0$, we have, according to (4.4.4), $|F(z)| \le C|z|$ for all $z \in \mathbf{C}$, so that $\|F(u)\|_{L^2(\mathbf{R}^d)} \le C\|u\|_{L^2(\mathbf{R}^d)}$. Next, still by (4.4.4),

$$\iint_{\mathbf{R}^d\times\mathbf{R}^d} \frac{|F(u(x)) - F(u(y))|^2}{|x-y|^{d+2s}}\, dx\, dy \le C^2 \iint_{\mathbf{R}^d\times\mathbf{R}^d} \frac{|u(x)-u(y)|^2}{|x-y|^{d+2s}}\, dx\, dy.$$

We deduce that $F(u) \in H^s(\mathbf{R}^d)$ and the inequality (4.5.1). □

The function $F(z) = \frac{z}{1+|z|^2}$ is an example of a function satisfying the hypotheses of the theorem.

We will show later a continuity result for these functions.

Even when the function F is not Lipschitz, one can prove results analogous to that of the theorem. Here is an example.

Example 4.4.5 Let $s \in]0,1[$ and v an element of $H^s(\mathbf{R}^d) \cap L^\infty(\mathbf{R}^d)$ with real values such that $\|v\|_{L^\infty(\mathbf{R}^d)} < 1$. Then the function $F = \frac{v}{1+v}$ belongs to $H^s(\mathbf{R}^d)$ and we have

$$\|F\|_{H^s} \leq \frac{1}{(1 - \|v\|_{L^\infty})^2} \|v\|_{H^s}. \tag{4.4.6}$$

For this, we will use (4.4.3). Indeed, since $1 + v(x) \geq 1 - \|v\|_{L^\infty} > 0$, we have first $|F(x)| \leq \frac{|v(x)|}{1-\|v\|_{L^\infty}}$, hence

$$\|F\|_{L^2} \leq \frac{1}{1 - \|v\|_{L^\infty}} \|v\|_{L^2}. \tag{4.4.7}$$

Next, $F(x) - F(y) = \frac{v(x)-v(y)}{(1+v(x))(1+v(y))}$ so that $|F(x) - F(y)|^2 \leq \frac{|v(x)-v(y)|^2}{(1-\|v\|_{L^\infty})^4}$. We deduce that

$$\iint \frac{|F(x) - F(y)|^2}{|x - y|^{d+2s}} \, dx \, dy \leq \frac{1}{(1 - \|v\|_{L^\infty})^4} \iint \frac{|v(x) - v(y)|^2}{|x - y|^{d+2s}} \, dx \, dy. \tag{4.4.8}$$

Then (4.4.6) follows from (4.4.3), (4.4.7), (4.4.8) since $0 < 1 - \|v\|_{L^\infty} < 1$.

An Improvement of Inequality (4.4.1)

Proposition 4.4.6 *Let $s \geq 0$. Then the space $H^s(\mathbf{R}^d) \cap L^\infty(\mathbf{R}^d)$ is an algebra and there exists a constant $C > 0$ such that for all $u, v \in H^s(\mathbf{R}^d) \cap L^\infty(\mathbf{R}^d)$*

$$\|uv\|_{H^s} \leq C\big(\|u\|_{H^s}\|v\|_{L^\infty} + \|u\|_{L^\infty}\|v\|_{H^s}\big). \tag{4.4.9}$$

This proposition implies Proposition 4.4.1 since when $s > \frac{d}{2}$ the space $H^s(\mathbf{R}^d)$ is continuously embedded in $L^\infty(\mathbf{R}^d)$.

Before giving the proof of the above proposition let us state some corollaries.

Corollary 4.4.7 *Let $s \geq 0$ and P a polynomial of degree $m \geq 1$ such that $P(0) = 0$. There exists a polynomial Q on $\mathbf{R}$, of degree $m - 1$ such that*

$$\|P(u)\|_{H^s} \leq Q(\|u\|_{L^\infty})\|u\|_{H^s} \tag{4.4.10}$$

for all $H^s(\mathbf{R}^d) \cap L^\infty(\mathbf{R}^d)$.

Proof We deduce from Proposition 4.4.6 and by induction on $n \geq 1$ that

$$\|u^n\|_{H^s} \leq C_n \|u\|_{L^\infty}^{n-1} \|u\|_{H^s}.$$

The corollary follows immediately. □

Corollary 4.4.8 *Let $s \geq 0$ and P a polynomial of degree $m \geq 1$ such that $P(0) = 0$. For $R > 0$ let $B_\infty(0, R) = \{u \in L^\infty(\mathbf{R}^d) : \|u\|_{L^\infty} \leq R\}$. Then there exists $C > 0$ such that*

$$\|P(u) - P(v)\|_{H^s} \leq C\Big(\sum_{\ell=0}^{m-1} R^\ell\Big)\|u - v\|_{H^s} \quad \forall u, v \in B_\infty(0, R) \cap H^s(\mathbf{R}^d).$$

Proof It is analogous to that of Corollary 4.4.3 starting from Corollary 4.4.2. □

Proof (*Proposition* 4.4.6) The case $s = 0$ being trivial, we may assume $s > 0$. According to (4.3.11) we can write

$$\begin{aligned} uv &= \sum_{j=-1}^{+\infty} \sum_{k=-1}^{+\infty} \Delta_j u \Delta_k v, \\ &= \sum_{j=-1}^{+\infty} \sum_{k=-1}^{j-3} \Delta_j u \Delta_k v + \sum_{k=1}^{+\infty} \sum_{j=-1}^{k-3} \Delta_j u \Delta_k v + \sum_{|j-k|\leq 2} \Delta_j u \Delta_k v, \\ &= \sum_{j=-1}^{+\infty} S_{j-2}(v)\Delta_j u + \sum_{k=-1}^{+\infty} S_{k-2}(u)\Delta_k v + \sum_{|j-k|\leq 2} \Delta_j u \Delta_k v = U_1 + U_2 + R. \end{aligned}$$

To begin let us consider the term U_1. It can be written as $U_1 = \sum_{j=-1}^{+\infty} w_j$ where $w_j = S_{j-2}(v)\Delta_j u$. We have $\widehat{w}_j = \widehat{S_{j-2}(v)} * \widehat{\Delta_j u}$. Then, according to (4.3.7) and (4.3.8),

$$\operatorname{supp} \widehat{w}_j \subset \operatorname{supp} \widehat{S_{j-2}(v)} + \operatorname{supp} \widehat{\Delta_j u} \subset \{|\xi| \leq 2^{j-2}\} + C_j \subset \widetilde{C}_j(\frac{1}{4}, \frac{9}{4}),$$

because if $\xi = \eta + \zeta$ with $|\eta| \leq 2^{j-2}$ and $\frac{1}{2}2^j \leq |\zeta| \leq 2\,2^j$ we will have

$$|\xi| \leq |\eta| + |\zeta| \leq (\frac{1}{4} + 2)2^j \quad \text{and} \quad |\xi| \geq |\zeta| - |\eta| \geq (\frac{1}{2} - \frac{1}{4})2^j.$$

On the other hand

$$\|w_j\|_{L^2} \leq \|S_{j-2}(v)\|_{L^\infty} \|\Delta_j u\|_{L^2}.$$

Next, using (4.3.10) it follows that

$$\begin{aligned}\left|S_{j-2}(v)(x)\right| &= \left|\left[2^{(j-2)d}h(2^{(j-2)d}\cdot)\right]*u(x)\right|,\\ &\leq 2^{(j-2)d}\int_{\mathbf{R}^d}|h(2^{(j-2)d}(y))||v(x-y)|\,dy \leq \|h\|_{L^1}\|v\|_{L^\infty}.\end{aligned}$$

We deduce from (4.3.12) that

$$\|w_j\|_{L^2} \leq \|h\|_{L^1}\|v\|_{L^\infty}\|\Delta_j u\|_{L^2} \leq C(h)\|v\|_{L^\infty}c_j 2^{-js},$$

where $\sum_j c_j^2 \leq C\|u\|_{H^s}^2$. Using (4.3.13) and (4.3.14) we deduce that we will have $U_1 = \sum_j w_j \in H^s(\mathbf{R}^d)$ and

$$\|U_1\|_{H^s} \leq C\Big(\sum_{j=-1}^{+\infty} 2^{2js}\|w_j\|_{L^2}^2\Big)^{\frac{1}{2}} \leq C\|v\|_{L^\infty}\|u\|_{H^s}. \tag{4.4.11}$$

By swapping u and v we obtain

$$\|U_2\|_{H^s} \leq C\|u\|_{L^\infty}\|v\|_{H^s}. \tag{4.4.12}$$

Let us estimate the term R. It can be written as $R = \sum_{\alpha=-2}^{2}\sum_{k=-1}^{+\infty}\Delta_{k+\alpha}u\Delta_k v$. We have

$$\begin{aligned}\operatorname{supp}(\widehat{\Delta_{k+\alpha}u\Delta_k v}) &= \operatorname{supp}(\widehat{\Delta_{k+\alpha}u}*\widehat{\Delta_k v}) \subset \operatorname{supp}(\widehat{\Delta_{k+\alpha}u}) + \operatorname{supp}(\widehat{\Delta_k v}),\\ &\subset \mathcal{C}_{\alpha+k} + \mathcal{C}_k \subset B(0, C2^k).\end{aligned}$$

Next, for $-2 \leq \alpha \leq 2$, using (4.3.10) we can write

$$\begin{aligned}\|\Delta_{k+\alpha}u\Delta_k v\|_{L^2} &\leq \|\Delta_{k+\alpha}u\|_{L^2}\|\Delta_k v\|_{L^\infty} \leq \|\Delta_{k+\alpha}u\|_{L^2}\|2^{kd}\left[\theta(2^k\cdot)\right]*v\|_{L^\infty}\\ &\leq \|\Delta_{k+\alpha}u\|_{L^2}\|2^{kd}\left[\theta(2^k\cdot)\right]\|_{L^1}\|v\|_{L^\infty} \leq C2^{-ks}c_k\|v\|_{L^\infty}\end{aligned}$$

where $\sum_k c_k^2 \leq C\|u\|_{H^s}^2$. Since $s > 0$ we deduce from point (3) of Theorem 4.3.2 that

$$\|R\|_{H^s} \leq C\|u\|_{H^s}\|v\|_{L^\infty}. \tag{4.4.13}$$

The proposition then follows from (4.4.11), (4.4.12) and (4.4.13). □

Remark 4.4.9 1. Corollaries 4.4.2 and 4.4.7 extend to C^∞ functions.

2. We have shown in these corollaries that polynomial functions operate on the spaces $H^s(\mathbf{R}^d)$ for $s > \frac{d}{2}$ and on $H^s(\mathbf{R}^d) \cap L^\infty(\mathbf{R}^d)$ for $s \geq 0$. One could ask what happens for $s < \frac{d}{2}$. An answer was given by the Swedish mathematician, Dahlberg. He proved that the only functions F that operate by composition on $H^s(\mathbf{R}^d)$ for integer s and $2 \leq s < \frac{d}{2}$ are the linear functions $F(z) = \lambda z$.

4.5 A Continuity Result in Sobolev Spaces

Although it applies to all situations (composition or flow), the following result is especially interesting for flows of partial differential equations.

For $s \in \mathbf{R}$, $r > 0$ and $u_0 \in H^s(\mathbf{R}^d)$ we set

$$B_s(u_0, r) = \{u \in H^s(\mathbf{R}^d) : \|u - u_0\|_{H^s} < r\}.$$

Theorem 4.5.1 *Let* $\Phi : B_s(u_0, r) \to H^s(\mathbf{R}^d), \quad u \mapsto \Phi(u)$. *We assume that*

1. *Φ is bounded on $B_s(u_0, r)$ i.e.*

$$\exists C > 0 : \|\Phi(u)\|_{H^s} \leq C, \quad \forall u \in B_s(u_0, r). \tag{4.5.1}$$

2. *Φ is Lipschitz in small norm i.e.*

$$\exists s_0 < s,\ \exists C > 0 : \|\Phi(u) - \Phi(v)\|_{H^{s_0}} \leq C\|u - v\|_{H^{s_0}}, \quad \forall u, v \in B_s(u_0, r). \tag{4.5.2}$$

3. *Propagation of regularity i.e.*

$$\exists s_1 > s,\ \exists C > 0 : \|\Phi(u)\|_{H^{s_1}} \leq C\|u\|_{H^{s_1}}, \quad \forall u \in B_s(u_0, r) \cap H^{s_1}(\mathbf{R}^d). \tag{4.5.3}$$

Then the map Φ is continuous.

Remark 4.5.2 The expression "propagation of regularity" is used in condition 3. to indicate that we require the quantity $\Phi(u)$ (which is initially assumed to belong to the space H^s if $u \in H^s$) to belong to a better space H^{s_1} (since $s_1 > s$) if $u \in H^{s_1}$.

Corollary 4.5.3 *Let $s \in]0, 1[$ and $F \in Lip(\mathbf{C}, \mathbf{C})$ such that $F(0) = 0$.*

Then the map $u \mapsto F(u) = F \circ u$ sends $H^s(\mathbf{R}^d)$ to $H^s(\mathbf{R}^d)$ and is continuous.

Proof The fact that F maps $H^s(\mathbf{R}^d)$ into itself was shown in Theorem 4.4.4. The inequality (4.4.5) shows that condition 1. of Theorem 4.5.1 is satisfied. For condition 2., we will take $s_0 = 0$. Indeed, the fact that F is Lipschitz implies that

$$\|F(u) - F(v)\|_{L^2} \leq C\|u - v\|_{L^2}.$$

Finally, condition 3. is satisfied for s_1 such that $s < s_1 < 1$ thanks to inequality (4.4.5). The mapping F is therefore continuous from $B_s(u_0, r)$ into $H^s(\mathbf{R}^d)$, for any $u_0 \in H^s(\mathbf{R}^d)$ and any $r > 0$. It is therefore continuous from $H^s(\mathbf{R}^d)$ into itself. □

Idea of the Proof of Theorem 4.5.1

The notations Δ_j and S_n were introduced in (4.3.7), (4.3.8).

Let $(u_k)_{k\in\mathbf{N}} \subset H^s(\mathbf{R}^d)$ which converges to u in $H^s(\mathbf{R}^d)$. To show that $(\Phi(u_k))_{k\in\mathbf{N}}$ converges to $\Phi(u)$ in $H^s(\mathbf{R}^d)$, we fix $\varepsilon > 0$ and will successively show that

1. we can choose n and k_0 large enough so that

$$\|\Phi(u) - \Phi(S_n(u))\|_{H^s} \le \varepsilon/3 \quad \text{and} \quad \|\Phi(u_k) - \Phi(S_n(u_k))\|_{H^s} \le \varepsilon/3 \text{ for } k \ge k_0,$$

2. with n fixed, $\lim_{k\to+\infty} \|\Phi(S_n(u_k)) - \Phi(S_n(u))\|_{H^s} = 0$.

To begin the proof of Theorem 4.5.1, we introduce the following notion.

We will denote by ℓ^2_+ the set of sequences $(c_j)_{j\in\mathbf{N}}$ of positive real numbers such that $\sum_{j\in\mathbf{N}} c_j^2 < +\infty$.

Frequency Envelopes

Definition 4.5.4 A sequence $(c_j)_{j\in\mathbf{N}} \in \ell^2_+$ is called an s-frequency envelope of a function $u \in H^s(\mathbf{R}^d)$ if it has the following properties:

1. Energy bound: we have

$$\|\Delta_j u\|_{H^s} \le c_j, \quad \forall j \in \mathbf{N}. \tag{4.5.4}$$

2. Slow variation: there exists $\delta > 0$ such that

$$c_j \le C\, 2^{\delta|k-j|} c_k, \quad \forall j, k \in \mathbf{N}. \tag{4.5.5}$$

Moreover, this s-envelope will be called precise if there exists $C > 0$ such that

$$\sum_{j\in\mathbf{N}} c_j^2 \le C\, \|u\|^2_{H^s}.$$

Lemma 4.5.5 *A precise s-envelope always exists, for any $\delta > 0$.*

Proof Indeed, set $\widetilde{c}_m = \|\Delta_m u\|_{H^s}$ then,

$$c_i = \sup_m 2^{-\delta|j-m|} \widetilde{c}_m.$$

Condition (4.5.4) is satisfied since $c_j \ge \widetilde{c}_j$. Since $|k-m| \le |k-j| + |j-m|$, we have

$$c_j \le \sup_m 2^{\delta|k-j|} 2^{-\delta|m-k|} \widetilde{c}_m \le 2^{\delta|k-j|} c_k.$$

Finally, since $c_j^2 \leq \sup_m 2^{-2\delta|j-m|}\widetilde{c}_m^2 \leq \sum_{m\in\mathbf{N}} 2^{-2\delta|j-m|}\widetilde{c}_m^2$, we have, by definition of $\widetilde{c}_m$,

$$\sum_{j\in\mathbf{N}} c_j^2 \leq \sum_{m\in\mathbf{N}}\Big(\sum_{j\in\mathbf{N}} 2^{-2\delta|j-m|}\Big)\widetilde{c}_m^2 \leq C\sum_{m\in\mathbf{N}} \|\Delta_m u\|_{H^s}^2 \leq C'\sum_{m\in\mathbf{N}} 2^{2ms}\|\Delta_m u\|_{L^2}^2$$
$$\leq C''\|u\|_{H^s}^2,$$

according to (4.3.12). □

Proposition 4.5.6 *Let $(f_n)_{n\in\mathbf{N}}$ be a sequence in $H^s(\mathbf{R}^d)$ which converges to f in $H^s(\mathbf{R}^d)$. Then for any $\delta > 0$ the family of s-envelopes defined by*

$$c_{j,n} = \sup_m 2^{-\delta|j-m|}\widetilde{c}_{m,n}, \quad \widetilde{c}_{m,n} = \|\Delta_m f_n\|_{H^s}$$

converges in ℓ^2_+ to the sequence (c_j) which is the s-envelope of f defined analogously.

Proof First, note that the sequences $(\widetilde{c}_m)$ and $(\widetilde{c}_{m,n})$, for fixed n, tend to zero as m tends to $+\infty$, since they belong to ℓ^2_+. We have

$$|c_{j,n} - c_j|^2 = |\sup_m 2^{-\delta|j-m|}\widetilde{c}_{m,n} - \sup_m 2^{-\delta|j-m|}\widetilde{c}_m|^2 = \sup_m 2^{-2\delta|j-m|}|\widetilde{c}_{m,n} - \widetilde{c}_m|^2,$$
$$\leq \sum_m 2^{-2\delta|j-m|}|\widetilde{c}_{m,n} - \widetilde{c}_m|^2.$$

We deduce that

$$\sum_j |c_{j,n} - c_j|^2 \leq \sum_m\Big(\sum_j 2^{-2\delta|j-m|}\Big)|\widetilde{c}_{m,n} - \widetilde{c}_m|^2 \leq C\sum_m |\widetilde{c}_{m,n} - \widetilde{c}_m|^2,$$
$$\leq C\sum_m \big|\|\Delta_m f_n\|_{H^s} - \|\Delta_m f\|_{H^s}\big|^2 \leq C\sum_m \|\Delta_m(f_n - f)\|_{H^s}^2,$$
$$\leq C\sum_m 2^{2ms}\|\Delta_m(f_n - f)\|_{L^2}^2 \leq C\|f_n - f\|_{H^s}^2,$$

according to (4.3.12). □

Regularization Procedure

Let $u \in H^s(\mathbf{R}^d)$ and $(c_j)_{j\in\mathbf{N}}$ an s-frequency envelope of u corresponding to $\delta > 0$ such that $s_1 - s - \delta > 0$. Let S_j, Δ_j be the operators defined in (4.3.7) and (4.3.8).

Proposition 4.5.7 *We have the following properties.*

$$(i) \quad \|\Delta_k S_n(u)\|_{H^s} \leq c_k,$$

$$(ii) \quad \|S_n(u)\|_{H^{s_1}} \leq C 2^{n(s_1-s)} c_n,$$

$$(iii) \quad \|S_{n+1}(u) - S_n(u)\|_{H^{s_0}} \leq C 2^{-n(s-s_0)} c_n,$$

$$(iv) \quad \lim_{n\to+\infty} \|S_n(u) - u\|_{H^s} = 0.$$

Proof We have

$$(i) \quad \|\Delta_k S_n(u)\|_{H^s}^2 = \|S_n(\Delta_k u)\|_{H^s}^2 = \int_{\mathbf{R}^d} \langle\xi\rangle^{2s} |\psi(2^{-n}\xi)\varphi(2^{-k}\xi)\widehat{u}(\xi)|^2 \, d\xi,$$
$$\leq \int_{\mathbf{R}^d} \langle\xi\rangle^{2s} |\varphi(2^{-k}\xi)\widehat{u}(\xi)|^2 \, d\xi = \|\Delta_k u\|_{H^s}^2 \leq c_k^2.$$

$$(ii) \quad \|S_n(u)\|_{H^{s_1}} \leq \sum_{k=-1}^{n-1} \|\Delta_k u\|_{H^{s_1}} \leq C \sum_{k=-1}^{n-1} 2^{k(s_1-s)} \|\Delta_k u\|_{H^s},$$
$$\leq C \sum_{k=-1}^{n-1} 2^{k(s_1-s)} c_k.$$

The sequence (c_k) is slowly varying, with $\delta > 0$ such that $s_1 - s - \delta > 0$ that is, $c_k \leq 2^{\delta|n-k|} c_n$. Consequently,

$$\|S_n(u)\|_{H^{s_1}} \leq \sum_{k=-1}^{n} 2^{k(s_1-s)} 2^{\delta(n-k)} c_n \leq \sum_{k=-1}^{n} 2^{k(s_1-s-\delta)} 2^{\delta n} c_n,$$
$$\leq C 2^{n(s_1-s-\delta)} 2^{\delta n} c_n \leq C 2^{n(s_1-s)} c_n.$$

$$(iii) \quad \|S_{n+1}(u) - S_n(u)\|_{H^{s_0}} = \|\Delta_n u\|_{H^{s_0}} \leq 2^{-n(s-s_0)} \|\Delta_n u\|_{H^s} \leq 2^{-n(s-s_0)} c_n.$$

(iv) Using (4.3.8) we see that

$$\|S_n(u) - u\|_{H^s}^2 = \int_{\mathbf{R}^d} \langle\xi\rangle^{2s} |\psi(2^{-n}\xi) - 1|^2 |\widehat{u}(\xi)|^2 \, d\xi,$$

and the result follows from the dominated convergence theorem. □

It follows from point (iv) of the above proposition that for $u \in B_s(u_0, r)$ and n large enough, $S_n(u)$ belongs to $B_s(u_0, r)$. We can then define $\Phi(S_n u)$.

Proposition 4.5.8

$$\begin{aligned}
&(i) \;\; \exists C > 0 : \|\Phi(S_n u)\|_{H^{s_1}} \leq C\, 2^{n(s_1-s)} c_n, \quad \forall n \geq n_0, \\
&(ii) \;\; \exists C > 0 : \|\Phi(u) - \Phi(S_n(u))\|_{H^{s_0}} \leq C\, 2^{-n(s-s_0)} c_n \quad \forall n \geq n_0, \\
&(iii) \;\; \lim_{n\to+\infty} \|\Phi(u) - \Phi(S_n(u))\|_{H^s} = 0.
\end{aligned}$$

Proof (i) It follows from (4.5.3) and (ii) of Proposition 4.5.7 that

$$\|\Phi(S_n u)\|_{H^{s_1}} \leq C\|S_n(u)\|_{H^{s_1}} \leq C' 2^{n(s_1-s)} c_n.$$

(ii) We deduce from (4.5.2) and (iii) of Proposition 4.5.7 that

$$\begin{aligned}
\|\Phi(S_{n+1}(u)) - \Phi(S_n(u))\|_{H^{s_0}} &\leq C\|S_{n+1}(u) - S_n(u)\|_{H^{s_0}}, \\
&\leq C\, 2^{-n(s-s_0)} c_n.
\end{aligned} \tag{4.5.6}$$

On the other hand, from (4.5.2), Φ is continuous on $H^{s_0}(\mathbf{R}^d)$ and point (iv) of Proposition 4.5.7 shows that $(S_n(u))$ converges to u in $H^{s_0}(\mathbf{R}^d)$. Thus, $\Phi(S_n(u))$ converges to $\Phi(u)$ in $H^{s_0}(\mathbf{R}^d)$. From inequality (4.5.6), the telescoping series $\sum_{m\geq n}\big(\Phi(S_{m+1}(u)) - \Phi(S_m(u))\big)$ converges in $H^{s_0}(\mathbf{R}^d)$ and we have

$$\Phi(u) - \Phi(S_n(u)) = \sum_{m\geq n}\big(\Phi(S_{m+1}(u)) - \Phi(S_m(u))\big).$$

We deduce from (4.5.6) that

$$\begin{aligned}
\|\Phi(u) - \Phi(S_n(u))\|_{H^{s_0}} &\leq C\sum_{m\geq n} 2^{-m(s-s_0)} c_m \leq C\sum_{m\geq n} 2^{-m(s-s_0-\delta)} 2^{-n\delta} c_n, \\
&\leq C 2^{-n(s-s_0)} c_n,
\end{aligned}$$

which proves point (ii).

(iii) Suppose $m \leq n$. We can write, according to (4.5.6),

$$\begin{aligned}
\|\Delta_m\big(\Phi(S_{n+1}(u)) - \Phi(S_n(u))\|_{H^s} &\leq 2^{m(s-s_0)}\|\Delta_m\big(\Phi(S_{n+1}(u)) - \Phi(S_n(u))\|_{H^{s_0}}, \\
&\leq 2^{(m-n)(s-s_0)} c_n.
\end{aligned}$$

Next, if $m \geq n$ we use point (i) of Proposition 4.5.8. It follows

$$\begin{aligned}
\|\Delta_m\big(\Phi(S_{n+1}(u)) - \Phi(S_n(u))\|_{H^s} &\leq C2^{-m(s_1-s)}\|\Delta_m\big(\Phi(S_{n+1}(u)) - \Phi(S_n(u))\|_{H^{s_1}}, \\
&\leq C2^{(n-m)(s_1-s)} c_n.
\end{aligned}$$

Therefore, for all m, n we have

$$\|\Delta_m\big(\Phi(S_{n+1}(u)) - \Phi(S_n(u))\|_{H^s} \leq C2^{-\kappa|m-n|}c_n, \tag{4.5.7}$$

where $\kappa = \min(s - s_0, s_1 - s)$.

We have $S_n(u) \to u$ in $H^{s_0}(\mathbf{R}^d)$ so $\Phi(S_n(u)) \to \Phi(u)$ in $H^{s_0}(\mathbf{R}^d)$ by (4.5.2). For fixed m we have

$$\|\Delta_m(\Phi(S_n(u)) - \Phi(u))\|_{H^s} \leq 2^{m(s-s_0)}\|\Delta_m(\Phi(S_n(u)) - \Phi(u))\|_{H^{s_0}} \to 0.$$

Thus, in $H^s(\mathbf{R}^d)$ we can write, using (4.5.7),

$$\begin{aligned}\|\Delta_m\big(\Phi(u) - \Phi(S_n(u))\big)\|_{H^s} &\leq C\sum_{p\geq n}\|\Delta_m(\Phi(S_{p+1}(u)) - \Phi(S_p(u)))\|_{H^s},\\ &\leq C\sum_{p\geq n}2^{-\kappa|m-p|}c_p \leq C\sum_{p\in\mathbf{Z}}2^{-\kappa|m-p|}1_{p\geq n}c_p.\end{aligned}$$

where $1_{p\geq n}$ denotes the characteristic function of the set $\{p \in \mathbf{Z} : p \geq n\}$.

The right-hand side of the above inequality is the discrete convolution on $\mathbf{Z}$ of the sequences $(\alpha_p = 2^{-\kappa|p|})$ and $(d_p = 1_{p\geq n}c_p)$. Young's theorem implies that

$$\|\alpha * d\|_{\ell^2(\mathbf{Z})} \leq C\|\alpha\|_{\ell^1(\mathbf{Z})}\|d\|_{\ell^2(\mathbf{Z})}.$$

Since $\|\alpha\|_{\ell^1(\mathbf{Z})} = \sum_{p\in\mathbf{Z}} 2^{-\kappa|p|} < +\infty$ we obtain

$$\begin{aligned}\|\Phi(u) - \Phi(S_n(u))\|_{H^s}^2 &\leq \sum_m \|\Delta_m\big(\Phi(u) - \Phi(S_n(u))\big)\|_{H^s}^2,\\ &\leq C\sum_{p\geq n} c_p^2 \to 0, \quad n \to +\infty,\end{aligned} \tag{4.5.8}$$

since the right-hand side is the remainder of a convergent series as $(c_p) \in \ell^2_+$. Point (iii) is thus proved.

Note that this proves that $(\Phi(S_n(u)))$ converges to $\Phi(u)$ in $H^s(\mathbf{R}^d)$. □

Proof *(End of the proof of Theorem* 4.5.1) Let $(u_k)_{k\in\mathbf{N}} \subset H^s(\mathbf{R}^d)$ which converges in $H^s(\mathbf{R}^d)$ to $u \in B_s(u_0, r)$. Let $(c_{p,k})_p$ be an s-frequency envelope of u_k which satisfies the conclusion of Proposition 4.5.6. For k large enough we have $u_k \in B_s(u_0, r)$. We can thus apply (4.5.8) and write

$$\|\Phi(u_k) - \Phi(S_n(u_k))\|_{H^s}^2 \leq C\sum_{p\geq n} c_{p,k}^2, \qquad \|\Phi(u) - \Phi(S_n(u))\|_{H^s}^2 \leq C\sum_{p\geq n} c_p^2.$$

According to Proposition 4.5.6, for any $\varepsilon > 0$ we can choose n and k_0 large enough so that for $k \geq k_0$ we have

$$\|\Phi(u_k) - \Phi(S_n(u_k))\|_{H^s} \leq \varepsilon/3 \qquad \|\Phi(u) - \Phi(S_n(u))\|_{H^s} \leq \varepsilon/3. \tag{4.5.9}$$

We thus fix the integer n. We then write

$$\begin{aligned}\|\Phi(u_k) - \Phi(u)\|_{H^s} \leq \|\Phi(u_k) - \Phi(S_n(u_k))\|_{H^s} + \|\Phi(S_n(u_k)) - \Phi(S_n(u))\|_{H^s} \\ + \|\Phi(S_n(u)) - \Phi(u)\|_{H^s} = (1) + (2) + (3).\end{aligned}$$

The terms (1) and (3) are bounded by $\varepsilon/3$ according to (4.5.9). Let us estimate term (2). Since $s \in]s_0, s_1[$ we can write $s = \theta s_0 + (1-\theta)s_1$ where $0 < \theta < 1$. We then have the inequality

$$\|v\|_{H^s} \leq \|v\|_{H^{s_0}}^{\theta} \|v\|_{H^{s_1}}^{1-\theta}. \tag{4.5.10}$$

Indeed, we write

$$\langle\xi\rangle^{2s} |\widehat{v}(\xi)|^2 = \langle\xi\rangle^{2\theta s_0} |\widehat{v}(\xi)|^{2\theta} \langle\xi\rangle^{2(1-\theta)s_1} |\widehat{v}(\xi)|^{2(1-\theta)},$$

then we use Hölder's inequality with $p = \frac{1}{\theta}, q = \frac{1}{1-\theta}$.

$$\int_{\mathbf{R}^d} \langle\xi\rangle^{2s} |\widehat{v}(\xi)|^2 \, d\xi \leq \Big(\int_{\mathbf{R}^d} \langle\xi\rangle^{2s_0} |\widehat{v}(\xi)|^2 \, d\xi\Big)^{\theta} \Big(\int_{\mathbf{R}^d} \langle\xi\rangle^{2s_1} |\widehat{v}(\xi)|^2 \, d\xi\Big)^{1-\theta}.$$

□

Using inequality (4.5.10) we can write

$$\begin{aligned}(2) &\leq \|\Phi(S_n(u_k)) - \Phi(S_n(u))\|_{H^{s_0}}^{\theta} \|\Phi(S_n(u_k)) - \Phi(S_n(u))\|_{H^{s_1}}^{1-\theta}, \\ &\leq \|\Phi(S_n(u_k)) - \Phi(S_n(u))\|_{H^{s_0}}^{\theta} \big(\|\Phi(S_n(u_k))\|_{H^{s_1}} + \|\Phi(S_n(u))\|_{H^{s_1}}\big)^{1-\theta}.\end{aligned} \tag{4.5.11}$$

According to hypothesis (4.5.2) we have

$$\|\Phi(S_n(u_k)) - \Phi(S_n(u))\|_{H^{s_0}}^{\theta} \leq C\|S_n(u_k) - S_n(u)\|_{H^{s_0}}^{\theta} \leq C\|u_k - u\|_{H^{s_0}}^{\theta},$$

so that

$$\lim_{k\to+\infty} \|\Phi(S_n(u_k)) - \Phi(S_n(u))\|_{H^{s_0}}^{\theta} = 0. \tag{4.5.12}$$

Then, according to hypothesis (4.5.3) we have

$$\begin{aligned}&\|\Phi(S_n(u_k))\|_{H^{s_1}} \leq C\|S_n(u_k)\|_{H^{s_1}} \leq C2^{n(s_1-s)}\|u_k\|_{H^s} \leq C2^{n(s_1-s)}R, \\ &\|\Phi(S_n(u))\|_{H^{s_1}} \leq \|S_n(u)\|_{H^{s_1}} \leq C2^{n(s_1-s)}\|u\|_{H^s} \leq C2^{n(s_1-s)}R.\end{aligned}$$

It follows from (4.5.11) and (4.5.12) that

$$\lim_{k\to+\infty} (2) = 0. \tag{4.5.13}$$

We deduce from (4.5.9) and (4.5.13) that

$$\lim_{k\to+\infty} \|\Phi(u_k) - \Phi(u)\|_{H^s} = 0.$$

This proves the continuity of Φ on $B_s(u_0, R)$ and completes the proof of Theorem 4.5.1.

4.6 A Continuity Result in L^p Spaces

Let $(X, \mathcal{A}, \mu)$ be a measured space where μ is σ-finite that is, $X = \cup_{n\in\mathbf{N}} X_n$ where $X_n \in \mathcal{A}$, $X_n \subset X_{n+1}$, and $\mu(X_n) < +\infty$, $\forall n \in \mathbf{N}$. We aim, in what follows, to show the following result.

Theorem 4.6.1 *Let $g : \mathbf{C} \to \mathbf{C}$ be a continuous function. Suppose there exist $C > 0$ and $k \in [1, +\infty[$ such that*

$$(\star) \quad |g(z)| \leq C|z|^k, \quad \forall z \in \mathbf{C}.$$

If $u : X \to \mathbf{C}$ is measurable, we define $G(u) : X \to \mathbf{C}$ by $G(u)(x) = g(u(x))$. Then, for all $p \geq k$, G is a continuous operator from $L^p(X)$ into $L^{\frac{p}{k}}(X)$.

The rest of this paragraph is devoted to the proof of this result which will require several steps.

Lemma 4.6.2 *G maps $L^p(X)$ into $L^{\frac{p}{k}}(X)$.*

Proof This follows from $(*)$, since $|G(u(x))|^{\frac{p}{k}} \leq C|u(x)|^p$. □

Lemma 4.6.3 *For all z, z' in $\mathbf{C}$ we have*

$$(|g(z)| + |g(z')|)^{\frac{p}{k}} \leq 2^{\frac{p}{k}}(1 + 2^p)|z|^p + 2^{\frac{p}{k}+p}|z - z'|^p.$$

Proof Recall that for all $a > 0, b > 0, \alpha > 0$ we have $(a + b)^\alpha \leq 2^\alpha(a^\alpha + b^\alpha)$.
Using this inequality with $\alpha = \frac{p}{k}$ and assumption $(*)$, we can write

$$\begin{aligned}
\left(|g(z)| + |g(z')|\right)^{\frac{p}{k}} &\leq 2^{\frac{p}{k}}\left(|g(z)|^{\frac{p}{k}} + |g(z')|^{\frac{p}{k}}\right) \leq 2^{\frac{p}{k}}(|z|^p + |z'|^p), \\
&\leq 2^{\frac{p}{k}}(|z|^p + (|z - z'| + |z|)^p) \leq 2^{\frac{p}{k}}(|z|^p + 2^p(|z - z'|^p + |z|^p)), \\
&\leq 2^{\frac{p}{k}}(1 + 2^p)|z|^p + 2^{\frac{p}{k}+p}|z - z'|^p.
\end{aligned}$$

□

Lemma 4.6.4 *Let $f \in L^1(X)$.*

(i) For any $\varepsilon > 0$ there exists $n \in \mathbf{N}$ such that

$$\int_{X_n^c} |f(x)| d\mu \leq \varepsilon,$$

where X_n^c denotes the complement of X_n.

(ii) For any $\varepsilon > 0$ there exists $C_0 > 0$ such that

$$\int_{\{x \in X : |f(x)| > C_0\}} |f(x)| d\mu \leq \varepsilon.$$

Proof (i) Let us set $f_n(x) = 1_{X_n^c} |f(x)|$. We have $0 \leq f_n(x) \leq |f(x)|$. On the other hand, if $x_0 \in X$, there exists n_0 such that $x_0 \in X_{n_0}$. Then $x_0 \in X_n$ for all $n \geq n_0$ and thus we have $f_{n_0}(x_0) = 0$, which shows that $\lim_{n \to +\infty} f_n(x_0) = 0$. The dominated convergence theorem implies that $\lim_{n \to +\infty} \int_{X_n^c} |f(x)| \, d\mu = 0$.

(ii) Let us set $f_n(x) = 1_{\{x \in X : |f(x)| \geq n\}} |f(x)|$. We have $0 \leq f_n(x) \leq |f(x)|$. Since $f \in L^1(X)$, we have $|f(x)| < +\infty$ almost everywhere. For almost every fixed $x_0 \in X$ and $n < |f(x_0)|$ we have $f_n(x_0) = 0$ and thus $\lim_{n \to +\infty} f_n(x_0) = 0$. The dominated convergence theorem implies that

$$\lim_{n \to +\infty} \int_{\{x \in X : |f(x)| \geq n\}} |f(x)| \, d\mu = 0.$$

□

Lemma 4.6.5 *Let $u \in L^p(X)$, $\varepsilon > 0$, $\delta > 0$ and let $v \in L^p(X)$ which is such that $\|u - v\|_{L^p(X)} \leq \delta$.*

(i) There exists $n_0 \in \mathbf{N}$ such that

$$\int_{X_{n_0}^c} |g(u(x)) - g(v(x))|^{\frac{p}{k}} d\mu \leq \frac{1}{10} \varepsilon^{\frac{p}{k}} + 2^{\frac{p}{k} + p} \delta^p.$$

We fix n_0 in this way.

(ii) There exists $B_0 > 0$ such that

$$\int_{\{x \in X_{n_0} : |u(x)| > B_0\}} |g(u(x)) - g(v(x))|^{\frac{p}{k}} d\mu \leq \frac{1}{10} \varepsilon^{\frac{p}{k}} + 2^{\frac{p}{k} + p} \delta^p.$$

We fix B_0 in this way.

Proof Using Lemma 4.6.3 we get

$$|g(u(x)) - g(v(x))|^{\frac{p}{k}} \leq 2^{\frac{p}{k}} (1 + 2^p) |u(x)|^p + 2^{\frac{p}{k} + p} |u(x) - v(x)|^p. \qquad (4.6.1)$$

According to Lemma 4.6.4 (i) there exists n_0 such that

$$\int_{X_{n_0}^c} |u(x)|^p \, d\mu \leq \frac{1}{10} \frac{\varepsilon^{\frac{p}{k}}}{2^{\frac{p}{k}}(1+2^p)} \leq \frac{\varepsilon^{\frac{p}{k}}}{10}.$$

On the other hand,

$$2^{p+\frac{p}{k}} \int_{X_{n_0}^c} |u(x) - v(x)|^p \, d\mu \leq 2^{p+\frac{p}{k}} \|u - v\|_{L^p(X)}^p \leq 2^{p+\frac{p}{k}} \delta^p.$$

Point (i) then follows from these three inequalities.

Next, since $u \in L^p(X)$, Lemma 4.6.4 (ii) shows that there exists $B_0 > 0$ such that

$$\int_{\{x \in X_{n_0} : |u(x)| > B_0\}} |u(x)|^p \, d\mu \leq \frac{1}{10} \varepsilon^{\frac{p}{k}}.$$

It is then enough to use (4.6.1) to deduce point (ii) of the Lemma. □

Lemma 4.6.6 *Let $u \in L^p(X)$ and $\varepsilon > 0$. There exists $\delta > 0$ such that for all v in $L^p(X)$ satisfying $\|u - v\|_{L^p(X)} \leq \delta$, if we set*

$$A = \{x \in X_{n_0} : |u(x)| \leq B_0, |u(x) - v(x)| \geq \sqrt{\delta}\},$$

then

(i) $\mu(A) \leq \delta^{\frac{p}{2}}$,

(ii) $\displaystyle\int_A |g(u(x)) - g(v(x))|^{\frac{p}{k}} d\mu \leq \frac{1}{10} \varepsilon^{\frac{p}{k}} + 2^{\frac{p}{k}+p} \delta^p.$

Proof (i) Recall that if $f \in L^1(X)$, for any $\alpha > 0$ there exists $\eta > 0$ such that

$$A \in \mathcal{A}, \quad \mu(A) \leq \eta \quad \Longrightarrow \int_A |f(x)| d\mu \leq \alpha.$$

We take $f = |u|^p$, $\alpha = \frac{1}{10} \frac{\varepsilon^{\frac{p}{k}}}{2^{\frac{p}{k}}(1+2^p)}$ and $\delta > 0$ such that $\delta^{\frac{p}{2}} < \eta$. We can write

$$\delta^p \geq \int_A |u(x) - v(x)|^p \geq \delta^{\frac{p}{2}} \mu(A),$$

which proves the first point.

Next, since $\mu(A) \leq \eta$ we have $\int_A |u(x)|^p \, d\mu \leq \alpha$. It is enough to use (4.6.1) to prove point (ii). □

Lemma 4.6.7 *Let $\varepsilon > 0$. There exists $\delta > 0$ such that if we set*

$$\mathcal{B} = \{x \in X_{n_0} : |u(x)| \leq B_0, |u(x) - v(x)| \leq \sqrt{\delta}\},$$

then

$$\int_B |g(u(x)) - g(v(x))|^{\frac{p}{k}} d\mu \leq \frac{1}{10}\varepsilon^{\frac{p}{k}}.$$

Proof In $\mathcal{B}$ we have $|u(x)| \leq B_0$ and $|v(x)| \leq |u(x)| + \sqrt{\delta} \leq B_0 + 1$ if $\delta \leq 1$. In the ball of radius $B_0 + 1$ the function g is uniformly continuous, that is

$$\forall \alpha > 0 \quad \exists \gamma : |z - z'| \leq \gamma \Longrightarrow |g(z) - g(z')| \leq \alpha.$$

We take $\alpha = \frac{\varepsilon}{\left(10\mu(X_{n_0})\right)^{\frac{k}{p}}}$ and $\delta > 0$ such that $\sqrt{\delta} \leq \gamma$ then

$$|g(u(x)) - g(v(x))| \leq \frac{\varepsilon}{\left(10\mu(X_{n_0})\right)^{\frac{k}{p}}},$$

which implies that

$$\int_{\mathcal{B}} |g(u(x)) - g(v(x))|^{\frac{p}{k}} \, d\mu \leq \frac{\varepsilon^{\frac{p}{k}}}{10\mu(X_{n_0})}\mu(\mathcal{B}) \leq \frac{\varepsilon^{\frac{p}{k}}}{10},$$

because $\mathcal{B} \subset X_{n_0}$. □

4.6.1 End of the Proof of Theorem 4.6.1

Set $F(x) = |g(u(x)) - g(v(x))|^{\frac{p}{k}}$. With the choices of δ made in Lemmas 4.6.6 and 4.6.7, that is, $\delta \leq 1$, $\delta^{\frac{p}{2}} \leq \eta$, $\sqrt{\delta} \leq \gamma$ we write

$$\begin{aligned}
\int_X F(x) \, d\mu &= \int_{X_{n_0}^c} F(x) \, d\mu + \int_{X_{n_0}} F(x) \, d\mu, \\
&\leq \frac{\varepsilon^{\frac{p}{k}}}{10} + 2^{p+\frac{p}{k}}\delta^p + \int_{\{x \in X_{n_0} : |u(x)| > B_0\}} F(x) \, d\mu, \\
&\leq \frac{\varepsilon^{\frac{p}{k}}}{5} + 2^{p+1+\frac{p}{k}}\delta^p + \int_A F(x) \, d\mu + \int_B F(x) \, d\mu, \\
&\leq \frac{4\varepsilon^{\frac{p}{k}}}{5} + 2^{p+2+\frac{p}{k}}\delta^p.
\end{aligned}$$

By further choosing δ such that $2^{p+2+\frac{p}{k}}\delta^p \leq \frac{\varepsilon^{\frac{p}{k}}}{5}$ we obtain

$$\int_X |g(u(x)) - g(v(x))|^{\frac{p}{k}}\, d\mu \leq \varepsilon^{\frac{p}{k}},$$

which proves the announced continuity.

The purpose of the following result is to prove that condition $(*)$ of Theorem 4.6.1 is necessary for the operator G defined by $G(u)(x) = g(u(x))$ to map from $L^p(\mathbf{R}^d)$ to $L^{\frac{p}{k}}(\mathbf{R}^d)$.

Theorem 4.6.8 *Let $g : \mathbf{C} \to \mathbf{C}$ be a continuous function such that $g(0) = 0$. Suppose that G operates from $L^p(\mathbf{R}^d)$ into $L^{\frac{p}{k}}(\mathbf{R}^d)$, for $p \geq k \geq 1$ then*

$$(**)\quad \exists C > 0 : \quad |g(z)| \leq C|z|^k, \quad \forall z \in \mathbf{C} \setminus \{0\}.$$

Proof We proceed by contradiction. If $(**)$ is false, for every $n \in \mathbf{N} \setminus \{0\}$ there exists $z_n \in \mathbf{C} \setminus \{0\}$ such that

$$|g(z_n)| > n^k |z_n|^k. \tag{4.6.2}$$

Let (B_n) be a sequence of measurable **disjoint** sets in $\mathbf{R}^d$ such that

$$|z_n|^p |B_n| = \frac{1}{n^{p+1}}, \quad n \geq 1. \tag{4.6.3}$$

Set

$$u(x) = \sum_{n\geq 1} z_n \mathbf{1}_{B_n}(x),$$

where $\mathbf{1}_{B_n}$ is the characteristic function of B_n. Then $u \in L^p(\mathbf{R}^d)$.

Indeed, set $A_N = \cup_{n=1}^N B_n$ and $A_\infty = \cup_{n\geq 1} B_n$. Then $u \equiv 0$ on $\mathbf{R}^d \setminus A_\infty$ so

$$\begin{aligned}\int_{\mathbf{R}^d} |u(x)|^p\, dx &= \int_{A_\infty} |u(x)|^p\, dx = \sum_{n\geq 1} \int_{B_n} |u(x)|^p\, dx = \sum_{n\geq 1} \int_{B_n} |z_n|^p\, dx,\\ &= \sum_{n\geq 1} |z_n|^p |B_n| = \sum_{n\geq 1} \frac{1}{n^{p+1}} < +\infty,\end{aligned}$$

by (4.6.3). On the other hand, since $u(x) = z_n$ on B_n, $u(x) = 0$ if $x \notin B_n$ for all n and $g(0) = 0$ we have $g(u(x)) = \sum_{n\geq 1} g(z_n)\mathbf{1}_{B_n}$. Using (4.6.2), the same reasoning as above shows that

$$\int_{A_N} |g(u(x))|^{\frac{p}{k}}\, dx \geq \sum_{n=1}^N n^p |z_n|^p |B_n| = \sum_{n=1}^N \frac{1}{n} \to +\infty, \quad N \to +\infty.$$

In summary, there exists $u \in L^p(\mathbf{R}^d)$ such that $g(u(x)) \notin L^{\frac{p}{k}}(\mathbf{R}^d)$. □

4.7 Appendix

Proposition 4.7.1 *For $0 < s < 1$ the quantity,*

$$N(u) = \Big(\|u\|^2_{L^2(\mathbf{R}^d)} + \iint_{\mathbf{R}^d\times\mathbf{R}^d} \frac{|u(x)-u(y)|^2}{|x-y|^{d+2s}}\,dx\,dy\Big)^{\frac{1}{2}}$$

is a norm on the space $H^s(\mathbf{R}^d)$ equivalent to the usual norm $\|u\|_{H^s}$.

Proof In the integral in x set $x - y = z$ we get

$$\iint_{\mathbf{R}^d\times\mathbf{R}^d} \frac{|u(x)-u(y)|^2}{|x-y|^{d+2s}}\,dx\,dy = \iint_{\mathbf{R}^d\times\mathbf{R}^d} \frac{|u(y+z)-u(y)|^2}{|z|^{d+2s}}\,dz\,dy.$$

In the integral in y we will use Parseval's identity.
Since $\mathcal{F}_{y\to\xi}(u(y+z)) = e^{i\langle z,\xi\rangle}\widehat{u}(\xi)$ we obtain

$$\iint_{\mathbf{R}^d\times\mathbf{R}^d} \frac{|u(y+z)-u(y)|^2}{|z|^{d+2s}}\,dz\,dy = (2\pi)^{-d}\int_{\mathbf{R}^d_z}\int_{\mathbf{R}^d_\xi} \frac{|e^{i\langle z,\xi\rangle}-1|^2}{|z|^{d+2s}}|\widehat{u}(\xi)|^2\,d\xi\,dz.$$

□

Lemma 4.7.2 *We have*

$$F(\xi) := \int_{\mathbf{R}^d_z} \frac{|e^{i\langle z,\xi\rangle}-1|^2}{|z|^{d+2s}}\,dz = C_{s,d}|\xi|^{2s},$$

where $C_{s,d}$ is a positive constant depending only on s and the dimension d.

Proof Notice that the function F is well defined. Indeed, the integral over the set $\{z : |z| \geq 1\}$ is convergent since $d + 2s > d$. Next, on the set $\{z : |z| < 1\}$ we have $|e^{i\langle z,\xi\rangle} - 1|^2 \leq \langle z, \xi\rangle^2 \leq |z|^2|\xi|^2$ and we use the fact that $d + 2s - 2 < d$.

We show that the function F is radial, i.e., if A is an orthogonal matrix, we have $F(A\xi) = F(\xi)$. Indeed, denoting by ${}^tA$ the transpose matrix, we have

$$F(A\xi) = \int_{\mathbf{R}^d} \frac{|e^{i\langle {}^tAz,\xi\rangle}-1|^2}{|z|^{d+2s}}\,dz.$$

We set $y = {}^tAz$ so that $dy = dz$ (since $|\det A| = 1$) and $|{}^tAy| = |y|$ (because A is orthogonal). We then obtain $F(A\xi) = F(\xi)$.

On the other hand, for any fixed $\xi \in \mathbf{R}^d$, $\xi \neq 0$, there exists an orthogonal matrix A such that $A\xi = (0, \ldots, 0, |\xi|)$ (this is a rotation of the coordinates). From the above, we have

$$F(\xi) = \int_{\mathbf{R}^d} \frac{|e^{i|\xi|z_d} - 1|^2}{|z|^{d+2s}}\, dz.$$

We then switch to spherical coordinates that is, $z = r\omega$ where $r \in (0, +\infty)$ and $\omega \in S^{d-1}$. Then, $|z| = r$ and $dz = r^{d-1}\, dr\, d\omega$. We deduce

$$F(\xi) = \int_{S^{d-1}} \int_0^{+\infty} \frac{|e^{ir|\xi|\omega_d} - 1|^2}{r^{2s+1}}\, dr\, d\omega.$$

Let us set, in the integral in r, $\lambda = r|\xi|$. Then $dr = \frac{1}{|\xi|}\, d\lambda$ and

$$F(\xi) = \Big(\int_{S^{d-1}} \int_0^{+\infty} \frac{|e^{i\lambda\omega_d} - 1|^2}{\lambda^{2s+1}}\, d\lambda\, d\omega \Big) |\xi|^{2s} = C_{s,d}|\xi|^{2s}.$$

Note that the integral in λ above is indeed meaningful. In fact, for $\lambda \geq 1$ it converges because $2s + 1 > 1$ and for $0 < \lambda < 1$ we have $|e^{i\lambda\omega_d} - 1|^2 \leq \lambda^2$ and $2s - 1 < 1$. □

Proposition 4.7.1 follows immediately from Lemma 4.7.2.

Corollary 4.7.3 *Suppose that $s = k + \sigma$ where $k \in \mathbf{N}$ and $0 < \sigma < 1$. Then the quantity*

$$N_s(u) = \Big(\|u\|^2_{L^2(\mathbf{R}^d)} + \sum_{|\alpha|=k} \iint_{\mathbf{R}^d \times \mathbf{R}^d} \frac{|D^\alpha u(x) - D^\alpha u(y)|^2}{|x - y|^{d+2\sigma}}\, dx\, dy \Big)^{\frac{1}{2}}$$

is equivalent to the usual norm of H^s.

Proof According to Parseval's equality and Lemma 4.7.2 the quantity $N_s^2(u)$ is equal to

$$(2\pi)^{-d} \Big(\int_{\mathbf{R}^d} |\widehat{u}(\xi)|^2\, d\xi + C_{\sigma,d} \sum_{|\alpha|=k} \int_{\mathbf{R}^d} |\xi^\alpha|^2 |\xi|^{2\sigma} |\widehat{u}(\xi)|^2\, d\xi \Big).$$

Corollary 4.7.3 then follows from the fact that there exist two positive constants C_1, C_2 depending on k and d such that

$$C_1|\xi|^{2k} \leq \sum_{|\alpha|=k} |\xi^\alpha|^2 \leq C_2|\xi|^{2k}.$$

□

4.8 Comments

1. The reason why the continuity of composition operators on Sobolev spaces is less interesting is contained in Remark 4.4.9. Basically, this remark states that for $s > \frac{d}{2}$ all C^∞ functions operate on the spaces $H^s(\mathbf{R}^d)$ while for $2 \leq s < \frac{d}{2}$ (and s integer) only the functions $F(z) = cz$ operate.
2. For semilinear partial differential equations (for example $\partial_t^2 u - \Delta u = u^2$) the map $u_0 \mapsto u$ is Lipschitz, i.e., of the type $\|u - v\|_E \leq C\|u_0 - v_0\|_F$ and thus the continuity of the flow is trivial. This is no longer true for quasilinear equations (for example $\partial_t u + u\partial_x u = 0$) and Theorem 4.5.1 turns out to be essential in this case.

Chapter 5
The Burgers Equation

5.1 Prerequisites

Distributions. Bounded Lipschitz functions on $\mathbf{R}$. Differentiability. Sobolev spaces. Sobolev embeddings. Appendix of Chap. 4.

5.2 The Context and the Problem

The equation we will study in this chapter is a **nonlinear** partial differential equation. Here, we aim to solve the associated Cauchy problem, that is, the following problem:

$$\begin{cases} (\frac{\partial u}{\partial t} + u\frac{\partial u}{\partial x})(t, x) = 0, \quad t \in \mathbf{R}, \quad x \in \mathbf{R}, \\ u|_{t=0} = u_0. \end{cases} \tag{5.2.1}$$

where u_0 is assumed to be **real**-valued, which is essential.

This is an evolution problem, which we have already seen in Chap. 1, but which we will study in more detail in this chapter. This problem is interesting for several reasons. First, unlike linear equations, its solution does not always exist (as a C^1 function) for all times. Next, in the context of Sobolev spaces, its flow is continuous but not uniformly continuous, which can be seen as a characteristic property of quasi-linear equations (see Sect. 1.3.1 for terminology). Finally, it exhibits a regularity threshold for the initial data, below which the flow is no longer even continuous.

The properties of the solutions depend strongly on the regularity of the initial data u_0. Here, we will successively treat the cases where u_0 is Lipschitz, then where u_0 belongs to a Sobolev space $H^s(\mathbf{R})$ for various values of s. In the first case, we will use the method of characteristics described in Chap. 1. For data belonging to Sobolev spaces, we will use, for the continuity of the flow, a time-dependent version of the

C. Zuily, *Selected Topics in Partial Differential Equations*, Universitext,
https://doi.org/10.1007/978-3-032-24082-8_5

flow continuity criterion developed in Chap. 4. Finally, the non-uniform continuity of the flow will result from a construction based on the characterization of the supports of the solutions.

5.3 The $W^{1,\infty}(\mathbf{R})$ Theory

5.3.1 Preliminaries

We set in what follows

$$W^{1,\infty}(\mathbf{R}) = \{u \in L^\infty(\mathbf{R}) : u' \in L^\infty(\mathbf{R})\},$$

which we equip with the norm $\|u\|_{W^{1,\infty}(\mathbf{R})} = \|u\|_{L^\infty(\mathbf{R})} + \|u'\|_{L^\infty(\mathbf{R})}$.

Let us recall, without proof, some classical properties of this space.

1. The space $C_b^1(\mathbf{R})$ of C^1 functions on $\mathbf{R}$ such that u and u' are bounded on $\mathbf{R}$ is strictly contained in $W^{1,\infty}(\mathbf{R})$. Indeed, the function $u(x) = e^{-|x|}$ belongs to $W^{1,\infty}(\mathbf{R})$ but not to $C_b^1(\mathbf{R})$.
2. $W^{1,\infty}(\mathbf{R})$ is a Banach space consisting of continuous functions.
3. $W^{1,\infty}(\mathbf{R})$ coincides with the space Lip($\mathbf{R}$) defined by

   $$\left\{u \in L^\infty(\mathbf{R}) : \exists C > 0 : |u(x) - u(y)| \leq C|x-y|, \quad \forall x, y \in \mathbf{R}, |x-y| \leq 1\right\}.$$

   (See Theorem 5.9.3 in the appendix of this chapter).
4. If $u \in W^{1,\infty}(\mathbf{R})$ then u is differentiable almost everywhere.
5. If $u \in W^{1,\infty}(\mathbf{R})$ and if $u'(x) \geq 0$, almost everywhere, then u is increasing. This results from the fact that since $u' \in L^1_{\text{loc}}(\mathbf{R})$ we have

   $$u(y) - u(x) = \int_x^y u'(x)\, dx.$$

If $u \in W^{1,\infty}(\mathbf{R})$ we denote by $\inf_{\mathbf{R}} u'$ and $\sup_{\mathbf{R}} u'$ the essential infimum and essential supremum of the function $u' \in L^\infty$.

5.3.2 Existence and Uniqueness of a Solution

Let us consider **positive** times. We begin with a necessary condition for existence. Suppose that the problem (5.2.1) has a real C^1 solution in t and Lipschitz in x. Consider the **characteristic** of the equation, which is defined by

$$\dot{X}(t) = u(t, X(t)), \quad X(0) = x.$$

Then

$$\frac{d}{dt}\big[u(t, X(t))\big] = (\partial_t u + \dot{X}(t)\partial_x u)(t, X(t)) = (\partial_t u + u\partial_x u)(t, X(t)) = 0.$$

Thus, $u(t, X(t)) = u_0(x)$ so that $\dot{X}(t) = u_0(x)$ and $X(t) = x + tu_0(x)$. Consequently, we necessarily have

$$u(t, x + tu_0(x)) = u_0(x).$$

We can then state the following result.

Theorem 5.3.1 *Let $u_0 \in W^{1,\infty}(\mathbf{R})$. Then the problem* (5.2.1) *admits a unique maximal solution on* $[0, T[\times\mathbf{R}$ *for* $T \geq T^*$ *where*

$$T^* = +\infty \ \ \textit{if} \ \ \inf_{\mathbf{R}} u_0' \geq 0 \quad \textit{and} \quad T^* = -\frac{1}{\inf_{\mathbf{R}} u_0'} \ \ \textit{if} \ \ \inf_{\mathbf{R}} u_0' \in]-\infty, 0[.$$

Proof Let $F(t, x) = x + tu_0(x)$. This is a function differentiable almost everywhere (a.e.) and $\frac{\partial F}{\partial x}(t, x) = 1 + tu_0'(x)$ a.e. Then,

$$\text{if } \inf_{\mathbf{R}} u_0' \geq 0 \text{ we have } \frac{\partial F}{\partial x} > 0, \quad a.e.\ x \in \mathbf{R}, \quad \forall t \geq 0,$$
$$\text{if } -\infty < \inf_{\mathbf{R}} u_0'(x) < 0, \ \text{ we have } \ \frac{\partial F}{\partial x} > 0, \ a.e.\ x \in \mathbf{R}, \ \text{ for } 0 \leq t < -\frac{1}{\inf_{\mathbf{R}} u_0'}. \tag{5.3.1}$$

Set

$$T^* = +\infty \quad \text{if } \inf_{\mathbf{R}} u_0' \geq 0, \qquad T^* = -\frac{1}{\inf_{\mathbf{R}} u_0'} \quad \text{if } \inf_{\mathbf{R}} u_0' \in]-\infty, 0[. \tag{5.3.2}$$

Note that for $t \in [0, T^*[$ we have

$$1 + tu_0'(x) \geq 1 + t \inf_{\mathbf{R}} u_0' > 0. \tag{5.3.3}$$

According to (5.3.1) and (5.3.2), for all $t \in [0, T^*[$, the mapping $x \mapsto x + tu_0(x)$ is continuous and strictly increasing. On the other hand, since

$$x - t\|u_0\|_{L^\infty(\mathbf{R})} \leq F(t, x) \leq x + t\|u_0\|_{L^\infty(\mathbf{R})},$$

we have $\lim_{x\to\pm\infty} F(t,x) = \pm\infty$. It follows that the mapping $x \to F(t,x)$ is invertible from $\mathbf{R}$ to $\mathbf{R}$ and we can write

$$x + tu_0(x) = y \in \mathbf{R} \Longleftrightarrow x = \kappa(t,y). \tag{5.3.4}$$

Here are some properties of κ.

Proposition 5.3.2 *1. For all $t \in [0, T^*[$ the function $y \mapsto \kappa(t,y)$ is strictly increasing.*

2. For all $t, t_0 \in [0, T^*[$ and all $y \in \mathbf{R}$ the function $t \mapsto \kappa(t,y)$ satisfies

$$|\kappa(t,y) - \kappa(t_0,y)| \le \frac{\|u_0\|_{L^\infty(\mathbf{R})}}{1 + t\inf_{\mathbf{R}} u_0'}|t - t_0|.$$

3. For $t \in [0, T^*[$ and all $y_1, y_2 \in \mathbf{R}$ such that $|y_1 - y_2| \le 1$ the function $y \mapsto \kappa(t,y)$ satisfies

$$|\kappa(t,y_2) - \kappa(t,y_1)| \le \frac{1}{1 + t\inf_{\mathbf{R}} u_0'}|y_2 - y_1|.$$

4. There exists a set $B \subset \mathbf{R}$ of measure zero such that the function $y \mapsto \kappa(t,y)$ is differentiable on $\mathbf{R} \setminus B$ and

$$\frac{\partial\kappa}{\partial y}(t,y) = \frac{1}{1 + tu_0'(\kappa(t,y))}. \tag{5.3.5}$$

5. There exists a set $B \subset \mathbf{R}$ of measure zero such that for $y \in \mathbf{R} \setminus B$ the function $t \mapsto \kappa(t,y)$ is differentiable and

$$\frac{\partial\kappa}{\partial t}(t,y) + u_0(\kappa(t,y))\frac{\partial\kappa}{\partial y}(t,y) = 0. \tag{5.3.6}$$

Proof 1. Indeed, let $y_2 > y_1$. Set $x_j = \kappa(t,y_j)$ so that $y_j = F(t,x_j)$, $j = 1,2$. We have seen that for $t \in [0, T^*[$ the function $x \mapsto F(t,x)$ is increasing. Thus, if $x_2 \le x_1$ we would have $y_2 = F(t,x_2) \le y_1 = F(t,x_1)$ which is a contradiction. Therefore, $x_2 = \kappa(t,y_2) > x_1 = \kappa(t,y_1)$.

2. Let $t_0, t \in [0, T^*[$ and $y \in \mathbf{R}$. Set $x_t = \kappa(t,y)$, $x_0 = \kappa(t_0,y)$. Thus we have $y = F(t,x_t)$ and $y = F(t_0,x_0)$, where $F(t,x) = x + tu_0(x)$. Then,

$$\begin{aligned} 0 = F(t,x_t) - F(t_0,x_0) &= F(t,x_t) - F(t,x_0) + F(t,x_0) - F(t_0,x_0), \\ &= \int_{x_0}^{x_t} \frac{\partial F}{\partial x}(t,x)\,dx + (t - t_0)u_0(x_0). \end{aligned}$$

When $t \in [0, T^*[$ we have $\frac{\partial F}{\partial x}(t,x) \ge 1 + t\inf_{\mathbf{R}} u_0' > 0$. Thus, if $x_t > x_0$ we write

$$-(t-t_0)u_0(x_0) = \int_{x_0}^{x_t} \frac{\partial F}{\partial x}(t,x)\,dx \geq (x_t - x_0)(1 + t \inf_{\mathbf{R}} u_0'),$$

and if $x_t < x_0$ we write

$$(t-t_0)u_0(x_0) = \int_{x_t}^{x_0} \frac{\partial F}{\partial x}(t,x)\,dx \geq (x_0 - x_t)(1 + t \inf_{\mathbf{R}} u_0').$$

In summary,

$$|\kappa(t,y) - \kappa(t_0,y)| = |x_t - x_0| \leq \frac{\|u_0\|_{L^\infty(\mathbf{R})}}{1 + t \inf_{\mathbf{R}} u_0'} |t - t_0|.$$

3. We may assume that $y_2 > y_1$. Set $x_j = \kappa(t, y_j)$ so that $y_j = x_j + tu_0(x_j)$. Then, according to point 1, we have $x_2 \geq x_1$. Then,

$$x_2 - x_1 = y_2 - y_1 - t(u_0(x_2) - u_0(x_1)) = y_2 - y_1 - t \int_{x_1}^{x_2} u_0'(s)\,ds. \tag{5.3.7}$$

We have $u_0'(s) \geq \inf_{\mathbf{R}} u_0'$, $t \geq 0$ and $x_2 \geq x_1$. It follows that we can write

$$t \int_{x_1}^{x_2} u_0'(s)\,ds \geq t(x_2 - x_1) \inf_{\mathbf{R}} u_0',$$

so that using (5.3.7) we obtain

$$x_2 - x_1 \leq y_2 - y_1 - t(x_2 - x_1) \inf_{\mathbf{R}} u_0',$$

hence

$$(1 + t \inf_{\mathbf{R}} u_0')(x_2 - x_1) \leq y_2 - y_1.$$

According to (5.3.3) we have $1 + t\ \inf_{\mathbf{R}} u_0' > 0$ so

$$x_2 - x_1 = \kappa(t, y_2) - \kappa(t, y_1) \leq \frac{1}{1 + t\ \inf_{\mathbf{R}} u_0'}(y_2 - y_1).$$

4. There exists a set $A \subset \mathbf{R}$ of measure zero such that for all $t \in (0, T^*)$ the function $x \mapsto g(x) = x + tu_0(x)$ is differentiable with respect to x on $\mathbf{R} \setminus A$ and its derivative is $g'(x) = 1 + tu_0'(x) > 0$. It follows that its inverse function κ is differentiable on $\mathbf{R} \setminus g(A)$. Indeed, let $y_0 \notin g(A)$. Set $\kappa(t, y) = x$, $\kappa(t, y_0) = x_0$, then $x_0 \notin A$ and we have $y = F(t, x)$, $y_0 = F(t, x_0)$ so that

$$\frac{\kappa(t,y) - \kappa(t,y_0)}{y - y_0} = \frac{x - x_0}{F(t,x) - F(t,x_0)}. \tag{5.3.8}$$

Since $y \mapsto \kappa(t, y)$ is continuous, according to point 3. if $y \to y_0$ then $x \to x_0$ and the right-hand side of (5.3.8) tends to $\frac{1}{\partial_x F(t,x_0)} = \frac{1}{1+tu_0'(x_0)}$. Moreover, since g is locally Lipschitz, $g(A)$ is of measure zero. Next, for $y_0 \in \mathbf{R} \setminus g(A)$ we have (5.3.5).

5. We keep the notations used in the previous points for A and B. Let $t_0 \in (0, T^*)$ and $y_0 \in \mathbf{R} \setminus B$. We have $\frac{\partial \kappa}{\partial t}(t_0, y_0) = \lim_{h\to 0} \frac{\kappa(t_0+h, y_0)-\kappa(t_0, y_0)}{h} =: \lim_{h\to 0} I_h$. Let us set

$$\begin{aligned} x = \kappa(t_0 + h, y_0) &\Longleftrightarrow y_0 = x + (t_0 + h)u_0(x), \\ x_0 = \kappa(t_0, y_0) &\Longleftrightarrow y_0 = x_0 + t_0 u_0(x_0). \end{aligned}$$

Then, $0 = x - x_0 + t_0(u_0(x) - u_0(x_0)) + hu_0(x)$ hence $\frac{1}{h} = -\frac{u_0(x)}{x-x_0+t_0(u_0(x)-u_0(x_0))}$. We deduce that

$$I_h = \frac{x - x_0}{h} = -\frac{u_0(x)}{1 + t_0 \frac{u_0(x)-u_0(x_0)}{x-x_0}}.$$

If $y_0 \notin B$ we have $x_0 \notin A$ so that the function $x \mapsto u_0(x)$ is differentiable at x_0. As $h \to 0$ we have $x \to x_0$, then $u_0(x) \to u_0(x_0)$. Therefore $\lim_{h\to 0} I_h = -\frac{u_0(x_0)}{1+t_0 u_0'(x_0)}$. Taking into account (5.3.5) we deduce that

$$\frac{\partial \kappa}{\partial t}(t_0, y_0) = -u_0(\kappa(t, y_0))\frac{\partial \kappa}{\partial y}(t_0, y_0),$$

which proves (5.3.6). □

Let us set

$$u(t, y) = u_0(\kappa(t, y)). \tag{5.3.9}$$

Then the function u defined by (5.3.9) is almost everywhere differentiable in (t, y) and it is the unique solution in $[0, T_0[\times\mathbf{R}$ of the problem (5.2.1).

Indeed, first we have $u(0, y) = u_0(\kappa(0, y)) = u_0(y)$. Next, let $y \notin B$ then $\kappa(t, y) \notin A$. We write

$$\frac{u_0(\kappa(t+h, y)) - u_0(\kappa(t, y))}{h} = \frac{u_0(\kappa(t+h, y)) - u_0(\kappa(t, y))}{\kappa(t+h, y) - \kappa(t, y)} \frac{\kappa(t+h, y) - \kappa(t, y)}{h}.$$

We have $\lim_{h\to 0} \kappa(t + h, y) = \kappa(t, y)$ so that, using the above equality, it follows that

$$\frac{\partial u}{\partial t}(t, y) = u_0'(\kappa(t, y))\frac{\partial \kappa}{\partial t}(t, y).$$

In the same way we have

$$\frac{\partial u}{\partial y}(t, y) = u_0'(\kappa(t, y))\frac{\partial \kappa}{\partial y}(t, y). \tag{5.3.10}$$

Using (5.3.6) we obtain

$$\begin{aligned}\frac{\partial u}{\partial t}(t,y) &= u_0'(\kappa(t,y))\frac{\partial \kappa}{\partial t}(t,y) = -u_0(\kappa(t,y))u_0'(\kappa(t,y))\frac{\partial \kappa}{\partial y}(t,y),\\ &= -u(t,y)\frac{\partial u}{\partial y}(t,y).\end{aligned} \tag{5.3.11}$$

Uniqueness follows immediately from the necessary condition described at the beginning of the paragraph.

We thus have a unique solution on the interval $[0, T^*)$ where T^* was defined above. The maximal solution then exists on the interval $[0, T)$ where $T \geq T^*$.

Note that using (5.3.4) we can write $\kappa(t, y) + tu_0(\kappa(t, y)) = y$ so that

$$\kappa(t,y) = y - tu(t,y). \tag{5.3.12}$$

□

Comments

1. We can give a geometric interpretation of the result obtained. Indeed, as above, the solution u is such that

$$u(t, y + tu_0(y)) = u_0(y).$$

This shows that the function $u = u(t, x)$ is constant and equal to $u_0(y)$ on the line D_y given by the equation $x = y + tu_0(y)$. If $\inf_{\mathbf{R}} u_0' \geq 0$ then u_0 is increasing so that for $y > y'$ the slope of the line D_y is greater than that of the line $D_{y'}$. These two lines will therefore never meet for positive t. On the contrary, if $\inf_{\mathbf{R}} u_0' < 0$ there exist points y and y' such that $y > y'$ but $u_0(y') > u_0(y)$ and in this case the two lines will meet at some point m_0. At this point, the solution u should have two different values $u_0(y)$ and $u_0(y')$, which is not possible.
2. In fact, the maximal time of existence is exactly $T^* = -\frac{1}{\inf_{\mathbf{R}} u_0'}$ in the second case. Indeed, from (5.3.5) we have

$$(1 + tu_0'(x))\partial_x\kappa(x + tu_0(x)) = 1. \tag{5.3.13}$$

On the other hand, it follows from (5.3.10) that

$$\partial_x u(t, x + tu_0(x)) = u_0'(x)\partial_x\kappa(t, x + tu_0(x)). \tag{5.3.14}$$

Multiplying (5.3.14) by $1 + tu_0'(x)$ and using (5.3.13) we obtain

$$(1 + tu_0'(x))\partial_x u(t, x + tu_0(x)) = u_0'(x). \tag{5.3.15}$$

Let x_0 such that $u_0'(x_0) < 0$. We deduce from (5.3.15) that for $tu_0'(x_0) > -1$, that is, for $0 \le t < T_{x_0}^* = -\frac{1}{u_0'(x_0)}$ we have

$$\partial_x u(t, x_0 + tu_0(x_0)) = \frac{1}{t + \frac{1}{u_0'(x_0)}}. \tag{5.3.16}$$

This equality shows that when t tends to $T_{x_0}^*$, $\partial_x u$ tends to $+\infty$ along the characteristic line, while u remains bounded since it is constant equal to $u_0(x_0)$. The maximal time of existence of the solution (as an element of $W^{1,\infty}(\mathbf{R})$) is therefore the smallest of these times as x_0 varies in $\mathbf{R}$, that is,

$$T^* = \inf_{\mathbf{R}} T_{x_0}^* = \inf_{\mathbf{R}} \Big(-\frac{1}{u_0'}\Big) = -\sup_{\mathbf{R}} \Big(\frac{1}{u_0'}\Big) = -\frac{1}{\inf_{\mathbf{R}} u_0'}$$

as announced.
Let us emphasize that at time T^* the solution u of (5.2.1) does not blow up (since it is constant along each characteristic line), but its derivative with respect to x blows up.
A completely similar discussion can be made for negative times t.

Let $u_0 \in W^{1,\infty}(\mathbf{R})$. If we take $T > 0$ such that $T\|u_0'\|_{L^\infty(\mathbf{R})} < 1$ then the solution will exist on $[0, T]$. Indeed, in the case where $\inf_{\mathbf{R}} u_0' \in]-\infty, 0[$ we have

$$T(-\inf_{\mathbf{R}} u_0') = T\sup_{\mathbf{R}}(-u_0') \le T\sup_{\mathbf{R}} |u_0'| = T\|u_0'\|_{L^\infty(\mathbf{R})} < 1,$$

and thus $T < T^*$ where T^* was defined in (5.3.2).

We will work in the interval $[0, T]$ where $T\|u_0'\|_{L^\infty(\mathbf{R})} < 1$.

Using (5.3.5), (5.3.6) and the estimate $1 + tu_0'(\kappa(t, y)) \ge 1 - T\|u_0'\|_{L^\infty} > 0$ we easily obtain the following result.

Proposition 5.3.3

$$(i) \qquad \sup_{(t,y)\in[0,T]\times\mathbf{R}} \left|\frac{\partial\kappa}{\partial y}(t, y)\right| \le \frac{1}{1 - T\|u_0'\|_{L^\infty(\mathbf{R})}},$$

$$(ii) \qquad \sup_{(t,y)\in[0,T]\times\mathbf{R}} \left|\frac{\partial\kappa}{\partial t}(t, y)\right| \le \frac{\|u_0\|_{L^\infty(\mathbf{R})}}{1 - T\|u_0'\|_{L^\infty(\mathbf{R})}}.$$

In particular, if $T\|u_0'\|_{L^\infty(\mathbf{R})} \le 1 - \delta$ with $\delta \in (0, 1)$ we have

$$\frac{1}{2-\delta} \le \frac{\partial\kappa}{\partial y}(t, y) \le \frac{1}{\delta}, \quad \forall y \in \mathbf{R}. \tag{5.3.17}$$

5.4 An Example

For $\lambda \geq 1$ and $\varepsilon \in (0, 1)$ we consider the function

$$u_0(x) = \lambda\big(e^{-|x+\varepsilon|} - e^{-|x-\varepsilon|}\big).$$

The function u_0 belongs to $W^{1,\infty}(\mathbf{R})$. Indeed, the function u_0 is continuous and bounded on $\mathbf{R}$ and differentiable on $\mathbf{R} \setminus \{\pm\varepsilon\}$. Since the function u_0 does not have any jump, its derivative in the sense of distributions (and almost everywhere) is given by the function

$$\begin{aligned} u_0'(x) &= \lambda\big(e^{\varepsilon} - e^{-\varepsilon}\big)e^{x} \quad \text{if} \quad x \in (-\infty, -\varepsilon), \\ u_0'(x) &= -\lambda e^{-\varepsilon}\big(e^{x} + e^{-x}\big) \quad \text{if} \quad x \in (-\varepsilon, \varepsilon), \\ u_0'(x) &= \lambda\big(e^{\varepsilon} - e^{-\varepsilon}\big)e^{-x} \quad \text{if} \quad x \in (\varepsilon, +\infty). \end{aligned} \tag{5.4.1}$$

It is clear that the function u_0' belongs to $L^\infty(\mathbf{R})$.

Since the initial data is in $W^{1,\infty}$, the problem (5.2.1) admits a unique maximal solution on the interval $[0, T^*)$. Let us compute T^*. On the intervals $(-\infty, -\varepsilon)$ and $(\varepsilon, +\infty)$ the function $u_0'(x)$ is positive. In the interval $(-\varepsilon, \varepsilon)$ the function u_0' is strictly negative. It increases from $-\lambda(1 + e^{-2\varepsilon})$ to $-2\lambda e^{-\varepsilon}$ between $-\varepsilon$ and 0 then decreases from $-2\lambda e^{-\varepsilon}$ to $-\lambda(1 + e^{-2\varepsilon})$ between 0 and ε. We deduce that

$$\inf_{x\in\mathbf{R}} u_0'(x) = \inf_{x\in(-\varepsilon,\varepsilon)} u_0'(x) = -\lambda(1 + e^{-2\varepsilon}).$$

From the general theory, we deduce that

$$T^* = \frac{1}{\lambda(1 + e^{-2\varepsilon})}.$$

5.5 The $H^s(\mathbf{R})$, $\frac{3}{2} < s < 2$ Theory

Let us first note that the Sobolev embedding theorem implies that for $s > \frac{3}{2}$ the space $H^s(\mathbf{R})$ is contained (with continuous embedding) in the space $W^{1,\infty}(\mathbf{R})$.

In what follows, we fix $s \in]\frac{3}{2}, 2[$ and $T > 0$. We consider the set

$$\mathcal{S}_T = \{u_0 \in H^s(\mathbf{R}) \,;\, T\|u_0'\|_{L^\infty} < 1\}, \tag{5.5.1}$$

equipped with the H^s norm.

The purpose of this paragraph is to prove the following result.

Theorem 5.5.1 *For all $u_0 \in \mathcal{S}_T$ the unique solution of the problem* (5.2.1) *given by Theorem 5.3.1, denoted by $\Phi(u_0)$, belongs to $C^0([0, T], H^s(\mathbf{R}))$.*

Moreover, the map $\Phi : \mathcal{S}_T \to C^0([0,T], H^s(\mathbf{R}))$ *is continuous.*

Remark 5.5.2 This result is also true for $s \geq 2$ but for convenience of calculations we will only consider the case $\frac{3}{2} < s < 2$.

Recall that if $u \in H^s(\mathbf{R})$ with $s > \frac{3}{2}$, then u and its derivative are continuous functions that tend to zero at infinity. We write $u, u' \in C^0_{\to 0}(\mathbf{R})$. In particular, u belongs to $W^{1,\infty}(\mathbf{R})$.

Since $H^s(\mathbf{R}) \subset W^{1,\infty}(\mathbf{R})$ we can apply Theorem 5.3.1 and deduce that the function u defined by $u(t,x) = u_0(\kappa(t,x))$ is necessarily the unique solution of the problem (5.2.1) on $[0, T]$.

It remains to prove that u belongs to $C^0([0,T], H^s(\mathbf{R}))$ and the continuity of the map Φ as stated in the theorem.

Before undertaking the proof of these assertions, we present a property of the solutions of the problem (5.2.1): the conservation of the L^2 norm.

Proposition 5.5.3 *Let* $u_0 \in \mathcal{S}_T$. *Then for all* $t \in [0, T]$ *the function* $x \mapsto u(t,x)$ *belongs to* $L^2(\mathbf{R})$ *and we have*

$$\int_{\mathbf{R}} u(t,x)^2\,dx = \int_{\mathbf{R}} u_0(x)^2\,dx. \tag{5.5.2}$$

Proof If $u_0 \in \mathcal{S}_T$ we have $u_0, u_0' \in C^0_{\to 0}(\mathbf{R})$. On the other hand, we have seen that for all $t \in [0, T]$ the map $x \mapsto \kappa(t,x)$ is a C^1-diffeomorphism from $\mathbf{R}$ to $\mathbf{R}$. Its inverse is $\kappa^{-1}(t,y) = y + tu_0(y)$ whose Jacobian is equal to $1 + tu_0'(y) > 0$.

According to the change of variables theorem, the function $x \mapsto u(t,x)^2$ belongs to $L^1(\mathbf{R})$ if and only if the function $y \mapsto u_0(y)^2(1 + tu_0'(y))$ belongs to $L^1(\mathbf{R})$, which is the case since $u_0 \in L^2(\mathbf{R})$ and $u_0' \in L^\infty(\mathbf{R})$. We then have

$$\int_{\mathbf{R}} u(t,x)^2\,dx = \int_{\mathbf{R}} u_0(y)^2(1 + tu_0'(y))\,dy. \tag{5.5.3}$$

We then note that, since $u_0 \in C^0_{\to 0}(\mathbf{R})$, we have

$$\int_{\mathbf{R}} u_0(y)^2 u_0'(y)\,dy = \frac{1}{3}\int_{\mathbf{R}} \left(u_0(y)^3\right)'\,dy = 0,$$

which, taking into account (5.5.3), gives (5.5.2). □

Let us return to the proof of Theorem 5.5.1.

If $u_0 \in \mathcal{S}_T$ there exists $\delta = \delta(u_0) > 0$ such that $T\|u_0'\|_{L^\infty} \leq 1 - \delta$. Let C_s be the norm of the embedding of $H^s(\mathbf{R})$ into $W^{1,\infty}(\mathbf{R})$. It is such that

$$\|u\|_{W^{1,\infty}} \leq C_s \|u\|_{H^s}, \quad \forall u \in H^s(\mathbf{R}).$$

Let $r = r(u_0) > 0$ such that

$$TC_s r \le \frac{1}{2}\delta. \tag{5.5.4}$$

Then,

$$\mathcal{S}_T = \cup_{u_0 \in \mathcal{S}_T} B_s(u_0, r), \tag{5.5.5}$$

where

$$B_s(u_0, r) = \{v_0 \in H^s(\mathbf{R}) : \|v_0 - u_0\|_{H^s} < r\}. \tag{5.5.6}$$

Indeed, if $v_0 \in B_s(u_0, r)$ we can write

$$\begin{aligned} T\|v_0'\|_{L^\infty} &\le T\|u_0'\|_{L^\infty} + T\|u_0' - v_0'\|_{L^\infty} \le 1 - \delta + TC_s\|u_0 - v_0\|_{H^s}, \\ &\le 1 - \delta + TC_s r \le 1 - \frac{\delta}{2}. \end{aligned} \tag{5.5.7}$$

Since continuity is a local property, it will suffice to prove the continuity of Φ on each ball $B_s(u_0, r)$.

5.5.1 An Abstract Result of Continuity in Sobolev Spaces

Here is a time-dependent version of Theorem 4.5.1 proved in Chap. 4.

For $s \in \mathbf{R}$, $u_0 \in H^s(\mathbf{R}^d)$ and $r > 0$ we set

$$B_s(u_0, r) = \{u \in H^s(\mathbf{R}^d) : \|u - u_0\|_{H^s} < r\}.$$

Let $T > 0$ and two real numbers s_0, s_1 such that

$$s_0 < s < s_1.$$

Theorem 5.5.4 *Let* $\Phi : B_s(u_0, r) \to L^\infty(0, T), H^{s_0}(\mathbf{R}^d))$ $u \mapsto \Phi(u)$. *Assume that*

1. *Φ is Lipschitz in the s_0 norm, i.e.*

$$\exists C > 0 : \|\Phi(u) - \Phi(v)\|_{L^\infty((0,T),H^{s_0})} \le C\|u - v\|_{H^{s_0}}, \quad \forall u, v \in B_s(u_0, r). \tag{5.5.8}$$

2. *Φ is bounded in the s_1 norm, i.e.*

$$\exists C > 0 : \|\Phi(u)\|_{L^\infty((0,T),H^{s_1})} \le C\|u\|_{H^{s_1}}, \quad \forall u \in B_s(u_0, r) \cap H^{s_1}(\mathbf{R}^d). \tag{5.5.9}$$

Then,

(i) interpolation : for all $v_0 \in B_s(u_0, r)$ *we have* $\Phi(v_0) \in L^\infty((0,T), H^s(\mathbf{R}^d))$.

(ii) continuity of Φ *: the map* $\Phi : B_s(u_0, r) \to L^\infty((0,T), H^s(\mathbf{R}^d))$ *is continuous.*

(iii) continuity in time : if for all $v_0 \in B_s(u_0, r) \cap H^{s_1}(\mathbf{R}^d)$ *we have* $\Phi(v_0) \in C^0([0,T], H^{s_0}(\mathbf{R}^d))$ *then* $\Phi(u_0) \in C^0([0,T], H^s(\mathbf{R}^d))$.

Remark 5.5.5 The constants C appearing in the assumptions 1. and 2. of the above theorem may depend on u, v but are bounded by constants depending only on u_o and r when u, v belong to $B_s(u_0, r)$.

The proof of this theorem will be given later, in Sect. 5.6. Before that, let us show how it implies Theorem 5.5.1.

Choice of Φ, s_0, s_1, r

The map Φ is the map that associates to u_0 the solution u of the Burgers equation. Taking into account (5.5.5) it is well defined on the ball $B_s(u_0, r)$ introduced in (5.5.6).

We shall take $s_0 = 0$, $s_1 = 2$.

1. Φ maps $B_s(u_0, r)$ into $L^\infty((0,T), L^2(\mathbf{R}))$

Let $v_0 \in B_s(u_0, r)$. Since $\Phi(v_0) = v_0 \circ \kappa$, by setting $y = \kappa(t,x) \Leftrightarrow x = y + tv_0(y)$ in the integral below, we can write, for $t \in (0,T)$,

$$\int_{\mathbf{R}} |\Phi(v_0)(t,x)|^2\,dx = \int_{\mathbf{R}} |v_0(y)|^2(1 + tv_0'(y))\,dy \le (1 + T\|v_0'\|_{L^\infty})\|v_0\|_{L^2}^2 \le 2\|v_0\|_{L^2}^2$$

according to (5.5.7).

We now need to verify that conditions 1. and 2. of Theorem 5.5.4 are satisfied.

2. Condition 1. is satisfied

Proposition 5.5.6 *Let* $s > \frac{3}{2}$. *Let* $T > 0, \delta > 0$ *and* $u_0 \in H^s(\mathbf{R})$ *such that* $T\|u_0'\|_{L^\infty(\mathbf{R})} \le 1 - \delta$. *Let* $v_0, w_0 \in B_s(u_0, r)$ *where* r *satisfies* (5.5.4). *Let* $\Phi(v_0), \Phi(w_0)$ *be the corresponding solutions of problem* (5.2.1) *on* $[0,T]$. *Then,*

$$\|\Phi(v_0) - \Phi(w_0)\|_{L^\infty((0,T),L^2(\mathbf{R}))}^2 \le 4\big(1 + \frac{4}{\delta^2}\big)\|u_0 - v_0\|_{L^2}^2.$$

Proof We have

$$\Phi(v_0)(t,x) = v_0(\kappa_1(t,x)), \quad \kappa_1(t,x) = y \Longleftrightarrow x = y + tv_0(y),$$
$$\Phi(w_0)(t,x) = w_0(\kappa_2(t,x)), \quad \kappa_2(t,x) = y \Longleftrightarrow x = y + tw_0(y).$$

We then write

$$(\Phi(v_0) - \Phi(w_0))(t,x) = A(t,x) + B(t,x),$$
$$A(t,x) = v_0(\kappa_1(t,x)) - w_0(\kappa_1(t,x)), \quad B(t,x) = w_0(\kappa_1(t,x)) - w_0(\kappa_2(t,x)).$$

By setting $y = \kappa_1(t,x)$ in the integral, we can write

$$\int_{\mathbf{R}} |A(t,x)|^2\,dx = \int_{\mathbf{R}} |v_0(y) - w_0(y)|^2 (1 + t v_0'(y))\,dy, \leq (1 + T\|v_0'\|_{L^\infty})\|v_0 - w_0\|_{L^2}^2,$$

whence, according to (5.5.7),

$$\int_{\mathbf{R}} |A(t,x)|^2\,dx \leq 2\|v_0 - w_0\|_{L^2}^2. \tag{5.5.10}$$

Next, by the mean value theorem,

$$\int_{\mathbf{R}} |B(t,x)|^2\,dx \leq \|w_0'\|_{L^\infty}^2 \int_{\mathbf{R}} |\kappa_1(t,x) - \kappa_2(t,x)|^2\,dx.$$

We will set in the integral $y = \kappa_1(t,x)$, which is equivalent to $x = y + t v_0(y)$. It follows that

$$\int_{\mathbf{R}} |B(t,x)|^2\,dx \leq \|w_0'\|_{L^\infty}^2 (1 + T\|v_0'\|_{L^\infty}) \int_{\mathbf{R}} |y - \kappa_2(t, y + t v_0(y))|^2\,dy.$$

We then note that $\kappa_2(t, y + t w_0(y)) = y$ and we use point 3. of Proposition 5.3.2. We obtain

$$|\kappa_2(t, y + t w_0(y)) - \kappa_2(t, y + t v_0(y))| \leq \frac{T}{1 - T\|w_0'\|_{L^\infty}} |w_0(y) - v_0(y)|.$$

Consequently,

$$\int_{\mathbf{R}} |B(t,x)|^2\,dx \leq T^2 \|w_0'\|_{L^\infty}^2 \frac{1 + T\|v_0'\|_{L^\infty}}{(1 - T\|w_0'\|_{L^\infty})^2} \|w_0 - v_0\|_{L^2}^2.$$

Next, from (5.5.7) we have $T\|w_0'\|_{L^\infty} \leq 1 - \frac{\delta}{2}$, so that $(1 - T\|w_0'\|_{L^\infty})^{-2} \leq \frac{4}{\delta^2}$. We thus obtain

$$\int_{\mathbf{R}} |B(t,x)|^2\,dx \leq \frac{8}{\delta^2}. \tag{5.5.11}$$

Proposition 5.5.6 then follows from the inequality $(a+b)^2 \leq 2(a^2 + b^2)$, from (5.5.10) and from (5.5.11). □

3. Condition 2. is satisfied

Lemma 5.5.7 *Let $u_0 \in H^s(\mathbf{R})$, $T > 0$, $\delta > 0$ such that $T\|u_0'\|_{L^\infty} \le 1 - \delta$. There exists a positive constant C depending on δ such that*

$$\|\Phi(v_0)\|_{L^\infty((0,T),H^2)} \le C\|v_0\|_{H^2}, \tag{5.5.12}$$

for any $v_0 \in B_s(u_0, r) \cap H^2(\mathbf{R})$.

Proof Recall (see 5.5.7) that $T\|v_0'\|_{L^\infty} \le 1 - \frac{\delta}{2}$. We have

$$\Phi(v_0)(t,x) = v_0(\kappa(t,x)) \text{ where } \kappa(t,x) = y \Longleftrightarrow x = y + tv_0(y).$$

By making the change of variables $y = \kappa(t,x)$ in the integrals, we deduce as previously,

$$\|\Phi(v_0)(t,\cdot)\|_{L^2}^2 \le (1 + T\|v_0'\|_{L^\infty})\|v_0\|_{L^2}^2 \le 2\|v_0\|_{L^2}^2. \tag{5.5.13}$$

Next,

$$\partial_x(\Phi(v_0))(t,x) = v_0'(\kappa(t,x))\partial_x\kappa(t,x).$$

Using (5.3.5) and the same change of variables as above, we get

$$\|\partial_x(\Phi(v_0))(t,\cdot)\|_{L^2}^2 \le \frac{1 + T\|v_0'\|_{L^\infty}}{(1 - T\|v_0'\|_{L^\infty})^2}\|v_0'\|_{L^2}^2 \le \frac{8}{\delta^2}\|v_0'\|_{L^2}^2. \tag{5.5.14}$$

Finally,

$$\partial_x^2(\Phi(v_0))(t,x) = v_0''(\kappa(t,x))(\partial_x\kappa(t,x))^2 + v_0'(\kappa(t,x))\partial_x^2\kappa(t,x) = (1) + (2).$$

As above, using (5.3.5) we have

$$\|(1)(t,\cdot)\|_{L^2}^2 \le \frac{1 + T\|v_0'\|_{L^\infty}}{(1 - T\|v_0'\|_{L^\infty})^4}\|v_0''\|_{L^2}^2 \le \frac{32}{\delta^4}\|v_0''\|_{L^2}^2. \tag{5.5.15}$$

Using the expression for $\partial_x\kappa$ given by (5.3.5) we obtain

$$\partial_x^2\kappa(t,x) = -\frac{tv_0''(\kappa(t,x))}{((1 + tv_0'(\kappa(t,x)))^3}, \tag{5.5.16}$$

whence we deduce,

$$(2) = -\frac{tv_0'(\kappa(t,x))v_0''(\kappa(t,x))}{((1 + tv_0'(\kappa(t,x)))^3}.$$

Consequently,

$$\|(2)\|_{L^2}^2 \leq C \frac{T^2 \|v_0'\|_{L^\infty}^2}{(1 - T\|v_0'\|_{L^\infty})^6} (1 + T\|v_0'\|_{L^\infty}) \|v_0''\|_{L^2}^2 \leq \frac{128}{\delta^6} \|v_0''\|_{L^2}^2. \tag{5.5.17}$$

Condition 2. then follows from (5.5.13), (5.5.14), (5.5.15), and (5.5.17). □

4. Continuity in time

We need to verify the condition appearing in (iii) of Theorem 5.5.4.

Lemma 5.5.8 *Let $u_0 \in H^s(\mathbf{R})$, $T > 0$, $\delta > 0$ such that $T\|u_0'\|_{L^\infty} \leq 1 - \delta$. For all v_0 belonging to $B_s(u_0, r) \cap H^2(\mathbf{R})$ we have*

$$\Phi(v_0) \in C^0([0, T], L^2(\mathbf{R})). \tag{5.5.18}$$

Proof Let $\varepsilon > 0$. Let $t_0 \in [0, T]$ and (t_k) a sequence tending to t_0. We may assume k large enough so that $|t_k - t_0| \leq 1$. Since $v_0 \in L^2(\mathbf{R})$ there exists R_0 large enough so that

$$(1 + (t_0 + 1)\|v_0'\|_{L^\infty(\mathbf{R})}) \int_{|y| > R_0} |v_0(y)|^2 \, dy \leq \frac{\varepsilon^2}{6}.$$

Let $R = R_0 + (t_0 + 1)\|v_0\|_{L^\infty}$. Then,

$$I_1 + I_2 := \int_{|x| > R} |v_0(\kappa(t_k, x))|^2 \, dx + \int_{|x| > R} |v_0(\kappa(t_0, x))|^2 \, dx \leq \frac{\varepsilon^2}{3}.$$

Indeed, in I_1 we can set $\kappa(t_k, x)) = y$, which is equivalent to $x = y + t_k v_0(y)$. So we have $dx = (1 + t_k v_0'(y)) \, dy$. Next, since $t_k \leq t_0 + 1$ we have

$$|x| \leq |y| + (1 + t_0)\|v_0\|_{L^\infty} \leq |y| + (1 + t_0)\|v_0\|_{L^\infty},$$

so, since $|x| > R$,

$$|y| > R - (1 + t_0)\|v_0\|_{L^\infty} = R_0.$$

Therefore,

$$I_1 \leq (1 + (t_0 + 1)\|v_0'\|_{L^\infty}) \int_{|y| > R_0} |v_0(y)|^2 \, dy \leq \frac{\varepsilon^2}{6}.$$

By exactly the same reasoning we have

$$I_2 \leq (1 + (t_0 + 1)\|v_0'\|_{L^\infty}) \int_{|y| > R_0} |v_0(y)|^2 \, dy \leq \frac{\varepsilon^2}{6}.$$

We deduce that

$$\int_{|x|>R} |v_0(\kappa(t_k, x)) - v_0(\kappa(t_0, x))|^2 \, dx \leq \frac{2\varepsilon^2}{3}. \tag{5.5.19}$$

It remains to examine the quantity

$$J = \int_{|x|<R} |v_0(\kappa(t_k, x)) - v_0(\kappa(t_0, x))|^2 \, dx.$$

The mean value theorem and Propositions 5.3.2 and 5.3.3 imply that for $k \geq k_0$,

$$\begin{aligned} J &\leq |B(0, R)| \|v_0'\|_{L^\infty}^2 |\kappa(t_k, x) - \kappa(t_0, x)|^2, \\ &\leq |B(0, R)| \|v_0')\|_{L^\infty}^2 \Big(\sup_{t \in [0,T]} |\partial_t \kappa(t, x)|\Big)^2 |t_k - t_0|^2, \\ &\leq |B(0, R)| \|v_0'\|_{L^\infty}^2 \Big(\frac{\|v_0\|_{L^\infty}}{1 - T\|v_0'\|_{L^\infty}} \Big)^2 |t_k - t_0|^2. \end{aligned}$$

It is enough to take $|t_k - t_0|$ small enough to have $J \leq \frac{\varepsilon^2}{3}$, which, together with (5.5.19), proves (5.5.18). □

This completes the proof of Theorem 5.5.1.

5.6 Proof of Theorem 5.5.4

The proof of points (i) and (ii) of this result was given in Chap. 4 (see Theorem 4.5.1.) It is enough to replace H^s by $L^\infty((0, T), H^s)$. Only the proof of point (iii) remains, which we will now detail.

For n large enough, $S_n(u) \in B_s(u_0, r) \cap H^{s_1}$ (see Proposition 4.5.7), so by hypothesis,

$$\Phi(S_n(u)) \in C^0([0, T], H^{s_0}(\mathbf{R}^d)). \tag{5.6.1}$$

Let us show that $\Phi(S_n(u)) \in C^0([0, T], H^s(\mathbf{R}^d))$.

We use the following inequality on Sobolev spaces. For $s_0 < s < s_1$ we have

$$\|v\|_{H^s} \leq \|v\|_{H^{s_0}}^A \|v\|_{H^{s_1}}^B \quad A = \frac{s_1 - s}{s_1 - s_0}, \quad B = \frac{s - s_0}{s_1 - s_0}.$$

To show this inequality, write $s = \lambda s_0 + (1 - \lambda)s_1$, $0 < \lambda < 1$ then

$$\langle \xi \rangle^{2s} |\widehat{v}(\xi)|^2 = \langle \xi \rangle^{2\lambda s_0} |\widehat{v}(\xi)|^{2\lambda} \langle \xi \rangle^{2(1-\lambda)s_1} |\widehat{v}(\xi)|^{2(1-\lambda)}$$

and use Hölder's inequality with $\alpha = \frac{1}{\lambda} > 1$ and $\beta = \frac{1}{1-\lambda} > 1$. Now set $\theta_n = \Phi(S_n(u))$. We write, with the above notations,

$$\|\theta_n(t,\cdot)-\theta_n(t',\cdot)\|_{H^s} \le \|\theta_n(t,\cdot)-\theta_n(t',\cdot)\|^A_{H^{s_0}} \|\theta_n(t,\cdot)-\theta_n(t',\cdot)\|^B_{H^{s_1}},$$
$$\le \|\theta_n(t,\cdot)-\theta_n(t',\cdot)\|^A_{H^{s_0}} \left(2\|\theta_n\|_{L^\infty([0,T],H^{s_1})}\right)^B$$

On the other hand we have

$$\|\theta_n\|_{L^\infty([0,T],H^{s_1})} \le C\|S_n(u)\|_{H^{s_1}} \le C2^{n(s_1-s_0)}\|u\|_{H^{s_0}}.$$

Using (5.6.1) we obtain the desired result.

Convergence in $L^\infty((0,T),H^s(\mathbf{R}^d))$ is uniform convergence in $t\in[0,T]$, with values in H^s. Since $\Phi(S_n(u))$ converges to $\Phi(u)$ in $L^\infty((0,T),H^s(\mathbf{R}^d))$, the limit $\Phi(u)$ belongs to $C^0([0,T],H^s(\mathbf{R}^d))$.

5.7 Non-Uniform Continuity of the Flow

We have seen previously that for any $u_0\in H^s(\mathbf{R})$ where $\frac{3}{2}<s<2$ and for any $T>0$ such that $T\|u_0'\|_{L^\infty(\mathbf{R})}<1$, the problem (5.2.1) admits a unique solution $u\in C^0([0,T],H^s(\mathbf{R}))$ and the map $u_0\mapsto u$ is continuous from $H^s(\mathbf{R})$ into $C^0([0,T],H^s(\mathbf{R}))$.

In this section, we aim to show the following result, where $u(t)$ denotes the function $x\mapsto u(t,x)$.

Theorem 5.7.1 *For all $t\in(0,T]$ the map from $H^s(\mathbf{R})$ to $H^s(\mathbf{R})$, $u_0\mapsto u(t)$, is not uniformly continuous.*

Proof It is enough to find $\varepsilon_0>0$ and two sequences of initial data (u_n^0) and (v_n^0) in $H^s(\mathbf{R})$ such that

$$\lim_{n\to+\infty}\|u_n^0-v_n^0\|_{H^s(\mathbf{R})}=0,$$

and if u_n and v_n denote the corresponding solutions of problem (5.2.1),

$$\|u_n(t)-v_n(t)\|_{H^s(\mathbf{R})}\ge\varepsilon_0.$$

Construction of the sequences (u_n^0) *and* (v_n^0).

Let $\chi\in C_0^\infty(\mathbf{R})$, $\chi(x)=1$ if $|x|\le\frac{1}{2}$, $\chi(x)=0$ if $|x|\ge 1$. Let (λ_n), (ε_n) be two sequences of positive real numbers such that

$$\lim_{n\to+\infty}\lambda_n=+\infty,\quad \lim_{n\to+\infty}\varepsilon_n=0,\quad \lim_{n\to+\infty}\lambda_n\varepsilon_n=+\infty. \tag{5.7.1}$$

Set

$$u_n^0(x)=\lambda_n^{\frac{1}{2}-s}\chi(\lambda_n x),\quad v_n^0(x)=u_n^0(x)+\varepsilon_n\chi(x). \tag{5.7.2}$$

Lemma 5.7.2 *For n large enough, we have the following estimates.*

$$(i)\quad \|u_n^0\|_{H^s} \le \|\chi\|_{H^s}, \quad \|v_n^0\|_{H^s} \le (1+\varepsilon_n)\|\chi\|_{H^s} \le 2\|\chi\|_{H^s},$$
$$(ii)\quad \|\partial_x u_n^0\|_{L^\infty} \le \lambda_n^{\frac{3}{2}-s}\|\chi'\|_{L^\infty} \quad \|\partial_x v_n^0\|_{L^\infty} \le (\lambda_n^{\frac{3}{2}-s}+\varepsilon_n)\|\chi'\|_{L^\infty},$$

Proof (i) Indeed we have

$$\widehat{\chi(\lambda_n\cdot)}(\xi) = \frac{1}{\lambda_n}\widehat{\chi}\Big(\frac{\xi}{\lambda_n}\Big),$$

so that

$$\|\chi(\lambda_n\cdot)\|_{H^s}^2 = \frac{1}{\lambda_n^2}\int_{\mathbf{R}}(1+|\xi|^2)^s\left|\widehat{\chi}\Big(\frac{\xi}{\lambda_n}\Big)\right|^2 d\xi = \frac{1}{\lambda_n}\int_{\mathbf{R}}(1+\lambda_n^2|\eta|^2)^s\,|\widehat{\chi}(\eta)|^2\,d\eta,$$

hence, for n large enough,

$$\|\chi(\lambda_n\cdot)\|_{H^s}^2 \le \lambda_n^{2s-1}\|\chi\|_{H^s}^2,$$

which proves (i). The assertion (ii) is obvious. □

In what follows, we fix $T > 0$. Recall the following facts.

Lemma 5.7.2 (ii) shows that $T\|\partial_x u_n^0\|_{L^\infty} + T\|\partial_x v_n^0\|_{L^\infty} \to 0$. Therefore, the solutions u_n and v_n with initial data u_n^0, v_n^0 at $t = 0$, exist on $[0, T]$ for n large enough. Moreover, according to (5.3.5),

$$\partial_y\kappa_1(t,y) = \frac{1}{1+t\partial_x u_n^0(\kappa_1(t,y))}, \quad \partial_y\kappa_2(t,y) = \frac{1}{1+t\partial_x v_n^0(\kappa_2(t,y))}, \tag{5.7.3}$$

and we have, for n large enough,

$$\frac{1}{2} \le \partial_y\kappa_j(t,y) \le 2 \quad j = 1, 2. \tag{5.7.4}$$

Next, since $u_n^0 - v_n^0 = \varepsilon_n\chi$, $\chi \in C_0^\infty(\mathbf{R})$, we obviously have

$$\lim_{n\to+\infty}\|u_n^0 - v_n^0\|_{H^s(\mathbf{R})} = 0. \tag{5.7.5}$$

The next step is to show that there exists $\varepsilon_0 > 0$ such that

$$\|u_n(t,\cdot) - v_n(t,\cdot)\|_{H^s} \ge \varepsilon_0. \tag{5.7.6}$$

According to (5.3.9) we have

$$u_n(t,y)=u_n^0(\kappa_1(t,y)),\quad v_n(t,y)=v_n^0(\kappa_2(t,y)).$$

We write

$$\begin{cases} v_n(t,y)-u_n(t,y)=A_n+B_n\,, \\ A_n=(v_n^0-u_n^0)(\kappa_2(t,y))=\varepsilon_n\chi(\kappa_2(t,y)), \\ B_n=u_n^0(\kappa_2(t,y))-u_n^0(\kappa_1(t,y)). \end{cases} \tag{5.7.7}$$

Estimate of A_n.

We will estimate the H^2 norm of A_n and use the fact that

$$\|A_n(t,\cdot)\|_{H^s}\leq\|A_n(t,\cdot)\|_{H^2}. \tag{5.7.8}$$

First,

$$\|A_n(t,\cdot)\|_{L^2}^2=\varepsilon_n^2\int_{\mathbf{R}}|\chi(\kappa_2(t,y))|^2\,dy.$$

We have, setting $\kappa_2(t,y)=Y\Leftrightarrow y=Y+tv_n^0(Y)$, so $dy=(1+t\partial_x v_n^0(Y))\,dY$,

$$\|A_n(t,\cdot)\|_{L^2}^2\leq\varepsilon_n^2\int_{\mathbf{R}}|\chi(Y)|^2\,(1+t\partial_x v_n^0(Y))\,dx\leq C(T,\chi)\varepsilon_n^2, \tag{5.7.9}$$

Next,

$$\|\partial_y A_n(t,\cdot)\|_{L^2}^2=\varepsilon_n^2\int_{\mathbf{R}}|\chi'(\kappa_2(t,y)|^2|\partial_y\kappa_2(t,y)|^2\,dy,$$

so that using (5.7.4) and the same change of variables as above we get

$$\|\partial_y A_n(t,\cdot)\|_{L^2}^2\leq C(T,\chi)\varepsilon_n^2. \tag{5.7.10}$$

Finally,

$$\begin{aligned} \partial_y^2 A_n(t,y)&=\varepsilon_n\chi''(\kappa_2(t,y))(\partial_y\kappa_2(t,y))^2+\varepsilon_n\chi'(\kappa_2(t,y))\partial_y^2\kappa_2(t,y) \\ &=(1)+(2). \end{aligned} \tag{5.7.11}$$

Using (5.7.4) and the same change of variables as above we get

$$\|(1)\|_{L^2}^2\leq C(T,\chi)\varepsilon_n^2. \tag{5.7.12}$$

According to (5.7.3) we can write

$$\partial_y^2\kappa_2(t,y)=-\frac{\partial_x^2 v_n^0(\kappa_2(t,y))\partial_y\kappa_2(t,y)}{(1+t\partial_x v_n^0(\kappa_2(t,y)))^2}=-\frac{\partial_x^2 v_n^0(\kappa_2(t,y))}{(1+t\partial_x v_n^0(\kappa_2(t,y)))^3}.$$

Recall that $v_n^0(x)=\lambda^{\frac{1}{2}-s}\chi(\lambda_n x)+\varepsilon_n\chi(x)$. Then,

$$\partial_x^2 v_n^0(\kappa_2(t,y)) = \lambda^{\frac{5}{2}-s}\chi''(\lambda_n\kappa_2(t,y)) + \varepsilon_n\chi''(\kappa_2(t,y)).$$

We deduce that

$$(2) = -\varepsilon_n \frac{\chi'(\kappa_2(t,y))\lambda_n^{\frac{5}{2}-s}\chi''(\lambda_n\kappa_2(t,y))}{(1+t\partial_x v_n^0(\kappa_2(t,y)))^3} - \varepsilon_n^2 \frac{\chi'(\kappa_2(t,y))\chi''(\kappa_2(t,y))}{(1+t\partial_x v_n^0(\kappa_2(t,y)))^3}.$$

The first term on the right-hand side of the above equality is zero for n large enough. Indeed, on the support of $\chi''(\lambda_n\kappa_2(t,y))$ we have $|\kappa_2(t,y))| \le \frac{1}{\lambda_n} \le \frac{1}{4}$. But then $\chi'(\kappa_2(t,y)) = 0$ since $\chi(x) = 1$ for $|x| \le \frac{1}{2}$. Therefore, using (5.7.3) we get

$$\|(2)\|_{L^2}^2 \le C\varepsilon_n^4 \int_{\mathbf{R}} |\chi'(\kappa_2(t,y))\chi''(\kappa_2(t,y))|^2\, dy.$$

Using the same change of variables as above, we finally obtain

$$\|(2)\|_{L^2}^2 \le C(T,\chi)\varepsilon_n^4. \tag{5.7.13}$$

It follows from (5.7.11) to (5.7.13) that

$$\|A_n(t,\cdot)\|_{H^2} \le C(T,\chi)\varepsilon_n. \tag{5.7.14}$$

Estimate of the term B_n. Recall that

$$B_n(t,y) = u_n^0(\kappa_2(t,y)) - u_n^0(\kappa_1(t,y)).$$

The crucial point of the proof is the following.

Lemma 5.7.3 *Let $t > 0$. There exist $n_0 \ge 1$ and $C > 0$ such that for $n \ge n_0$ we have*

$$\begin{aligned} &(i)\quad \lambda_n|\kappa_1(t,y)| \le 1 \Longrightarrow \lambda_n|\kappa_2(t,y)| \ge C\varepsilon_n\lambda_n,\\ &(ii)\quad |x| \le \frac{1}{\lambda n} \Longleftrightarrow |\kappa_1(t,x)| \le \frac{1}{\lambda n}. \end{aligned} \tag{5.7.15}$$

Corollary 5.7.4 *We have, for fixed t and n large enough,*

$$\operatorname{supp} \partial_x u_n^0(\kappa_1(t,\cdot)) \cap \operatorname{supp} \partial_x u_n^0(\kappa_2(t,\cdot)) = \emptyset, \tag{5.7.16}$$

where supp *denotes the support in x.*

Proof of the Corollary As $\partial_x u_n^0(x) = \lambda^{\frac{3}{2}-s}\chi'(\lambda_n x)$, if $y \in \operatorname{supp} \partial_x u_n^0(\kappa_1(t,\cdot))$ we have $y \in \operatorname{supp} \chi'(\lambda_n\kappa_1(t,\cdot))$ so $\lambda_n|\kappa_1(t,y)| \le 1$. It follows from Lemma 5.7.3 that for n large enough, we have $\lambda_n|\kappa_2(t,y)| \ge 2$, since $\varepsilon_n\lambda_n \to +\infty$.

Therefore, $y \notin \operatorname{supp} \partial_x u_n^0(\kappa_2(t,\cdot))$. □

Proof of the Lemma (i) First recall that

$$\kappa_1(t, z + tu_n^0(z)) = z, \quad \kappa_2(t, z + tv_n^0(z)) = z,$$

so that, setting $z = \kappa_1(t, y) \iff y = z + tu_n^0(z)$, we can write

$$\begin{aligned}\kappa_2(t, y) &= \kappa_2(t, z + tu_n^0(z)) = \kappa_2(t, z + tv_n^0(z)) + t(u_n^0(z) - v_n^0(z))\partial_y\kappa_2(t, y^*),\\ &= z + t(u_n^0(z) - v_n^0(z))\partial_y(t, y^*) = z + t\varepsilon_n\chi(z)\partial_y\kappa_2(t, y^*).\end{aligned}$$

Then,

$$\lambda_n\kappa_2(t, y) = \lambda_n z + t\lambda_n\varepsilon_n\chi(z)\partial_y\kappa_2(t, y^*).$$

If $z = \kappa_1(t, y)$ then $\lambda_n|z| \leq 1$ by hypothesis. Thus $|z| \leq \frac{1}{\lambda_n} \leq \frac{1}{2}$ if n is large enough, hence $\chi(z) = 1$. Since $\partial_y\kappa_2(t, y^*) \geq \frac{1}{2}$ it follows that

$$\lambda_n|\kappa_2(t, y)| \geq \frac{1}{2}t\lambda_n\varepsilon_n - \lambda_n|z| \geq \frac{1}{2}t\lambda_n\varepsilon_n - 1 \geq \frac{1}{4}t\lambda_n\varepsilon_n,$$

for $n \geq n_0(t)$, since by hypothesis $\lambda_n\varepsilon_n \to +\infty$ and $t > 0$.

Let us show (ii). The map $y \mapsto \kappa(t, y)$ is strictly increasing. It is therefore a bijection from $[-\frac{1}{\lambda_n}, \frac{1}{\lambda_n}]$ onto its image. We will show that $\kappa(t, \pm\frac{1}{\lambda_n}) = \pm\frac{1}{\lambda_n}$. Indeed, we have

$$\kappa(t, \pm\frac{1}{\lambda_n}) = X \Leftrightarrow \pm\frac{1}{\lambda_n} = \kappa^{-1}(t, X) = X + tu_n^0(X).$$

But $\kappa^{-1}(t, \pm\frac{1}{\lambda_n}) = \pm\frac{1}{\lambda_n} + tu_n^0(\pm\frac{1}{\lambda_n}) = \pm\frac{1}{\lambda_n}$ since $u_n^0(+\frac{1}{\lambda_n}) = \lambda_n^{\frac{1}{2}-s}\chi(\pm 1) = 0$. Therefore, $X = \pm\frac{1}{\lambda_n}$. □

Let us return to the estimate of the term B_n. The goal is to prove that there exists $c_0 > 0$ such that

$$\|B_n(t, \cdot)\|_{H^s} \geq c_0. \tag{5.7.17}$$

To do this, we write

$$\|B_n(t, \cdot)\|_{H^s} \geq \|\partial_x B_n(t, \cdot)\|_{H^{s-1}}, \quad \frac{1}{2} < s - 1 < 1.$$

We have

$$\begin{aligned}\partial_x B_n(t, y) &= \partial_y\kappa_2(t, y)\partial_x u_n^0(\kappa_2(t, y)) - \partial_y\kappa_1(t, y)\partial_x u_n^0(\kappa_1(t, y)),\\ &:= F_2(t, y) - F_1(t, y).\end{aligned}$$

By the characterization of the spaces H^σ for $0 < \sigma < 1$ recalled in Proposition 5.9.1, we can write

$$\|\partial_x B_n(t,\cdot)\|_{H^{s-1}} \geq \iint_{\mathbf{R}\times\mathbf{R}} \frac{|F_2(t,x)-F_1(t,x)-F_2(t,y)+F_1(t,y)|^2}{|x-y|^{2s-1}}\,dx\,dy,$$

since $1+2(s-1)=2s-1$. Set $E=\{x\in\mathbf{R}:\lambda_n|\kappa_1(t,x)|<1\}$. Then, according to Lemma 5.7.3, (ii) we have

$$E=\Big(-\frac{1}{\lambda_n},\frac{1}{\lambda_n}\Big).$$

Moreover, according to Corollary 5.7.4, $F_2(t,\cdot)=0$ on E so that

$$\|\partial_x B_n(t,\cdot)\|_{H^{s-1}} \geq \int_{-\frac{1}{\lambda_n}}^{\frac{1}{\lambda_n}}\int_{-\frac{1}{\lambda_n}}^{\frac{1}{\lambda_n}} \frac{|F_1(t,y)-F_1(t,x)|^2}{|x-y|^{2s-1}}\,dx\,dy.$$

According to (5.7.3) we have

$$F_1(t,y)=\frac{\partial_x u_n^0(\kappa_1(t,y))}{1+t\partial_x u_n^0(\kappa_1(t,y))},$$

so that

$$F_1(t,y)-F_1(t,x)=\frac{\partial_x u_n^0(\kappa_1(t,y))-\partial_x u_n^0(\kappa_1(t,x))}{(1+t\partial_x u_n^0(\kappa_1(t,x))(1+t\partial_x u_n^0(\kappa_1(t,y)))}.$$

Using (5.7.3), (5.7.4) it follows that

$$\|\partial_x B_n(t,\cdot)\|_{H^{s-1}} \geq C\int_{-\frac{1}{\lambda_n}}^{\frac{1}{\lambda_n}}\int_{-\frac{1}{\lambda_n}}^{\frac{1}{\lambda_n}} \frac{|\partial_x u_n^0(\kappa_1(t,y))-\partial_x u_n^0(\kappa_1(t,x))|^2}{|x-y|^{2s-1}}\,dx\,dy.$$

Set $\kappa_1(t,y)=Y,\ \kappa_1(t,x)=X \Leftrightarrow y=Y+tu_n^0(Y),\ x=X+tu_n^0(X)$. We have

$$|u_n^0(Y)-u_n^0(X)|\leq \|\partial_x u_n^0\|_{L^\infty}|X-Y|\leq \lambda_n^{\frac{3}{2}-s}\|\chi'\|_{L^\infty}|X-Y|.$$

Since $\lambda_n^{\frac{3}{2}-s}\to 0$ we have $|x-y|\geq \frac{1}{2}|X-Y|$ for n large enough.

On the other hand, $dx=(1+t\partial_x u_n^0(X))\,dX,\ dy=(1+t\partial_x u_n^0(Y))\,dY$. Using (5.7.16) (ii) we obtain

$$\begin{aligned}\|\partial_x B_n(t,\cdot)\|_{H^{s-1}} \geq{}& C'\int_{-\frac{1}{\lambda_n}}^{\frac{1}{\lambda_n}}\int_{-\frac{1}{\lambda_n}}^{\frac{1}{\lambda_n}} \frac{|\partial_x u_n^0(Y)-\partial_x u_n^0(X)|^2}{|X-Y|^{2s-1}}\,dX\,dY,\\ \geq{}& C'\lambda^{3-2s}\int_{-\frac{1}{\lambda_n}}^{\frac{1}{\lambda_n}}\int_{-\frac{1}{\lambda_n}}^{\frac{1}{\lambda_n}} \frac{|\chi'(\lambda_n Y)-\chi'(\lambda_n X)|^2}{|X-Y|^{2s-1}}\,dX\,dY.\end{aligned}$$

Setting finally $\lambda_n X = a$, $\lambda_n Y = b$ we get

$$\|\partial_x B_n(t,\cdot)\|_{H^{s-1}} \geq C \int_{-1}^{1}\int_{-1}^{1} \frac{|\chi'(a)-\chi'(b)|^2}{|b-a|^{2s-1}}\,da\,db =: c_0.$$

The quantity c_0 is strictly positive and finite. Indeed writing

$$|\chi'(a)-\chi'(b)|^2 \leq \|\chi''\|_{L^\infty}^2 |b-a|^2,$$

it is bounded above by

$$C_1 \int_{-1}^{1}\int_{-1}^{1} \frac{1}{|b-a|^{2s-3}}\,da\,db,$$

which is finite since $0 < 2s-3 < 1$.

This completes the proof of (5.7.17). Using (5.7.7), (5.7.14), and (5.7.17), we obtain eventually that there exists $\varepsilon_0 > 0$ such that for n large enough,

$$\|u_n(t,\cdot) - v_n(t,\cdot)\|_{H^s(\mathbf{R})} \geq \varepsilon_0,$$

which ends the proof of Theorem 5.7.1. □

5.8 The Case $u_0 \in H^s(\mathbf{R})$, $s < \frac{3}{2}$

We aim to show the following result.

Theorem 5.8.1 *Let $1 \leq s < \frac{3}{2}$. For any $\delta \in (0,1)$ there exists $u_0 \in H^s(\mathbf{R})$ satisfying $\|u_0\|_{H^s} < \delta$ such that the problem (5.2.1) admits a solution u which belongs to $C^0([0,T_0], H^s(\mathbf{R}))$, for some $T_0 > 0$, and satisfies $\|u(T_0)\|_{H^s} \geq \frac{1}{\delta}$.*

This result implies that for $s < \frac{3}{2}$ the flow, that is, the map $u_0 \mapsto u$, is not continuous from $H^s(\mathbf{R})$ into $C^0([0,T_0], H^s(\mathbf{R}))$, contrary to the case $s > \frac{3}{2}$ that we saw previously. The following is devoted to the proof of this theorem.

We will take up the example from Sect. 5.4. For $\lambda \geq 1$ and $\varepsilon \in (0,1)$ we considered the function

$$u_0(x) = \lambda\big(e^{-|x+\varepsilon|} - e^{-|x-\varepsilon|}\big).$$

λ and ε will be chosen later on depending on s and δ appearing in the statement of Theorem 5.8.1.

Lemma 5.8.2 *1. The function u_0 belongs to $W^{1,\infty}(\mathbf{R})$.*
2. For all $s < \frac{3}{2}$ we have $u_0 \in H^s(\mathbf{R})$ and there exists $C = C(s) \geq 1$, such that

$$C^{-1}\lambda\varepsilon^{\frac{3}{2}-s} < \|u_0\|_{H^s} \leq C\lambda\varepsilon^{\frac{3}{2}-s}. \tag{5.8.1}$$

Proof 1. This was shown in Sect. 5.4.

2. Consider the function $v(x) = e^{-|x|} \in L^1(\mathbf{R})$. We have $\widehat{v}(\xi) = \frac{2}{1+\xi^2}$. Indeed,

$$\begin{aligned}\widehat{v}(\xi) &= \int_{\mathbf{R}} e^{-ix\xi-|x|}\,dx = \int_{-\infty}^{0} e^{-(i\xi-1)x}\,dx + \int_{0}^{+\infty} e^{-(i\xi+1)x}\,dx,\\ &= -\frac{1}{i\xi-1} + \frac{1}{i\xi+1} = \frac{2}{1+\xi^2}.\end{aligned}$$

Then,

$$\mathcal{F}(e^{-|x\pm\varepsilon|})(\xi) = \int_{\mathbf{R}} e^{-ix\xi}e^{-|x\pm\varepsilon|}\,dx = e^{\pm i\varepsilon}\widehat{v}(\xi) = e^{\pm i\varepsilon}\frac{2}{1+\xi^2}.$$

We deduce that

$$\widehat{u_0}(\xi) = \lambda\frac{4i\sin(\varepsilon\xi)}{1+\xi^2}.$$

The function $(1+\xi^2)^{\frac{s}{2}}\widehat{u_0}(\xi)$ belongs to $L^2(\mathbf{R})$ for all $s < \frac{3}{2}$. Let us show (5.8.1). We have

$$\begin{aligned}\|u_0\|_{H^s}^2 &= 16\lambda\int_{\mathbf{R}}(1+\xi^2)^{s-2}\sin^2(\varepsilon\xi)\,d\xi = 32\lambda\int_0^{+\infty}(1+\xi^2)^{s-2}\sin^2(\varepsilon\xi)\,d\xi,\\ &:= 32\lambda\, I_\varepsilon.\end{aligned}$$

In the integral I_ε set $\eta = \varepsilon\xi$. We get

$$I_\varepsilon = \frac{1}{\varepsilon}\int_0^{+\infty}\left(1+\frac{\eta^2}{\varepsilon^2}\right)^{s-2}\sin^2\eta\,d\eta = \varepsilon^{3-2s}\int_0^{+\infty}\frac{\sin^2\eta}{(\varepsilon^2+\eta^2)^{2-s}}\,d\eta.$$

Since $0 < \varepsilon < 1$ we have

$$\varepsilon^{3-2s}\int_0^{+\infty}\frac{\sin^2\eta}{(1+\eta^2)^{2-s}}\,d\eta \le I_\varepsilon \le \varepsilon^{3-2s}\int_0^{+\infty}\frac{\sin^2\eta}{\eta^{4-2s}}\,d\eta.$$

Since $s < \frac{3}{2}$ we have $4-2s > 1$. The integrals on the left and right converge at $+\infty$. On the other hand, near $\eta = 0$ we have $\frac{\sin^2\eta}{\eta^{4-2s}} \sim \frac{1}{\eta^{2-2s}}$ and $2-2s < 1$ since $s \ge 1$. The integral on the right therefore converges at zero. This completes the proof of (5.8.1). □

Let us return to the proof of Theorem 5.8.1. The initial data being in $W^{1,\infty}$, problem (5.2.1) admits a unique maximal solution on the interval $[0, T^*)$ where (see Sect. 5.4)

$$T^* = \frac{1}{\lambda(1+e^{-2\varepsilon})}.$$

Lemma 5.8.3 *Let* $\varepsilon \in (0,1)$. *Set* $T_0 = \frac{(1-\varepsilon^2)e^{\varepsilon}}{2\lambda}$. *Then,*
(*i*) $0 < T_0 < T^*$,
(*ii*) $\|u(T_0)\|_{H^1} \geq 2e^{-1}\varepsilon^{-\frac{1}{2}}\lambda$.

Proof (*i*) First, let ε_0 be such that $\cosh \varepsilon_0 = 2$, where cosh denotes the hyperbolic cosine. Then, $\varepsilon_0 > 1$. For $\varepsilon \in (0,1)$ we have $\cosh \varepsilon \leq 1 + \varepsilon^2$. Indeed let us set $f(x) = \cosh x - 1 - x^2$. We have $f'(x) = \sinh x - 2x$ and $f''(x) = \cosh x - 2$. Thus, for $x \in (0, \varepsilon_0)$ we have $f''(x) < 0$ so f' is decreasing and since $f'(0) = 0$ we have $f'(x) < 0$, so f is decreasing, hence $f(x) < 0$ since $f(0) = 0$.

Since $1 + \varepsilon^2 < \frac{1}{1-\varepsilon^2}$, we deduce that $\cosh \varepsilon < \frac{1}{1-\varepsilon^2}$. It follows that we will have $\frac{T_0}{T^*} = (1-\varepsilon^2)\cosh \varepsilon < 1$.

(*ii*) We can write

$$\|u(T_0)\|^2_{H^1} \geq \int_{\mathbf{R}} |(\partial_x u)(T_0, x)|^2 \, dx.$$

We have $u(t,x) = u_0(\kappa(t,x))$ where $\kappa(t,x) = y \iff x = y + tu_0(y)$ and according to (5.3.10),

$$(\partial_x u)(t,x) = (\partial_x \kappa)(t,x) u_0'(\kappa(t,x)).$$

Then, according to (5.3.5) we have $(\partial_x \kappa)(t,x) = \frac{1}{1+tu_0'(y)}$. We deduce that

$$\|u(T_0)\|^2_{H^1} \geq \int_{\mathbf{R}} (\partial_x \kappa)(T_0, x) |u_0'(\kappa(T_0, x))|^2 (\partial_x \kappa)(T_0, x) \, dx.$$

Setting in the integral $y = \kappa(T_0, x)$ we have $dy = (\partial_x \kappa)(T_0, x)\, dx$ so that

$$\|u(T_0)\|^2_{H^1} \geq \int_{\mathbf{R}} \frac{1}{1 + T_0 u_0'(y)} |u_0'(y)|^2 \, dy \geq \int_0^{\varepsilon} \frac{1}{1 + T_0 u_0'(y)} |u_0'(y)|^2 \, dy. \quad (5.8.2)$$

According to the monotonicity of u_0' on the interval $(0, \varepsilon)$ we have

$$-\lambda(1 + e^{-2\varepsilon}) = u_0'(\varepsilon) \leq u_0'(y) \leq u_0'(0) = -2\lambda e^{-\varepsilon} < 0.$$

Thus $|u_0'(y)|^2 \geq 4\lambda^2 e^{-2\varepsilon}$ and

$$1 + T_0 u_0'(y) \leq 1 - 2\lambda e^{-\varepsilon} \frac{(1-\varepsilon^2)e^{\varepsilon}}{2\lambda} = \varepsilon^2,$$

$$1 + T_0 u_0'(y) \geq 1 - \lambda(1 + e^{-2\varepsilon}) \frac{(1-\varepsilon^2)e^{\varepsilon}}{2\lambda} = 1 - (1-\varepsilon^2)\cosh \varepsilon > 0.$$

We deduce from (5.8.2) that

$$\|u(T_0)\|_{H^1}^2 \geq \int_0^{\varepsilon} 4\varepsilon^{-2}\lambda^2 e^{-2\varepsilon}\, dy \geq 4e^{-2}\varepsilon^{-1}\lambda^2.$$

□

End of the proof of Theorem 5.8.1 Let $\delta \in (0, 1)$. It remains to choose the parameters ε (small) and λ (large) as a function of δ. We take $\lambda \geq \frac{2}{\delta}$ so that $\frac{\delta}{\lambda} \leq \frac{1}{2}\delta^2$. Then we take $\varepsilon \leq \left(\frac{\delta}{C\lambda}\right)^{\frac{2}{3-2s}}$ so that, according to Lemma 5.8.2, we have $\|u_0\|_{H^s} \leq \delta$. Next, since $C \geq 1$ we have $\frac{\delta}{C\lambda} \leq \frac{\delta}{\lambda} \leq \frac{1}{2}\delta^2$ hence

$$\varepsilon \leq (\frac{1}{2}\delta^2)^{\frac{2}{3-2s}} \leq 1 \leq \frac{1}{4}(\lambda\delta)^2.$$

We deduce that $\lambda\varepsilon^{-\frac{1}{2}} \geq \frac{2}{\delta}$. Using Lemma 5.8.3 and the fact that $s \geq 1$ we obtain

$$\|u(T_0)\|_{H^s} \geq \|u(T_0)\|_{H^1} \geq 2e^{-1}\lambda\varepsilon^{-\frac{1}{2}} \geq \frac{4}{e\,\delta} \geq \frac{1}{\delta}.$$

Notice that $T_0 \leq C\delta$. □

5.9 Appendix

5.9.1 Characterization of the Spaces H^s for $0 < s < 1$

Recall the result proved in the Appendix of Chap. 4.

Proposition 5.9.1 *For $0 < s < 1$ the quantity*

$$N(u) = \left(\|u\|_{L^2(\mathbf{R}^d)}^2 + \iint_{\mathbf{R}^d \times \mathbf{R}^d} \frac{|u(x) - u(y)|^2}{|x - y|^{d+2s}}\, dx\, dy\right)^{\frac{1}{2}}$$

is a norm on the space $H^s(\mathbf{R}^d)$ equivalent to the usual norm $\|u\|_{H^s}$.

Corollary 5.9.2 *Suppose that $s = k + \sigma$ where $k \in \mathbf{N}$ and $0 < \sigma < 1$. Then the quantity*

$$N_s(u) = \left(\|u\|_{L^2(\mathbf{R}^d)}^2 + \sum_{|\alpha|=k} \iint_{\mathbf{R}^d \times \mathbf{R}^d} \frac{|D^\alpha u(x) - D^\alpha u(y)|^2}{|x - y|^{d+2\sigma}}\, dx\, dy\right)^{\frac{1}{2}}$$

is equivalent to the usual norm of H^s.

5.9.2 *Characterization of Bounded Lipschitz Functions on* $\mathbf{R}^d$

We denote by Lip $(\mathbf{R}^d)$ the space of functions $u \in L^\infty(\mathbf{R}^d)$ such that

$$\exists\, A > 0 : |u(x) - u(y)| \le A\,|x - y|, \quad \text{a.e } x, y \in \mathbf{R}^d\,. \tag{5.9.1}$$

The purpose of this paragraph is to show the following result.

Theorem 5.9.3 *The following two conditions are equivalent:*

$$(i) \quad u \in Lip\,(\mathbf{R}^d),$$

$$(ii) \quad u \in L^\infty(\mathbf{R}^d) \text{ and } \frac{\partial u}{\partial x_i} \in L^\infty(\mathbf{R}^d), \quad i = 1, \dots, d\,.$$

Proof $(i) \Longrightarrow (ii)$

Lemma 5.9.4 *Let* $T \in \mathcal{D}'(\mathbf{R}^d)$*. For* $h \in \mathbf{R}$*, we denote by* $\tau_h^i T$ *the distribution defined by*

$$\langle \tau_h^i T, \varphi\rangle = \langle T, \tau_{-h}^i \varphi\rangle, \quad \varphi \in C_0^\infty(\mathbf{R}^d),$$

where $\tau_{-h}^i\varphi(x) = \varphi(x_1, \dots, x_i - h, \dots, x_d)$*. Then, in* $\mathcal{D}'(\mathbf{R}^d)$*, we have*

$$\lim_{\substack{h\to 0\\ h\neq 0}} \frac{\tau_h^i T - T}{h} = \frac{\partial T}{\partial x_i}.$$

Proof We have

$$\Big\langle \frac{\tau_h^i T - T}{h}, \varphi\Big\rangle = -\langle T, \varphi_{-h}^i\rangle, \quad \text{where } \varphi_{-h}^i(x) = \frac{\tau_{-h}^i\varphi(x) - \varphi(x)}{-h}.$$

It is sufficient to show that (φ_h^i) converges to $\frac{\partial\varphi}{\partial x_i}$ in $C_0^\infty(\mathbf{R}^d)$. If supp φ is contained in $\{x : |x| \le M\}$ and $|h| \le \varepsilon$, then supp $\varphi_h^i \subset \{x : |x| \le M + \varepsilon\}$. Next, by the Taylor formula with integral remainder, we can write (denoting $e_i = (0, \dots, 1, \dots, 0)$)

$$\Big|\partial_x^\alpha\, \varphi(x - h\, e_i) - \partial_x^\alpha\, \varphi(x) + h\, \partial_x^\alpha \frac{\partial\varphi}{\partial x_i}(x)\Big| \le C\,|h|^2 \sum_{|\beta|\le|\alpha|+2} \sup_{\mathbf{R}^d} |\partial_x^\beta\, \varphi(x)|,$$

so $(\partial_x^\alpha\, \varphi_h^i)$ converges uniformly on $\mathbf{R}^d$ to $\partial_x^\alpha \frac{\partial\varphi}{\partial x_i}$, which proves the desired result. □

If $u \in$ Lip $(\mathbf{R}^d)$ and $h \neq 0$, from (5.9.1) we have

$$\Big|\frac{\tau_h^i u(x) - u(x)}{h}\Big| = \Big|\frac{u(x + h\, e_i) - u(x)}{h}\Big| < A.$$

Thus,

$$\left|\left\langle \frac{\tau_h^i u - u}{h}, \varphi\right\rangle\right| = \left|\int \frac{u(x + h\, e_i) - u(x)}{h}\, \varphi(x)\, dx\right| \le A\, \|\varphi\|_{L^1(\mathbf{R}^d)}\,.$$

Letting h tend to zero and using Lemma 5.9.4 we obtain

$$\left|\left\langle \frac{\partial u}{\partial x_i}, \varphi\right\rangle\right| \le A\, \|\varphi\|_{L^1(\mathbf{R}^d)}\,, \quad \forall \varphi \in C_0^\infty(\mathbf{R}^d)\,. \tag{5.9.2}$$

It follows from (5.9.2) that the map $\varphi \mapsto F(\varphi) = \left\langle \frac{\partial u}{\partial x_i}, \varphi\right\rangle$ is a continuous linear form on $C_0^\infty(\mathbf{R}^d)$ endowed with the L^1 norm. Since $C_0^\infty(\mathbf{R}^d)$ is dense in $L^1(\mathbf{R}^d)$, there exists a continuous linear form $\widetilde{F}$ on L^1 such that $\widetilde{F} = F$ on $C_0^\infty(\mathbf{R}^d)$. Since $\widetilde{F} \in (L^1(\mathbf{R}^d))' = L^\infty(\mathbf{R}^d)$, there exists $g \in L^\infty(\mathbf{R}^d)$ such that $\widetilde{F}(\varphi) = \int g\, \varphi\, dx$, for all $\varphi \in L^1(\mathbf{R}^d)$. If $\varphi \in C_0^\infty(\mathbf{R}^d)$, we have

$$\int g\, \varphi\, dx = \widetilde{F}(\varphi) = F(\varphi) = \left\langle \frac{\partial u}{\partial x_i}, \varphi\right\rangle,$$

that is, $\frac{\partial u}{\partial x_i} = g$ in $\mathcal{D}'$ and $\frac{\partial u}{\partial x_i} \in L^\infty(\mathbf{R}^d)$.

$(ii) \Longrightarrow (i)$.

Lemma 5.9.5 *Let $u \in L^\infty(\mathbf{R}^d)$ such that $\frac{\partial u}{\partial x_i} \in L^\infty(\mathbf{R}^d)$, $i = 1, \dots, n$. Suppose further that u is* **continuous.** *Let $\rho \in C_0^\infty(\mathbf{R}^d)$, $\rho \ge 0$, be such that $\int \rho(x)\, dx = 1$ and set $\rho_\varepsilon(x) = \varepsilon^{-d} \rho\left(\frac{x}{\varepsilon}\right)$. Set $u_\varepsilon = u * \rho_\varepsilon$. Then,*

$$(i) \quad |u_\varepsilon(x) - u_\varepsilon(x')| \le \sum_{i=1}^d \Big\| \frac{\partial u}{\partial x_i} \Big\|_{L^\infty(\mathbf{R}^d)} |x - x'|, \quad \forall\, x, x' \in \mathbf{R}^d,$$

$$(ii) \quad u \in Lip\,(\mathbf{R}^d).$$

Proof (i) Since u is continuous, we know that u_ε converges uniformly on every compact to u. Moreover, $u_\varepsilon \in C^\infty$ and $\frac{\partial u_\varepsilon}{\partial x_i} = \rho_\varepsilon * \frac{\partial u}{\partial x_i}$. We then have

$$\left\| \frac{\partial u_\varepsilon}{\partial x_i} \right\|_{L^\infty} \le \|\rho_\varepsilon\|_{L^1} \left\| \frac{\partial u}{\partial x_i} \right\|_{L^\infty} \le \left\| \frac{\partial u}{\partial x_i} \right\|_{L^\infty}, \tag{5.9.3}$$

since $\|\rho_\varepsilon\|_{L^1} = \|\rho\|_{L^1} = 1$.

Using the mean value inequality and (5.9.3), we can write

$$|u_\varepsilon(x) - u_\varepsilon(x')| \le \sum_{i=1}^d \Big\| \frac{\partial u}{\partial x_i} \Big\|_{L^\infty} |x - x'|.$$

(ii) It follows from (i), letting ε tend to zero, that

$$|u(x) - u(x')| \leq \sum_{i=1}^{n} \left\| \frac{\partial u}{\partial x_i} \right\|_{L^\infty} |x - x'|.$$

We deduce that $u \in \mathrm{Lip}\,(\mathbf{R}^d)$. □

To complete the implication $(ii) \Longrightarrow (i)$, it remains to prove, according to Lemma 5.9.5, that u is continuous.

Let $u \in L^\infty(\mathbf{R}^d)$ such that $\frac{\partial u}{\partial x_i} \in L^\infty(\mathbf{R}^d)$, $i = 1, \ldots, d$. Let $E = c_d\, |x|^{2-d}$ if $d \geq 3$ and $E = c \,\mathrm{Log}\, |x|$ if $d = 2$, a fundamental solution of Δ (the Laplacian).

Let $\theta \in C_0^\infty(\mathbf{R}^d)$, $\theta = 1$ for $|x| \leq 1$. We then have

$$\Delta(\theta\, E) = \delta_0 + \omega, \quad \omega \in C_0^\infty(\mathbf{R}^d). \tag{5.9.4}$$

Indeed, we know that $\Delta\, E = \delta_0$. Then $\Delta(\theta E) = \theta\, \Delta E + 2 \sum_{i=1}^{d} \frac{\partial \theta}{\partial x_i} \frac{\partial E}{\partial x_i} + (\Delta\, \theta)\, E$. Since $\theta = 1$ for $|x| \leq 1$, $\frac{\partial \theta}{\partial x_i}$ and $\Delta\theta$ vanish for $|x| \leq 1$ and $E \in C^\infty(\mathbf{R}^d \setminus 0)$ so $\frac{\partial \theta}{\partial x_i} \frac{\partial E}{\partial x_i}$ and $\Delta\theta \cdot E$ are in $C_0^\infty(\mathbf{R}^d)$. We then have $\theta\, \Delta\, E = \theta \delta_0 = \theta(0)\, \delta_0 = \delta_0$, which proves (5.9.4).

Set $E_i = \theta\, \frac{\partial E}{\partial x_i}$. Then, $E_i \in L^1(\mathbf{R}^d)$, $i = 1, \ldots, d$.

Indeed, we have $\frac{\partial E}{\partial x_i} = c'_d\, \frac{x_i}{|x|^d}$. Thus, $\frac{\partial E}{\partial x_i}$ is locally integrable, therefore $\theta\, \frac{\partial E}{\partial x_i}$ belongs to $L^1(\mathbf{R}^d)$. Now we can write

$$u = \sum_{i=1}^{d} \frac{\partial u}{\partial x_i} * E_i + \widetilde{\omega}, \quad \widetilde{\omega} \in C^\infty(\mathbf{R}^d). \tag{5.9.5}$$

Indeed, from (5.9.4) we have

$$\begin{aligned} u =& u * \delta_0 = u * \Delta(\theta E) - u * \omega = \sum_{i=1}^{d} \frac{\partial u}{\partial x_i} * \frac{\partial}{\partial x_i} (\theta E) - u * \omega, \\ =& \sum_{i=1}^{d} \frac{\partial u}{\partial x_i} * \big(\theta\, \frac{\partial E}{\partial x_i}\big) + \sum_{i=1}^{d} \frac{\partial u}{\partial x_i} * \big(\frac{\partial \theta}{\partial x_i}\, E\big) - u * \omega. \end{aligned}$$

Since $\frac{\partial \theta}{\partial x_i}\, E \in C_0^\infty(\mathbf{R}^d)$, we have $\frac{\partial u}{\partial x_i} * \big(\frac{\partial \theta}{\partial x_i}\, E\big) \in C^\infty(\mathbf{R}^d)$; similarly ω belongs to $C_0^\infty(\mathbf{R}^d)$, so $u * \omega$ belongs to $C^\infty(\mathbf{R}^d)$, which proves (5.9.5).

Let $x, x' \in \mathbf{R}^d$ and $(\tau_{x-x'} f)(z) = f(x - x' + z)$. We have

$$\left| \Big(\frac{\partial u}{\partial x_i} * E_i\Big)(x) - \Big(\frac{\partial u}{\partial x_i} * E_i\Big)(x') \right| \leq \| \tau_{x-x'}\, E_i - E_i \|_{L^1(\mathbf{R}^d)} \left\| \frac{\partial u}{\partial x_i} \right\|_{L^\infty(\mathbf{R}^d)}. \tag{5.9.6}$$

Indeed, let us denote $E_i = \theta\, \frac{\partial E}{\partial x_i}$. We have

$$\begin{aligned}\left|\left(\frac{\partial u}{\partial x_i} * E_i\right)(x) - \left(\frac{\partial u}{\partial x_i} * E_i\right)(x')\right| &= \left|\int \frac{\partial u}{\partial x_i}(y)\Big[E_i(x-y) - E_i(x'-y)\Big]\,dy\right|, \\ &\le \left\|\frac{\partial u}{\partial x_i}\right\|_{L^\infty} \int |E_i(x-x'+z) - E_i(z)|\,dz, \\ &\le \left\|\frac{\partial u}{\partial x_i}\right\|_{L^\infty} \|\tau_{x-x'} E_i - E_i\|_{L^1}.\end{aligned}$$

Now since $E_i \in L^1(\mathbf{R}^d)$ and translation is continuous in $L^1(\mathbf{R}^d)$ we deduce that the function $\frac{\partial u}{\partial x_i} * E_i$ is continuous. It follows from (5.9.5) that u is continuous, which completes the proof of Theorem 5.9.3. □

5.10 Comments

1. Johannes Martinus Burgers (1895–1981) was a Dutch physicist. The equation that bears his name and that he discussed in 1948 comes from fluid mechanics. It appears in various fields of applied mathematics, such as the modeling of gas dynamics, acoustics, or road traffic. It also appears in earlier works by mathematicians Andrew Forsyth and Harry Bateman.
2. Concerning Sobolev spaces, we have only considered here the cases $1 < s < \frac{3}{2}$ and $\frac{3}{2} < s < 2$. The latter choice was made so as not to obscure the text with long and tedious calculations, but it should be known that the results obtained (continuity and non-uniform continuity) are valid throughout the range $s > \frac{3}{2}$.
3. Another interesting case that we have not treated, but which is the subject of numerous works, is when the initial data is less regular than Lipschitz, for example only $L^\infty(\mathbf{R})$. Several new phenomena appear. First, the possible solutions are no longer regular enough for their derivatives to exist, so one is led to the notion of weak solutions, that is, "in the sense of distributions." But it turns out that these weak solutions are no longer unique, which leads to seeking a criterion allowing, among all these solutions, to distinguish one, which leads to the notion of "entropy" solution.
   We refer the reader to reference [7] for the study of this case.

Chapter 6
Mathematical Analysis of Surface Waves

6.1 Prerequisites

Differential equations. Local inversion theorem. Diffeomorphisms.

6.2 Context and Problem

6.2.1 The Problem

In 1813 the French Academy of Sciences opened a competition asking

"To determine the propagation of waves on the surface of a heavy fluid of indefinite depth".

Many famous mathematicians, such as Augustin-Louis Cauchy, Siméon Denis Poisson etc. participated in this competition. Cauchy, who was 26 years old at that time, was awarded the "Grand Prix". It is remarkable that Cauchy made use, in his memoir, of Fourier integrals even before Fourier had published his major work on the subject.

However, the mathematical study of waves began much earlier, in the mid-18th century, with the pioneering work of Leonhard Euler (c. 1752), Pierre-Simon de Laplace (c. 1775–1776), building upon D'Alembert's foundations and Joseph-Louis Lagrange (c. 1781). Later, the field was significantly advanced by figures such as George B. Airy (c. 1845), George Gabriel Stokes (c. 1847), and Joseph Boussinesq (c. 1872).

Since then and up to the present day, it has been an active field of study.

These studies were initially motivated by navigation problems, the prediction of tides, and more recently by issues of offshore installation safety, tsunami analysis, etc.

C. Zuily, *Selected Topics in Partial Differential Equations*, Universitext,
https://doi.org/10.1007/978-3-032-24082-8_6

The mathematical model that we will present here, aims to answer the question posed. It seeks to model the propagation of waves on the surface of a fluid occupying a domain whose upper boundary is variable. One can think of the moving surface of an ocean or a lake. In this model, the fluid will be assumed to be

- subject to the force of gravity,
- incompressible,
- irrotational.

These terms, as well as their mathematical modeling, will be explained later.

The point of view we will adopt here is essentially that which Augustin-Louis Cauchy developed in his 1814 memoir, in response to the question of the Academy.

Some Types of Waves

Water waves are waves that propagate on the surface of a fluid (a lake, an ocean, etc.) by *energy transfer*, originating from a phenomenon that may be the impact of an object on the surface of the fluid, the action of the wind (wind waves), the action of the moon and/or the sun (tides), an underwater volcanic eruption, an earthquake (tsunamis), etc.

The surface of the fluid is two-dimensional, but it is sometimes useful, as a first approach, to simplify the problem and assume that the spatial dimension is one (which would correspond to taking the trace of the waves on a plane perpendicular to the fluid). In this case, an (ideal) periodic wave is characterized by:

$$\lambda = \text{wavelength},\ A = \text{amplitude},\ T = \text{period},\ h = \text{depth}.$$

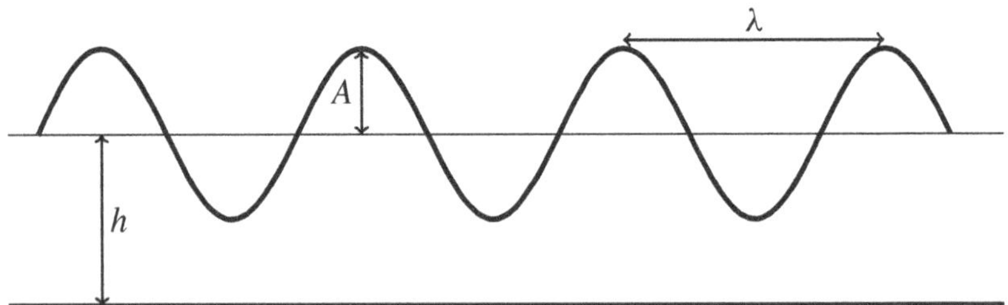

- wind waves: $\lambda \sim 1$ to 100 m, period $\sim$ seconds to minutes,
- tides: $\lambda \sim$ hundreds of km, period $\sim$ 12h,
- tsunamis: $\lambda \sim 100$ km, $\lambda >> h$, period $\sim$ minutes to hours,
- other types: breakers, rogue waves (poorly understood) etc.

6.3 Modeling

We will work in a basis (e_1, e_2, e_3) of $\mathbf{R}^3$ for which e_3 is a vertical vector.

The domain occupied by the fluid at time t will be denoted Ω_t. This will be a connected open subset of $\mathbf{R}^3$. It will be bounded at its upper part by the surface of the fluid denoted Σ_t and at its lower part by a bottom Γ. We will assume that Σ_t is a surface given by the graph of a function.

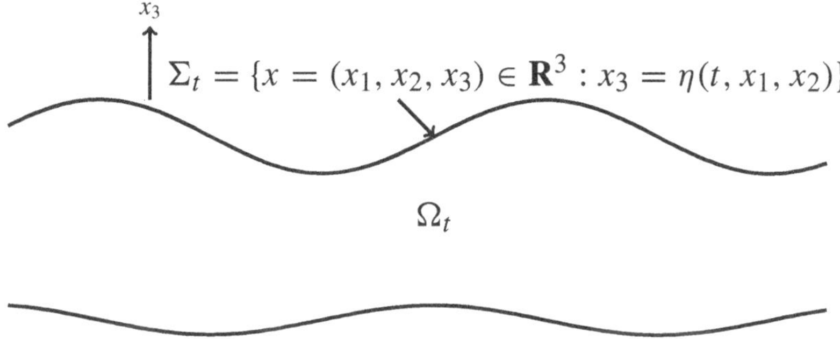

The surface $\Sigma_t = \{x = (x_1, x_2, x_3) \in \mathbf{R}^3 : x_3 = \eta(t, x_1, x_2)\}$ is an **unknown of the problem** and we will denote

$$\Omega = \{(t, x) : t \in \mathbf{R}^+, x \in \Omega_t\}, \quad \Sigma = \{(t, x) : t \in \mathbf{R}^+, x \in \Sigma_t\}.$$

6.3.1 *The Equations*

We shall denote by $X(t) = (X_1(t), X_2(t), X_3(t))$ the position of a fluid particle at time t and by $v(t, X) = (v_1(t, X), v_2(t, X), v_3(t, X))$ the velocity of the particle at time t and at point X. They are related by the equations

$$X_j'(t) = v_j(t, X(t)), \quad j = 1, 2, 3, \tag{6.3.1}$$

where X_j' denotes the derivative with respect to time of the function X_j.

We will assume v is of class C^1 so that we can easily solve this problem, and we will denote by $X(t, y)$ the solution of (6.3.1) such that $X(0) = y \in \mathbf{R}^3$.

The Equation of Motion

The first equation of the model expresses Newton's law, which states that the acceleration (which is equal to $X_j''(t)$, the second derivative) is proportional to the forces acting on the particles. Using (6.3.1) we see that

$$\begin{aligned} X_j''(t) &= \frac{\partial v_j}{\partial t}(t, X(t)) + \sum_{k=1}^{3} \frac{\partial v_j}{\partial x_k}(t, X(t))X_k'(t), \\ &= \Big(\frac{\partial v_j}{\partial t} + \sum_{k=1}^{3} v_k \frac{\partial v_j}{\partial x_k}\Big)(t, X(t)) = \Big(\frac{\partial v_j}{\partial t} + (v \cdot \nabla_x)v_j\Big)(t, X(t)), \end{aligned}$$

where we have denoted $(v \cdot \nabla_x)$ the operator $\sum_{k=1}^{3} v_k \frac{\partial}{\partial x_k}$.

In our model, the forces acting on the fluid particles are, on the one hand, the *gradient of pressure* due to the fluid and, on the other hand the *gravity*. Since gravity acts downward, we have

$$F = -\nabla_x P - g e_3$$

where P is the pressure, $\nabla_x P$ is the vector $(\frac{\partial P}{\partial x_1}, \frac{\partial P}{\partial x_2}, \frac{\partial P}{\partial x_3})$ and g is the acceleration due to gravity.

There are other models in which other forces are involved (surface tension, etc.). The first system of equations is therefore

$$\frac{\partial v}{\partial t} + (v \cdot \nabla_x)v = -\nabla_x P - g e_3 \quad \text{in } \Omega. \tag{6.3.2}$$

These are the Euler equations.

6.3.2 *Incompressibility*

This is a physical notion that expresses the fact that, infinitesimally, the volume of an open set occupied by the fluid (that is, its Lebesgue measure) does not change over time. We will show that mathematically this is expressed by the equation,

$$\operatorname{div} v = \sum_{j=1}^{3} \frac{\partial v_j}{\partial x_j} = 0 \quad \text{in } \Omega, \tag{6.3.3}$$

where $\operatorname{div} v$, defined above, is the divergence of the vector field v.

In what follows, we may take $n = 3$ but the result we are going to prove is true in any dimension.

Let us introduce some notations.

Let $M(n, \mathbf{R})$ be the algebra of $n \times n$ matrices with real coefficients, I an open interval of $\mathbf{R}$, and $k \in \mathbf{N}$.

We denote by $C^k(I, M(n, \mathbf{R}))$ the set of maps $A : I \to M(n, \mathbf{R}), t \mapsto A(t)$, where $A(t) = (a_{ij}(t))_{1\le i,j\le n}$, is such that the functions $t \mapsto a_{ij}(t)$ belong to $C^k(I)$.

If A belongs to $C^1(I, M(n, \mathbf{R}))$ we denote by $A'(t)$ the matrix $(a'_{ij}(t))_{1\le i,j\le n}$ (the matrix of derivatives).

If $A = (a_{ij})_{1\le i,j\le n} \in M(n, \mathbf{R})$ we denote by $\mathrm{Tr}(A) = \sum_{i=1}^n a_{ii}$ its trace and $\det A$ its determinant.

Lemma 6.3.1 *Let I be an open interval of* $\mathbf{R}$. *We consider*

$$A \in C^0(I, M(n, \mathbf{R})), \quad B \in C^1(I, M(n, \mathbf{R})), \quad t_0 \in I \text{ and } \quad \Delta(t) = \det B(t).$$

We assume that $B'(t) = A(t)B(t)$ for all $t \in I$. Then,

$$\Delta'(t) = \Delta(t) Tr\,(A(t)), \quad t \in I,$$

so that $\Delta(t) = \Delta(t_0) \exp\left(\int_{t_0}^t Tr\,(A(s))\,ds \right)$.

Proof We have $\Delta'(t) = \sum_{i=1}^n \det\big(l_1(t), \ldots, l'_i(t), \ldots, l_n(t)\big)$ where $l_i(t)$ denotes the ith row of the matrix $B(t) = (b_{ij}(t))$. Let us expand each determinant in the sum with respect to the ith row. Denoting by cof the cofactor, we obtain

$$\Delta'(t) = \sum_{i=1}^n \sum_{k=1}^n b'_{ik}(t)\,\mathrm{cof}(b_{ik}(t)). \tag{6.3.4}$$

We will show that

$$\Delta'(t) = \mathrm{Tr}\big(B'(t)\,\mathrm{adj}(B(t))\big), \tag{6.3.5}$$

where $\mathrm{adj}(B(t)) = {}^t(\mathrm{cof}(b_{ik}(t)))$ is the adjugate matrix of $B(t)$.

Indeed, if $c_{ij} = \mathrm{cof}(b_{ij})$ and $\mathrm{adj}(B) = (d_{ij})$ we have $d_{ij} = c_{ji} = \mathrm{cof}(b_{ji})$ so that

$$\big(B'(t)\,\mathrm{adj}(B(t))\big)_{ij} = \sum_{k=1}^n b'_{ik}(t)\,d_{kj}(t) = \sum_{k=1}^n b'_{ik}(t)\,\mathrm{cof}(b_{jk})(t).$$

Then,

$$\mathrm{Tr}\big(B'(t)\,\mathrm{adj}(B(t))\big) = \sum_{i=1}^n \sum_{k=1}^n b'_{ik}(t)\,\mathrm{cof}(b_{ik}(t))$$

so that (6.3.5) follows from (6.3.4). As by hypothesis $B'(t) = A(t)B(t)$ and since

$$B(t)\,\mathrm{adj}(B(t)) = \det B(t) Id = \Delta(t) Id,$$

it follows from (6.3.5) that

$$\Delta'(t) = \mathrm{Tr}\big(A(t)\, B(t)\, \mathrm{adj}(B(t))\big) = \Delta(t)\, \mathrm{Tr}(A(t)),$$

which, by integrating the differential equation, implies the lemma. □

Next, if we denote by $X(t, y)$ the position at time t of the particle originating at time $t = 0$ from the point $y = (y_1, y_2, y_3)$, i.e., $X(0, y) = y$, by differentiating equation (6.3.1) with respect to y_j and setting $B(t) = \left(\frac{\partial X_i}{\partial y_j}(t, y)\right)$, $A(t) = \left(\frac{\partial v_i}{\partial x_j}(t, X(t, y))\right)$, we easily see that

$$B'(t) = A(t)B(t), \quad B(0) = Id.$$

Since

$$\mathrm{Tr}(A(t)) = \sum_{j=1}^{3} \frac{\partial v_j}{\partial x_j}(t, X(t, y)) = \mathrm{div}\, v(t, X(t, y)), \tag{6.3.6}$$

it follows from Lemma 6.3.1 that

$$\det B(t) = \det\left(\frac{\partial X}{\partial y}(t, y)\right) = \exp\Big(\int_0^t \mathrm{div}\, v(s, X(s, y))\, ds\Big), \tag{6.3.7}$$

since $\frac{\partial X}{\partial y}(0, y)$ is the identity matrix.

It follows from (6.3.7) that

$$\text{the determinant of } B(t) = \big(\frac{\partial X}{\partial y}(t, y)\big) \text{ is always strictly positive.}$$

Corollary 6.3.2 *Let $t_0 \in \mathbf{R}$, O an open subset of $\mathbf{R}^d$, Φ the map $y \mapsto X(t_0, y)$ and $\Omega = \Phi(O)$. Then Φ is a diffeomorphism from O onto Ω.*

Proof According to (6.3.7) and the local inversion theorem, it suffices to show the injectivity. Assume $X(t_0, y) = X(t_0, y')$. Set $E = \{t \in [0, T] : X(t, y) = X(t, y')\}$. By hypothesis $t_0 \in E$. On the other hand, by continuity, E is closed. Let us show that it is open in $[0, T]$. Let $\bar{t} \in E$ and consider the differential systems

$$U'(t) = v(t, U(t)), \quad U(0) = X(\bar{t}, y), \tag{6.3.8}$$

$$V'(t) = v(t, V(t)), \quad V(0) = X(\bar{t}, y'). \tag{6.3.9}$$

Since $\bar{t} \in E$ we have $X(\bar{t}, y) = X(\bar{t}, y')$. The uniqueness of solutions to this system implies that there exists $\varepsilon > 0$ such that $U(t) = V(t)$ for $|t - \bar{t}| \leq \varepsilon$. Now $X(t, y)$ is the unique solution of (6.3.8) and $X(t, y')$ is the unique solution of (6.3.9). Thus, $X(t, y) = X(t, y')$ for $|t - \bar{t}| \leq \varepsilon$ which shows that $B(\bar{t}, \varepsilon) \subset E$. We deduce that $E = [0, T]$ so, $0 \in E$ and thus $y = y'$. □

Let us return to the problem of incompressibility.

Let O be an open set occupied by the fluid at time $t = 0$. At time $t > 0$ the fluid originating from O occupies the open set $O_t = \{X(t, y) : y \in O\}$ whose volume is $V(t) = \int_{O_t} dx$. According to Corollary 6.3.2 we can set, in the integral, $x = X(t, y)$ so that

$$dx = \left|\det\left(\frac{\partial X}{\partial y}(t, y)\right)\right| dy = \det\left(\frac{\partial X}{\partial y}(t, y)\right) dy = \Delta(t, y)\, dy,$$

because, according to (6.3.7), this determinant is positive. Thus,

$$V(t) = \int_O \Delta(t, y)\, dy.$$

According to Lemma 6.3.1 and (6.3.7) we have

$$V'(t) = \int_O \Delta'(t, y)\, dy = \int_O \operatorname{div} v(t, X(t, y))\Delta(t, y)\, dy. \tag{6.3.10}$$

Therefore, if $\operatorname{div} v \equiv 0$ we have $\mu(O_t) = V(t) = V(0) = \mu(O)$ (where μ is the Lebesgue measure).

Conversely, if there exists (t_0, x_0) such that $\operatorname{div} v(t_0, x_0) \neq 0$ there exists an open set $\Omega \subset \mathbf{R}^3$ such that $\operatorname{div} v(t_0, x)$ is nonzero in Ω. According to Corollary 6.3.2 there exists an open set O such that $\operatorname{div} v(t_0, X(t_0, y))$ is nonzero in O (for example > 0). Then,

$$\int_O \operatorname{div} v(t_0, X(t_0, y))\Delta(t_0, y)\, dy > 0.$$

It follows from (6.3.10) that $V'(t_0) > 0$, which shows that $\mu(O_{t_0+\varepsilon}) > \mu(O_{t_0})$ for small $\varepsilon > 0$.

6.3.3 *Irrotationality*

This term expresses the absence of vorticity in the fluid and is mathematically expressed by the formula

$$\operatorname{curl} v = 0 \quad \text{in } \Omega, \tag{6.3.11}$$

where

$$\operatorname{curl} v = (\partial_1 v_2 - \partial_2 v_1,\ \partial_1 v_3 - \partial_3 v_1,\ \partial_2 v_3 - \partial_3 v_2), \quad \partial_j = \frac{\partial}{\partial x_j}. \tag{6.3.12}$$

We will set in what follows

$$\omega_{ij} = \partial_i v_j - \partial_j v_i, \quad i, j = 1, 2, 3,$$

so that

$$\omega_{ji} = -\omega_{ij} \quad \text{and curl } v = (\omega_{12}, \omega_{13}, \omega_{23}).$$

We will show that assuming the fluid is irrotational does not add an equation on v but rather makes an irrotationality assumption on the initial condition. Indeed, we have the following result.

Proposition 6.3.3 *Let v be a solution of the Euler equation* (6.3.2). *Set* $v_0 = v|_{t=0}$. *If curl* $v_0 = 0$ *then for all* $t \geq 0$ *we have curl* $v(t, \cdot) = 0$.

Proof By differentiating the jth equation of (6.3.2) with respect to x_i and then the ith equation with respect to x_j and taking the difference we obtain

$$\partial_t \omega_{ij} + (v \cdot \nabla)\, \omega_{ij} + \sum_{k=1}^{3} \big(\partial_i v_k \partial_k v_j - \partial_j v_k \partial_k v_i\big) = 0.$$

On the other hand, we have

$$\begin{aligned} \partial_i v_k \partial_k v_j - \partial_j v_k \partial_k v_i &= \partial_i v_k (\partial_k v_j - \partial_j v_k) + \partial_j v_k (\partial_i v_k - \partial_k v_i) \\ &= \partial_i v_k\, \omega_{kj} + \partial_j v_k\, \omega_{ki}, \end{aligned}$$

so that, since $\omega_{ki} = -\omega_{ik}$, we obtain

$$\partial_t \omega_{ij} + (v \cdot \nabla)\, \omega_{ij} + \sum_{k=1}^{3} \big(\partial_i v_k\, \omega_{kj} - \partial_j v_k\, \omega_{ik}\big) = 0, \tag{6.3.13}$$

Let us denote by $\nabla v(t, \cdot)$ the matrix $\big(\partial_i v_j\big)_{1 \leq i, j \leq 3}$ and $A(t)$ the matrix $(\omega_{ij}(t, \cdot))_{1 \leq i, j \leq 3}$. Since $\omega_{jj} = 0$ and $\omega_{ji} = -\omega_{ij}$, if curl$v_0 = 0$ we have $A(0) \equiv 0$.

Using (6.3.13) and denoting by ${}^t(\nabla v)$ the transpose of the matrix ∇v, we obtain the matrix equation

$$\partial_t A(t) + (v \cdot \nabla) A(t) + \nabla v \cdot A(t) - A(t) \cdot {}^t(\nabla v) = 0.$$

Set $B(t) = A(t, X(t))$, $W(t) = \nabla v(t, X(t))$. We have $\partial_t B(t) = [\partial_t A + (v \cdot \nabla) A](t, X(t))$ so that the above equation can be written as

$$\partial_t B(t) = B(t)^t W(t) - W(t) B(t).$$

Consequently, since $B(0) = A(0, X(0)) = 0$, we have

$$B(t) = \int_0^t \big(B(s)^t W(s) - W(s) B(s)\big)\, ds,$$

so that, denoting $\|M\| = \sup_{1\le i,j\le 3}|m_{ij}|$, if $M = (m_{ij})$, we obtain

$$\|B(t)\| \le 6\int_0^t \|W(s)\|\|B(s)\|\,ds.$$

Gronwall's inequality (see Lemma 1.7.1 in the appendix of Chap. 1) shows that $B(t) = 0$ for all t and thus $A(t,x) = 0$ for all (t,x) according to Corollary 6.3.2. □

A Consequence of Irrotationality

We have the following result.

Proposition 6.3.4 *Let $x \mapsto v(x) = (v_1(x), v_2(x), v_3(x))$ be a C^1 vector field in $\mathbf{R}^3$ such that curl $v = 0$ in Ω. Then there exists a function $\phi \in C^2(\mathbf{R}^3)$ such that $v = \nabla\phi$.*

Proof From (6.3.12) we have the three equations in $\mathbf{R}^3$,

$$\partial_1 v_2 - \partial_2 v_1 = 0, \quad \partial_1 v_3 - \partial_3 v_1 = 0, \quad \partial_2 v_3 - \partial_3 v_2 = 0.$$

We will show that the following function is suitable. Set for $x = (x_1, x_2, x_3)$,

$$\phi(x) = \int_0^{x_1} v_1(t, x_2, x_3)\,dt + \int_0^{x_2} v_2(0, s, x_3)\,ds + \int_0^{x_3} v_3(0, 0, u)\,du.$$

First, we have $\partial_1\phi(x) = v_1(x)$. Next

$$\begin{aligned}\partial_2\phi(x) &= \int_0^{x_1} \partial_2 v_1(t, x_2, x_3)\,dt + v_2(0, x_2, x_3) = \int_0^{x_1} \partial_1 v_2(t, x_2, x_3)\,dt + v_2(0, x_2, x_3),\\ &= v_2(x) - v_2(0, x_2, x_3) + v_2(0, x_2, x_3) = v_2(x).\end{aligned}$$

Finally

$$\begin{aligned}\partial_3\phi(x) &= \int_0^{x_1} \partial_3 v_1(t, x_2, x_3)\,dt + \int_0^{x_2} \partial_3 v_2(0, t, x_3)\,dt + v_3(0, 0, x_3),\\ &= \int_0^{x_1} \partial_1 v_3(t, x_2, x_3)\,dt + \int_0^{x_2} \partial_2 v_3(0, t, x_3)\,dt + v_3(0, 0, x_3),\\ &= v_3(x) - v_3(0, x_2, x_3) + v_3(0, x_2, x_3) - v_3(0, 0, x_3) + v_3(0, 0, x_3) = v_3(x).\end{aligned}$$

□

Remark 6.3.5 We have an analogous result if $\mathbf{R}^3$ is replaced by a simply connected open set Ω (Poincaré's Lemma).

Corollary 6.3.6 *Let Ω be a connected open subset of $\mathbf{R}^3$. Let $x \mapsto v(x)$, where $v(x) = (v_1(x), v_2(x), v_3(x))$, be a C^1 vector field in Ω such that curl $v(x) = 0$ and div $v(x) = 0$ for all x in Ω. Let ϕ be the function given by Proposition 6.3.4. Then ϕ is harmonic in Ω, that is, $\Delta\phi = 0$ in Ω where $\Delta = \sum_{j=1}^3 \frac{\partial^2}{\partial x_j^2}$.*

Proof This follows from the fact that $0 = \operatorname{div} v = \operatorname{div}(\nabla\phi) = \sum_{j=1}^{3} \frac{\partial^2\phi}{\partial x_j^2}$. □

6.3.4 Boundary Conditions

As mentioned, the upper boundary Σ of the open set Ω is an unknown of the problem. We have assumed that this boundary is the graph of a function. We introduce the notation $x = (\widetilde{x}, x_3)$, $\widetilde{x} = (x_1, x_2) \in \mathbf{R}^2$ so that

$$\Sigma_t = \{x \in \mathbf{R}^3 : x_3 = \eta(t, \widetilde{x})\}, \quad \Sigma = \{(t, x) \in \mathbf{R} \times \mathbf{R}^3 : x \in \Sigma_t\}. \tag{6.3.14}$$

The kinematic condition.
It expresses the following fact:

"*any particle that at time* $t = 0$ *is on the surface remains on the surface for all later times*".

This is mathematically expressed by the equation

$$\partial_t \eta = \sqrt{1 + |\nabla_{\widetilde{x}}\eta|^2}\,(v \cdot n) \quad \text{on } \Sigma. \tag{6.3.15}$$

Here $|\nabla_{\widetilde{x}}\eta|^2 = \sum_{j=1}^{2} \left(\frac{\partial\eta}{\partial x_j}\right)^2$ and $n(t, \widetilde{x}) = (1 + |\nabla_{\widetilde{x}}\eta(t, \widetilde{x})|^2)^{-\frac{1}{2}}(1, -\nabla_{\widetilde{x}}\eta(t, \widetilde{x}))$ denotes the unit outward normal to Σ at the point $(t, \widetilde{x}, x_3 = \eta(t, \widetilde{x}))$. Moreover $v \cdot n$ denotes the scalar product in $\mathbf{R}^3$ of the vectors v and n.

Indeed, let us set $\Phi(t) = x_3(t) - \eta(t, \widetilde{x}(t))$ (where $\widetilde{x}(t) = (x_1(t), x_2(t))$ and also, $X(t) = (\widetilde{x}(t), x_3(t))$. Suppose that $X(0) \in \Sigma_0$. Then $\Phi(0) = 0$. Next, let us denote by $\widetilde{v} = (v_1, v_2)$ the horizontal components and v_3 the vertical component of the velocity. By differentiating $\Phi(t)$, we obtain

$$\begin{aligned} \Phi'(t) &= x_3'(t) - \partial_t\eta(t, \widetilde{x}(t)) - \nabla_{\widetilde{x}}\eta(t, \widetilde{x}(t)) \cdot \widetilde{x}'(t), \\ &= v_3\big(t, X(t)\big) - \partial_t\eta(t, \widetilde{x}(t)) - \nabla_{\widetilde{x}}\eta(t, \widetilde{x}(t)) \cdot \widetilde{v}(t, X(t)). \end{aligned}$$

Using the fact that $x_3(t) = \Phi(t) + \eta(t, \widetilde{x}(t))$ we get

$$\begin{aligned} \Phi'(t) = v_3\big(t, \widetilde{x}(t), \Phi(t) + \eta(t, \widetilde{x}(t))\big) - \partial_t\eta(t, \widetilde{x}(t)) \\ - \nabla_{\widetilde{x}}\eta(t, \widetilde{x}(t)) \cdot \widetilde{v}\big(t, \widetilde{x}(t), \Phi(t) + \eta(t, \widetilde{x}(t))\big) =: F(t, \Phi(t)). \end{aligned}$$

Since $n(t, \widetilde{x}(t)) = (1 + |\nabla_{\widetilde{x}}\eta(t, \widetilde{x}(t))|^2)^{-\frac{1}{2}}(1, -\nabla_{\widetilde{x}}\eta(t, \widetilde{x}(t)))$ and $v = (\widetilde{v}, v_3)$, if $\Phi \equiv 0$, we have

$$F(t, 0) = -\partial_t\eta(t, \widetilde{x}(t)) + \sqrt{1 + |\nabla_x\eta(t, \widetilde{x}(t))|^2}\big(v(t, \widetilde{x}(t), \eta(t, \widetilde{x}(t))) \cdot n(t, \widetilde{x}(t))\big),$$

so that if Eq. (6.3.15) is satisfied, we have $F(t, 0) = 0$. In summary,

$$\Phi'(t) = F(t, \Phi(t)), \quad \Phi(0) = 0, \quad F(t, 0) = 0.$$

Thus, $\Phi \equiv 0$ is the unique solution to this problem. This shows that as long as $X(t)$ exists, we have $x_3(t) = \eta(t, \widetilde{x}(t))$, that is, $X(t) \in \Sigma_t$. Conversely, if $\Phi(t) = 0$ for all $t \geq 0$, then $\Phi'(t) \equiv 0$ and by the calculation above, (6.3.15) is satisfied.

The dynamic condition.
It expresses the continuity of the pressure across the free surface.

$$P = P_0 \quad \text{on } \Sigma, \quad (\text{where } P_0 \text{ is the atmospheric pressure}).$$

However, since P appears in the equations only through its gradient, we may, by abuse of notation, still denote by P the quantity $P - P_0$ and assume that

$$P = 0 \quad \text{on } \Sigma. \tag{6.3.16}$$

The bottom condition.
We will assume that the fluid moves parallel to the bottom. When the bottom is regular, denoting by ν the unit normal to Γ, we will assume

$$v \cdot \nu = 0 \quad \text{on } \Gamma, \tag{6.3.17}$$

that is, the velocity is tangent to the bottom.

6.3.5 *The System in* (η, v)

In summary, the system in (η, v) is as follows:

$$\begin{cases} (i) \;\; \dfrac{\partial v}{\partial t} + (v \cdot \nabla_x) v = -\nabla_x P - g e_3 \quad \text{in } \Omega, \\ (ii) \;\; \partial_t \eta = \sqrt{1 + |\nabla_{\widetilde{x}} \eta|^2}\, (v \cdot n) \quad \text{on } \Sigma, \\ (iii) \;\; \operatorname{div} v = 0 \quad \text{in } \Omega, \\ (iv) \;\; P = 0 \quad \text{on } \Sigma, \\ (v) \;\; v \cdot \nu = 0 \quad \text{on } \Gamma. \end{cases} \tag{6.3.18}$$

The Cauchy problem consists in solving the system (6.3.18) with the initial conditions $v|_{t=0} = v_0$, $\eta|_{t=0} = \eta_0$ and moreover $\operatorname{curl} v_0 = 0$, if we are interested in irrotational fluids.

6.3.6 Conservation of Energy

Let (η, v) be a regular solution of (6.3.18), for simplicity in infinite depth, i.e.

$$\Omega_t = \{x = (\widetilde{x}, x_3) : \widetilde{x} \in \mathbf{R}^2, -\infty < x_3 < \eta(t, \widetilde{x})\}.$$

Assume that, for $t \geq 0$, we have $v_j(t, \cdot) \in L^2(\Omega_t)$, $j = 1, 2, 3$ and $\eta(t, \cdot) \in L^2(\mathbf{R}^2)$.

The energy of this solution at time $t > 0$ is, by definition, the quantity

$$E(t) = \frac{1}{2} \iint_{\Omega_t} |v(t, x)|^2 \, dx + \frac{g}{2} \int_{\mathbf{R}^2} \eta(t, \widetilde{x})^2 \, d\widetilde{x}. \tag{6.3.19}$$

We set

$$E(0) = \frac{1}{2} \iint_{\Omega_0} |v_0(x)|^2 \, dx + \frac{g}{2} \int_{\mathbf{R}^2} \eta_0(\widetilde{x})^2 \, d\widetilde{x}.$$

We will show that for suitable solutions, the quantity (6.3.19) is conserved over time.

To do this, recall that, according to Corollary 6.3.2 and (6.3.7) (since $\operatorname{div} v = 0$), we have

(i) $\Omega_t = \{x = X(t, y) : y \in \Omega_0\}$,

(ii) the map $\Omega_0 \to \Omega_t$, $y \mapsto x = X(t, y)$ is a diffeomorphism,

(iii) the Jacobian determinant of this map is equal to 1.

It follows from this, by setting $x = X(t, y)$ in the integral that

$$\iint_{\Omega_t} |v(t, x)|^2 \, dx = \iint_{\Omega_0} |v(t, X(t, y))|^2 \, dy. \tag{6.3.20}$$

We will make assumptions on v (and η) allowing us to differentiate under the integral sign the right-hand side of (6.3.20) (and the quantity $E(t)$).

Let $I = (0, +\infty)$ and introduce the following functional space:

$$\mathcal{E} = \left\{h : I \times \Omega_0 \to \mathbf{R} : \forall K \subset\subset I, \ \exists g_K \in L^2(\Omega_0) : \sup_{t \in K} |h(t, y)| \leq |g_K(y)|\right\}.$$

We will assume in what follows that

(H1) the function $(t, \widetilde{x}) \mapsto \eta(t, \widetilde{x})$ is C^1 on $I \times \mathbf{R}^2$.

(H1') $\forall K \subset\subset I, \exists \theta_K \in L^2(\mathbf{R}^2) : \sup_{t \in K} |\eta(t, \widetilde{x})| \leq |\theta_K(\widetilde{x})|$,

(H2) the functions $(t, y) \mapsto v_j(t, X(t, y))$ are C^1 on $I \times \Omega_0$,

(H2') the functions $(t, y) \mapsto v_j(t, X(t, y))$, $\partial_t v(t, X(t, y))$, $\partial_{x_k} v_j(t, X(t, y))$ belong to $\mathcal{E}$.

We then have the following result.

Proposition 6.3.7 *Let (η, v) be a regular solution of the surface wave system in infinite depth satisfying conditions (Hi), (Hi'), $i = 1, 2$. Then for all $t \geq 0$ we have*

$$E(t) = E(0).$$

Proof We will denote by $y = (y_1, y_2, y_3)$ the running point of Ω_0 and the point $x = (x_1, x_2, x_3)$ that of Ω_t. Using the hypotheses and (6.3.20) we can differentiate under the integral sign and write

$$\begin{aligned}
\frac{1}{2}\frac{d}{dt}\iint_{\Omega_t} |v(t,x)|^2\,dx &= \frac{1}{2}\frac{d}{dt}\iint_{\Omega_0} |v(t, X(t,y))^2\,dy, \\
&= \iint_{\Omega_0} \big(v\cdot(\partial_t v + (X'(t,y)\cdot\nabla_x)v)\big)(t, X(t,y))\,dy, \\
&= \iint_{\Omega_0} \big(v\cdot(\partial_t v + (v\cdot\nabla_x)v)\big)(t, X(t,y))\,dy.
\end{aligned}$$

Returning to Ω_t via the inverse diffeomorphism and using Eq. (6.3.18) (i), we have

$$\begin{aligned}
\frac{1}{2}\frac{d}{dt}\iint_{\Omega_t} |v(t,x)|^2\,dx &= \iint_{\Omega_t} \big(v\cdot(\partial_t v + (v\cdot\nabla_x)v)\big)(t,x)\,dx, \\
&= \iint_{\Omega_t} v(t,x)\cdot(-\nabla_x(P(t,x) + gx_3))\,dx.
\end{aligned}$$

By the divergence theorem (see Theorem 12.6.4 in Chap. 12) we can write

$$\frac{1}{2}\frac{d}{dt}\iint_{\Omega_t} |v(t,x)|^2\,dx = \iint_{\Omega_t} \operatorname{div} v(t,x)(P(t,x) + gx_3)\,dx - \int_{\Sigma_t} (v\cdot n)(P + gx_3)\,d\sigma,$$

where Σ_t is the boundary of Ω_t, n is the unit normal at a point on the boundary, and $d\sigma$ is the Lebesgue measure on Σ_t.

By (6.3.18) we have $\operatorname{div} v = 0$ in Ω_t and $P|_{\Sigma_t} = 0$. It follows that

$$\frac{1}{2}\frac{d}{dt}\iint_{\Omega_t} |v(t,x)|^2\,dx = -g\int_{\Sigma_t} (v\cdot n)x_3\,d\sigma.$$

Since $\Sigma_t = \{x = (\widetilde{x}, x_3) : x_3 = \eta(t,\widetilde{x})\}$ where $\widetilde{x} = (x_1, x_2)$ we have

$$d\sigma = \sqrt{1 + |\nabla_{\widetilde{x}}\eta(t,\widetilde{x})|^2}\,d\widetilde{x}.$$

Thus,

$$\frac{1}{2}\frac{d}{dt}\iint_{\Omega_t} |v(t,x)|^2\,dx = -g\int_{\mathbf{R}^2} \sqrt{1 + |\nabla_x\eta(t,\widetilde{x})|^2}(v\cdot n)(t,\widetilde{x},\eta(t,\widetilde{x}))\eta(t,\widetilde{x})\,d\widetilde{x}.$$

On the other hand, according to (6.3.18) (ii) we have

$$\partial_t \eta(t, \widetilde{x}) = \sqrt{1 + |\nabla_x \eta(t, \widetilde{x})|^2}\, (v \cdot n)(t, \widetilde{x}, \eta(t, \widetilde{x})),$$

so tha

$$\frac{1}{2}\frac{d}{dt}\iint_{\Omega_t} |v(t,x)|^2\, dx = -g \int_{\mathbf{R}^2} \partial_t \eta(t, \widetilde{x})\eta(t, \widetilde{x})\, d\widetilde{x} = -\frac{g}{2}\frac{d}{dt}\int_{\mathbf{R}^2} \eta(t, \widetilde{x})^2\, d\widetilde{x}.$$

It follows from (6.3.19) that $\frac{d}{dt}E(t) = 0$. □

6.3.7 The Case of Irrotational Fluids. The System in (η, ϕ)

We are interested in the following in the case where the fluid is *irrotational*, i.e., curl $v_0 = 0$. By Proposition 6.3.4, there exists a function ϕ such that $v = \nabla_x \phi$. We will translate the system (6.3.18) into equations for (η, ϕ).

For $j = 1, 2, 3$, we have

$$(v \cdot \nabla) v_j = \sum_{k=1}^{3} \frac{\partial \phi}{\partial x_k} \frac{\partial^2 \phi}{\partial x_k \partial x_j} = \frac{1}{2}\frac{\partial}{\partial x_j} \sum_{k=1}^{3} \Big(\frac{\partial \phi}{\partial x_k}\Big)^2,$$

so that equation (i) becomes

$$\frac{\partial}{\partial x_j}\Big(\frac{\partial \phi}{\partial t} + \frac{1}{2}\sum_{k=1}^{3}\Big(\frac{\partial \phi}{\partial x_k}\Big)^2\Big) = -\frac{\partial}{\partial x_j}\Big(P + g x_3\Big), \quad j = 1, 2, 3.$$

We can integrate both sides and since in the Euler equation the initial pressure P only appears through $\nabla_x P$, we may, for convenience, take the constant of integration to be zero. We deduce

$$\frac{\partial \phi}{\partial t} + \frac{1}{2}\sum_{k=1}^{3}\Big(\frac{\partial \phi}{\partial x_k}\Big)^2 = -P - g x_3.$$

Next, consider (ii). We have $v \cdot n = \nabla \phi \cdot n = \frac{\partial \phi}{\partial n}$ (the normal derivative of ϕ), so that (ii) becomes

$$\frac{\partial \eta}{\partial t} - \sqrt{1 + |\nabla_{\widetilde{x}} \eta|^2}\, \frac{\partial \phi}{\partial n} = 0 \quad \text{on } \Sigma.$$

Since $n = \frac{1}{\sqrt{1+|\nabla_{\widetilde{x}}\eta|^2}}(-\nabla_{\widetilde{x}}\eta, 1)$ we have

$$\frac{\partial}{\partial n} = \frac{1}{\sqrt{1 + |\nabla_{\widetilde{x}} \eta|^2}}\Big(\frac{\partial}{\partial x_3} - \nabla_x \eta \cdot \nabla_x\Big), \tag{6.3.21}$$

so that the above equation can be rewritten as

$$\frac{\partial \eta}{\partial t} - \frac{\partial \phi}{\partial x_3} + \sum_{j=1}^{2} \frac{\partial \eta}{\partial x_j} \frac{\partial \phi}{\partial x_j} = 0 \quad \text{on } \Sigma.$$

Since $\operatorname{div}(\nabla_x \phi) = \sum_{j=1}^{3} \frac{\partial^2 \phi}{\partial x_j^2} := \Delta_x \phi$, condition (iii) becomes

$$\Delta_x \phi = 0 \quad \text{in } \Omega.$$

Finally, condition (v) reads

$$\frac{\partial \phi}{\partial \nu} = 0 \quad \text{on } \Gamma,$$

where ν is the unit normal to Γ.

In summary, the system in (η, ϕ) is written as

$$\begin{cases} (i)' & \dfrac{\partial \phi}{\partial t} + \dfrac{1}{2} \displaystyle\sum_{k=1}^{3} \Big(\frac{\partial \phi}{\partial x_k}\Big)^2 = -P - g x_3 \quad \text{in } \Omega, \\ (ii)' & \dfrac{\partial \eta}{\partial t} - \dfrac{\partial \phi}{\partial x_3} + \displaystyle\sum_{j=1}^{2} \frac{\partial \eta}{\partial x_j} \frac{\partial \phi}{\partial x_j} \quad \text{on } \Sigma, \\ (iii)' & \Delta_x \phi = 0 \quad \text{in } \Omega, \\ (iv)' & P = 0 \quad \text{on } \Sigma, \\ (v)' & \dfrac{\partial \phi}{\partial \nu} = 0 \quad \text{on } \Gamma. \end{cases} \tag{6.3.22}$$

6.4 Linear Approximation

In order to understand the nature of these equations and to deduce properties of the waves, and following Cauchy etc. we begin by studying a model case.

We assume the bottom is flat and given by the equation $x_3 = -h_0$ and that the surface of the fluid is $x_3 = 0$ (while keeping η in the equations as an unknown of the problem, otherwise everything is trivial) so that

$$\Omega = \{(t, \tilde{x}, x_3) : t \geq 0, \tilde{x} \in \mathbf{R}^2, -h_0 < x_3 < 0\}.$$

Next, we will neglect the quadratic terms that appear in the equations (which corresponds to considering waves of small amplitude compared to their wavelengths and speeds, see Remark 6.4.1 below).

Take the restriction of equations $(i)'$ and $(iii)'$ on $\Sigma = \{(t, \widetilde{x}, x_3) : x_3 = 0\}$. Since $P = 0$ and $x_3 = \eta$ on Σ, we obtain

$$\begin{aligned}
&(i)' + (ii)' \Longrightarrow (1) \quad \frac{\partial \phi}{\partial t} + g\eta = 0 \text{ and } \frac{\partial \eta}{\partial t} - \frac{\partial \phi}{\partial x_3} = 0, \text{ if } x_3 = 0,\\
&(iii)' \Longrightarrow (2) \quad \sum_{j=1}^{3} \frac{\partial^2 \phi}{\partial x_j^2} = 0, \ -h_0 < x_3 < 0, \qquad (6.4.1)\\
&(v)' \Longrightarrow (3) \quad \frac{\partial \phi}{\partial x_3} = 0, \text{ if } x_3 = -h_0.
\end{aligned}$$

Differentiating the first equation of (1) with respect to t and using the second, we finally obtain the system of equations

$$\begin{aligned}
&(1)' \quad \frac{\partial \phi}{\partial t} + g\eta = 0 \quad \text{and} \quad \frac{\partial^2 \phi}{\partial t^2} + g\frac{\partial \phi}{\partial x_3} = 0, \text{ if } x_3 = 0,\\
&(2)' \quad \sum_{j=1}^{3} \frac{\partial^2 \phi}{\partial x_j^2} = 0 \text{ in } \Omega, \qquad (6.4.2)\\
&(3)' \quad \frac{\partial \phi}{\partial y}(t, \widetilde{x}, -h_0) = 0, \ \forall (t, \widetilde{x}) \in \mathbf{R}^+ \times \mathbf{R}^2.
\end{aligned}$$

We look for solutions of (6.4.2) having a wave-like behavior in (x_1, x_2).

For $k = (k_1, k_2) \in \mathbf{R}^2 \setminus \{0\}$, $\omega \in \mathbf{R}$, $t \in \mathbf{R}$, $\widetilde{x} \in \mathbf{R}^2$ we set

$$\eta(t, \widetilde{x}) = A_k \cos(k \cdot \widetilde{x} - \omega t), \quad \phi(t, x) = Y_k(x_3) \sin(k \cdot \widetilde{x} - \omega t), \quad A_k \in \mathbf{R}, \qquad (6.4.3)$$

where $k \cdot \widetilde{x} = k_1 x_1 + k_2 x_2$. We denote $|k| = \sqrt{k_1^2 + k_2^2}$.

Here, the surface of our fluid will be periodic with amplitude $|A_k|$.

We proceed by necessary conditions. We first suppose that η, ϕ are solutions of the system (6.4.2). This will give us necessary conditions on Y_k, ω. We then check that in this way we obtain a solution of (6.4.2).

Using $(2)'$ we obtain

$$\Big(\sum_{j=1}^{3} \frac{\partial^2 \phi}{\partial x_j^2}\Big)(t, x) = (Y_k''(x_3) - (k_1^2 + k_2^2) Y_k(x_3)) \sin(k \cdot \widetilde{x} - \omega t) = 0, \quad \forall (t, x) \in \Omega.$$

This implies that $Y_k''(x_3) - (k_1^2 + k_2^2) Y_k(x_3) = 0$ for all $x_3 \in \mathbf{R}$, from which we deduce

$$Y_k(x_3) = \alpha_k e^{|k| x_3} + \beta_k e^{-|k| x_3}.$$

The condition (3)' implies that $Y_k'(-h_0) = 0$, that is, $\alpha_k |k| e^{-|k| h_0} - \beta_k |k| e^{|k| h_0} = 0$, hence $\alpha_k = e^{2|k| h_0} \beta_k$ and

$$Y_k(x_3) = e^{|k|h_0}\beta_k(e^{|k|(h_0+x_3)} + e^{-|k|(h_0+x_3)}) = 2e^{|k|h_0}\beta_k\cosh(|k|(h_0+x_3)),$$
$$\phi(t,x) = 2e^{|k|h_0}\beta_k\cosh(|k|(h_0+x_3))\sin(k\cdot\widetilde{x} - \omega t),$$

where cosh denotes the hyperbolic cosine.

We then use the first equation of $(1)'$. We obtain

$$2\omega e^{|k|h_0}\beta_k\cosh(|k|h_0)\cos(k\cdot\widetilde{x} - \omega t) = gA_k\cos(k\cdot\widetilde{x} - \omega t),$$

which implies that $2e^{|k|h_0}\beta_k = \frac{gA_k}{\omega\cosh(|k|h_0)}$ hence

$$\begin{cases} \phi(t,x) = \dfrac{gA_k}{\omega\cosh(|k|h_0)}\cosh(|k|(x_3+h_0))\sin(k\cdot\widetilde{x} - \omega t), \\ \eta(t,\widetilde{x}) = A_k\cos(k\cdot\widetilde{x} - \omega t). \end{cases} \tag{6.4.4}$$

So far, we have not used the second equation of (1)'. It will give us an important relation between ω and k. Indeed, using (6.4.4) and (1)' we obtain

$$\omega^2 = g|k|\tanh(|k|h_0), \tag{6.4.5}$$

where tanh denotes the hyperbolic tangent function. This equality is called **the dispersion relation**.

We deduce that

$$\omega(k) = \pm\sqrt{g|k|\tanh(|k|h_0)},$$

hence finally,

$$\begin{cases} \phi(t,x) = \dfrac{gA_k}{\omega(k)\cosh(|k|h_0)}\cosh(|k|(x_3+h_0))\sin(k\cdot\widetilde{x} - \omega(k)t), \\ \eta(t,\widetilde{x}) = A_k\cos(k\cdot\widetilde{x} - \omega(k)t). \end{cases} \tag{6.4.6}$$

Conversely, we check that η, ϕ given by (6.4.6) where $\omega(k)$ satisfies (6.4.5) is indeed a solution of the system (6.4.1).

In infinite depth ($h_0 = +\infty$) these equalities reduce to

$$\begin{cases} \phi(t,x) = \dfrac{gA_k}{\omega(k)}e^{|k|x_3}\sin(k\cdot\widetilde{x} - \omega(k)t), \\ \eta(t,\widetilde{x}) = A_k\cos(k\cdot\widetilde{x} - \omega(k)t), \qquad \omega(k)^2 = g|k|. \end{cases} \tag{6.4.7}$$

To interpret this result, let us place ourselves in dimension 1 of space that is

$$x = (x_1, x_2) \in \mathbf{R}^2,\ x_1 \in \mathbf{R},\ k \in \mathbf{R}\setminus\{0\}.$$

This number k is called *the wavenumber*. It is related to the wavelength λ by

$$|k| = \frac{2\pi}{\lambda}. \tag{6.4.8}$$

The solution described in (6.4.6) concerns the case where the depth is finite. Let us rewrite it in the case $d = 1$. We have

$$\begin{cases} \phi(t, x) = \dfrac{g A_k}{\omega(k) \cosh(|k| h_0)} \cosh(|k|(x_2 + h_0)) \sin(k x_1 - \omega(k) t), \\ \eta(t, x) = A_k \cos(k x_1 - \omega(k) t), \qquad \omega(k) = \pm\sqrt{g|k| \tanh(|k| h_0)}. \end{cases} \tag{6.4.9}$$

Let us make a digression. Let $u : \mathbf{R} \to \mathbf{R}$ be a continuous function and $a, b \in \mathbf{R}$ with $a \neq 0, b \neq 0$. For $t \geq 0$ set $u_t(x_1) = u(ax_1 - bt)$ and examine for two times $t_1 > t_0$ the link between the graphs of u_{t_1} and u_{t_0}. Since

$$u_{t_1}(x_1) = u\big(a\big(x_1 - \frac{b}{a}(t_1 - t_0)\big) - bt_0\big) = u_{t_0}\big(x_1 - \frac{b}{a}(t_1 - t_0)\big) = u_{t_0}(X),$$

we see that the graph of u_{t_1} is the translate of that of u_{t_0} by taking as new origin the point $x_1 = \frac{b}{a}(t_1 - t_0)$ (this translation is to the right if $\frac{b}{a} > 0$ and to the left if $\frac{b}{a} < 0$). On the other hand, in the time $t_1 - t_0$ a point of the graph has traveled the distance $|\frac{b}{a}|(t_1 - t_0)$. The speed of travel is therefore $|\frac{b}{a}|$. Returning to (6.4.9), we see that the surface of the fluid is a periodic function of the above type, with $a = k, b = \omega(k)$, which, as time increases, undergoes a translation (to the left or to the right) at the speed

$$\begin{aligned} \left|\frac{\omega(k)}{k}\right| &= \sqrt{\frac{g \tanh(|k| h_0)}{|k|}} = \sqrt{\frac{g \tanh(|k| h_0)\lambda}{2\pi}} \quad \text{if } h_0 < +\infty, \\ \left|\frac{\omega(k)}{k}\right| &= \sqrt{\frac{g}{|k|}} = \sqrt{\frac{g\lambda}{2\pi}} \quad \text{if } h_0 = +\infty. \end{aligned} \tag{6.4.10}$$

The formulas (6.4.10) imply, in particular, that *the speed of a wave depends on its wavelength.*

"Two distinct waves propagate at distinct speeds."

This phenomenon is called **dispersion**.

In finite depth, the dispersion relation simplifies in two cases: (recall that $\lambda = \frac{2\pi}{|k|}$)

Case (i) the water is shallow compared to the wavelength that is, $\lambda >> h_0$ (which is equivalent to $|k| h_0 << 1$),

Case (ii) the water is very deep compared to the wavelength that is, $\lambda << h_0$ (which is equivalent to $|k| h_0 >> 1$).

Then we have

$$\text{Case } (i) : \tanh(|k| h_0) \sim |k| h_0, \qquad \left|\frac{\omega(k)}{k}\right| \sim \sqrt{g h_0}, \quad \text{no dispersion,}$$

$$\text{Case } (ii): \tanh(|k|h_0) \sim 1, \quad \left|\frac{\omega(k)}{k}\right| \sim \sqrt{\frac{g}{|k|}}.$$

Let us compute the wave speed in particular cases.

In what follows we will take $g \sim 10 \text{ ms}^{-2}$. Since 1m $= 10^{-3}$ km and 1s $= \frac{1}{3610^2}$h we will have

$$g \sim 10 \cdot 10^{-3}(36)^2 10^4 \text{ kmh}^{-2} \sim 13 \cdot 10^4 \text{ kmh}^{-2}.$$

Case (i) : Tsunamis : $\lambda \sim 100$ km , $h_0 = 3$ km
We have: speed $\sim \sqrt{gh_0} \sim 630$ km/h
Case (ii) : Wind waves: $\lambda \sim 100$ m $= 10^{-1}$km, $h_0 = 3$ km
We have: speed $\sim \sqrt{\frac{g\lambda}{2\pi}} \sim \sqrt{\frac{13\,10^4 \cdot 10^{-1}}{6}} \sim 45$ km/h.

Remark 6.4.1 We have $|\frac{\partial \eta}{\partial x}(t, x_1)| \leq |k||A_k| = \frac{2\pi |A_k|}{\lambda}$; similarly $|\frac{\partial \phi}{\partial x_j}|$, $j = 1, 2$ are bounded by $\frac{g|kA_k|}{|\omega(k)|} = \frac{g|A_k|}{|\text{speed}|}$. All these derivatives are therefore small if the amplitude A_k is small compared to the wavelength and the speed, which justifies a posteriori neglecting their squares in the equations.

6.4.1 The Motion of Fluid Particles

Recall that the velocity v is equal to $\nabla\phi$. In the case of waves, we can consider that the displacement perpendicular to the motion is negligible that is, $v_2 \sim 0$. This amounts in (6.4.7) to assuming that $k = (k_1, 0)$, $k_1 \neq 0$. In infinite depth, we thus have

$$\begin{cases} \phi(t, x) = \dfrac{gA_k}{\omega(k_1)} e^{|k_1|x_2} \sin(k_1 x_1 - \omega(k_1)t), \\ \eta(t, x) = A_k \cos(k_1 x_1 - \omega(k_1)t), \qquad \omega(k_1)^2 = gk_1. \end{cases} \tag{6.4.11}$$

Since $v = \nabla\phi$ and $X'(t) = v(t, x(t))$ where $X(t) = (x_1(t), x_2(t))$, we obtain

$$\begin{cases} x_1'(t) = \dfrac{gk_1 A_k}{\omega(k_1)} e^{|k_1|x_2(t)} \cos(k_1 x_1(t) - \omega(k_1)t), \\ x_2'(t) = \dfrac{gk_1 A_k}{\omega(k_1)} e^{|k_1|x_2(t)} \sin(k_1 x_1(t) - \omega(k_1)t). \end{cases} \tag{6.4.12}$$

Suppose the quantity $B_k := \frac{gk_1 A_k}{|\omega(k_1)|}$ is small, that is, $g|k_1 A_k| << |\omega(k_1)| = \sqrt{g|k_1|}$ or again $\sqrt{g|k_1|}|A_k| << 1$, which expresses the fact that the wave height is small compared to the square root of the wavelength. In this case, (6.4.12) shows that (since $x_2(t) < 0$ and thus $e^{|k_1|x_2(t)} \leq 1$) $|x_1'(t)| + |y'(t)| << 1$. Therefore, $x_1(t)$ and $x_2(t)$ vary little. We can thus, as a first approximation, replace $x_1(t)$ and $x_2(t)$ by their values at zero, $x_1(0)$ and $x_2(0)$. We thus have roughly

$$x_1'(t) = \frac{gk_1 A_k}{\omega(k_1)} e^{|k_1|x_2(0)} \cos(k_1 x_1(0) - \omega(k_1)t),$$
$$x_2'(t) = \frac{gk_1 A_k}{\omega(k_1)} e^{|k_1|x_2(0)} \sin(k_1 x_1(0) - \omega(k_1)t).$$

which gives, since $\omega(k_1)^2 = g|k_1|$ and $\frac{k_1}{|k_1|} = \pm 1$,

$$x_1(t) - a = \pm A_k e^{|k_1|x_2(0)} \sin(k_1 x_1(0) - \omega(k_1)t),$$
$$x_2(t) - b = \mp A_k e^{|k_1|x_2(0)} \cos(k_1 x_1(0) - \omega(k_1)t).$$

Then,

$$(x_1(t) - a)^2 + (x_2(t) - b)^2 = A_k^2 e^{2|k_1|x_2(0)},$$

which shows that the water "particles" approximately describe small circles centered at (a, b).

Remark 6.4.2 The previous discussion consisted in modeling the propagation of a wave (corresponding to a value of k). If we want to model a wave train with different values of k, we are led to look for solutions that are superpositions of solutions of the type (6.4.3), that is, for example, to look for η as the real part of

$$u(t, x) = \int_{\mathbf{R}^2} e^{i(k\cdot x + \psi(k)t)} A(k)\, dk + \int_{\mathbf{R}^2} e^{i(k\cdot x - \psi(k)t)} B(k)\, dk,$$

where $\psi(k) = \sqrt{g|k| \tanh(|k|h_0)}$.

6.5 Comments

The surface wave equations, also called water waves equations, have aroused and continue to arouse great interest both from a mathematical perspective and in the field of oceanography.

Numerous versions of these equations have been proposed, each based on a different physical assumption. Here are a few examples.

One may wish, in the model, to take into account surface tension. This is a force of attraction and cohesion between the water molecules at the surface, which explains, for example, the observable fact that small birds can stand, without sinking, on the surface of a body of water.

Other authors have proposed studying the case of waves on the surface of shallow water, or conversely, deep water, or even the case of small amplitude waves, etc. Each of these physical assumptions leads, by making certain quantities negligible in the equations, to simplified and different models.

Another very interesting case is that of solitary waves, which can travel long distances without changing shape, such as the now famous one observed in 1834 by John Scott Russell on the Glasgow-Edinburgh canal.

Among the cases that are still largely open, as far as mathematical modeling is concerned, are those of waves that become breaking waves as they approach the shore, as well as those of sudden and dangerous waves, which, in great depths, have very large amplitudes and are therefore called “rogue waves”.

Chapter 7
Continuity of the Dirichlet-Neumann Operator

7.1 Prerequisites

Space of Lipschitz functions on $\mathbf{R}^d$, identification with the space $W^{1,\infty}(\mathbf{R}^d)$. Sobolev spaces. Hypoellipticity of the Laplacian. Fourier transform. Representation of continuous linear forms on Hilbert spaces.

7.2 The Context

This chapter follows the previous chapter. The object we will introduce here, the Dirichlet-Neumann operator, will, as we shall see, simplify the wave propagation equations introduced in Chap. 6. It is an example of an operator that behaves like a first-order differential operator but is not one. It belongs to the class of what are called pseudo-differential operators.

In the course of the proof, we will state and prove a lemma (see Lemma 7.4.9) that rigorously justifies the existence of traces on the boundary $x_d = 0$ for elements of the space $H^1(\mathbf{R}^d_+)$ where $\mathbf{R}^d_+ = \{x = (x_1, \ldots, x_d) \in \mathbf{R}^d : x_d > 0\}$.

7.3 Reminders, Notations, Results

First recall that

$$W^{1,\infty}(\mathbf{R}^d) = \{u \in L^\infty(\mathbf{R}^d) : \partial_{x_j} u \in L^\infty(\mathbf{R}^d), 1 \leq j \leq d\},$$

C. Zuily, *Selected Topics in Partial Differential Equations*, Universitext,
https://doi.org/10.1007/978-3-032-24082-8_7

which is equipped with the norm

$$\|u\|_{W^{1,\infty}} = \|u\|_{L^\infty} + \sum_{j=1}^{d} \|\partial_{x_j} u\|_{L^\infty}.$$

This space is identified with the space of bounded and Lipschitz functions v, that is,

$$v \in L^\infty(\mathbf{R}^d) \quad \text{and} \quad |v(x) - v(y)| \le C|x - y|, \quad \text{if } |x - y| \le 1.$$

(See Section 5.9.2 of Chap. 5).

In what follows, we denote by $H^s(\mathbf{R}^d)$ the usual Sobolev space on $\mathbf{R}^d$,

$$H^s(\mathbf{R}^d) = \{u \in \mathcal{S}'(\mathbf{R}^d) : \|u\|_{H^s}^2 := \int_{\mathbf{R}^d} (1 + |\xi|^2)^s |\widehat{u}(\xi)|^2 \, d\xi < +\infty\},$$

and if Ω is an open subset of $\mathbf{R}^d$,

$$H^1(\Omega) = \{u \in L^2(\Omega) : \partial_{x_j} u \in L^2(\Omega), 1 \le j \le d\},$$

equipped with the norm

$$\|u\|_{H^1(\Omega)} = \Big(\|u\|_{L^2(\Omega)}^2 + \sum_{j=1}^{d} \|\partial_{x_j} u\|_{L^2(\Omega)}^2\Big)^{\frac{1}{2}}.$$

Let $d \ge 1$, $\eta \in W^{1,\infty}(\mathbf{R}^d)$ and $h > 0$ such that $\|\eta\|_{L^\infty(\mathbf{R}^d)} < \frac{h}{2}$. We set

$$\Omega = \{(x, y) \in \mathbf{R}^d \times \mathbf{R} : -h < y < \eta(x)\},$$
$$\Sigma = \{(x, y) \in \mathbf{R}^d \times \mathbf{R} : y - \eta(x) = 0\}, \quad \Gamma = \{(x, y) \in \mathbf{R}^d \times \mathbf{R} : y = -h\}.$$

In what follows, we denote by $\Delta_{x,y} = \partial_y^2 + \sum_{j=1}^d \partial_{x_j}^2$ the Laplacian in $\mathbf{R}^{d+1}$, ∇_x the gradient in $\mathbf{R}^d$ and $\nabla_{x,y} = (\nabla_x, \partial_y)$ the gradient in $\mathbf{R}^{d+1}$.

Moreover, n will denote the unit normal vector to Σ,

$$n(x) = \frac{1}{\sqrt{1 + |\nabla_x \eta(x)|^2}} (-\nabla_x \eta(x), 1) \tag{7.3.1}$$

and $\frac{\partial}{\partial n} = n \cdot \nabla_{x,y}$ the normal derivative to Σ.

Finally, in what follows, all functions will be real-valued.

7.3.1 The Plan

1. We will begin by showing that for any $\psi \in H^{\frac{1}{2}}(\mathbf{R}^d)$ the problem,

$$\Delta_{x,y} u = 0, \quad \text{in } \Omega, \quad u|_\Sigma = \psi, \quad \partial_y u|_{y=-h} = 0, \tag{7.3.2}$$

admits a unique solution in a certain space to be specified.

2. We will then define the **Dirichlet-Neumann operator** by

$$G(\eta)\psi = \sqrt{1 + |\nabla_x \eta(x)|^2}\left(\frac{\partial u}{\partial n}\right)|_\Sigma, \tag{7.3.3}$$

It is an operator which, given the Dirichlet problem data, ψ, associates the data of the Neumann problem, hence its name.

Taking into account (7.3.1) we have

$$G(\eta)\psi = \big(\partial_y u - \nabla_x \eta \cdot \nabla_x u\big)|_\Sigma \tag{7.3.4}$$

3. Finally, we will prove the following result.

Theorem 7.3.1 *There exists $C > 0$ depending only on d, h such that for all η in $W^{1,\infty}(\mathbf{R}^d)$ with $\|\eta\|_{L^\infty(\mathbf{R}^d)} \leq \frac{1}{2}h$ and all ψ in $H^{\frac{1}{2}}(\mathbf{R}^d)$, we have*

$$\|G(\eta)\psi\|_{H^{-\frac{1}{2}}(\mathbf{R}^d)} \leq C(1 + \|\eta\|^2_{W^{1,\infty}(\mathbf{R}^d)})\|\psi\|_{H^{\frac{1}{2}}(\mathbf{R}^d)}. \tag{7.3.5}$$

This result shows that the Dirichlet-Neumann operator behaves like an operator of order 1.

7.3.2 An Example

We will first consider the particular case where $\eta \equiv 0$ and completely determine the Dirichlet-Neumann operator. In this case, (7.3.2) reads

$$\Delta_{x,y} u = 0, \quad \text{in } \Omega = \{(x, y) \in \mathbf{R}^d \times \mathbf{R} : -h < y < 0\}, \quad u|_{y=0} = \psi, \quad \partial_y u|_{y=-h} = 0.$$

According to the hypoellipticity of the Laplacian, u must necessarily be C^∞ in Ω. We seek u which is also tempered in x. We can then use the Fourier transform in x. Denoting by $\widehat{u}(\xi, y)$ the partial Fourier transform of u, our problem is equivalent to

$$(i) \quad (\partial_y^2 - |\xi|^2)\widehat{u}(\xi, y) = 0, \quad (ii) \quad \widehat{u}(\xi, 0) = \widehat{\psi}(\xi), \quad (iii) \quad \partial_y \widehat{u}(\xi, -h) = 0,$$

for $\xi \in \mathbf{R}^d$.

The equation (i) implies that $\widehat{u}(\xi, y) = C_1(\xi)e^{y|\xi|} + C_2(\xi)e^{-y|\xi|}$. The equality (ii) implies that $C_1(\xi) + C_2(\xi) = \widehat{\psi}(\xi)$. Finally, (iii) implies that

$$|\xi|C_1(\xi)e^{-h|\xi|} - |\xi|C_2(\xi)e^{h|\xi|} = 0.$$

We deduce that

$$C_1(\xi) = \frac{e^{h|\xi|}\widehat{\psi}(\xi)}{e^{h|\xi|} + e^{-h|\xi|}}, \quad C_2(\xi) = \frac{e^{-h|\xi|}\widehat{\psi}(\xi)}{e^{h|\xi|} + e^{-h|\xi|}},$$

and thus,

$$\widehat{u}(\xi, y) = \frac{e^{(h+y)|\xi|} + e^{-(h+y)|\xi|}}{e^{h|\xi|} + e^{-h|\xi|}}\widehat{\psi}(\xi).$$

On the other hand, the unit normal to the set $\{y = 0\}$ being the vector $(0, 1)$, the normal derivative is equal to ∂_y. From above, we have

$$\partial_y\widehat{u}(\xi, y)|_{y=0} = |\xi|\frac{e^{h|\xi|} - e^{-h|\xi|}}{e^{h|\xi|} + e^{-h|\xi|}}\widehat{\psi}(\xi) = |\xi|\tanh(h|\xi|)\widehat{\psi}(\xi),$$

where tanh denotes the hyperbolic tangent.

The formula (7.3.3) then gives

$$\widehat{G(0)\psi}(\xi) = |\xi|\tanh(h|\xi|)\widehat{\psi}(\xi).$$

and thus,

$$G(0)\psi(x) = \mathcal{F}^{-1}\big(|\xi|\tanh(h|\xi|)\widehat{\psi}(\xi)\big).$$

This formula shows that the operator $G(0)$ is not a differential operator, since the quantity which multiplies $\widehat{\psi}(\xi)$ is not a polynomial.

On the other hand, denoting $\langle\xi\rangle = (1 + |\xi|^2)^{\frac{1}{2}}$,

$$\|G(0)\psi\|^2_{H^{-\frac{1}{2}}} = \int_{\mathbf{R}^d} \langle\xi\rangle^{-1}|\widehat{G(0)\psi}(\xi)|^2\, d\xi = \int_{\mathbf{R}^d} \langle\xi\rangle^{-1}|\xi|^2(\tanh(h|\xi|))^2|\widehat{\psi}(\xi)|^2\, d\xi.$$

Since the hyperbolic tangent is bounded by 1 and $|\xi|^2 \leq \langle\xi\rangle^2$, we see that the inequality (7.3.5) is indeed satisfied.

7.4 The General Case

The proof of points 1., 2., 3., described in the outline, requires several steps.

7.4.1 Step 1. Lifting the Trace

This consists in showing that, possibly by adding a right-hand side to the equation in (7.3.2), we can reduce to the case where $\psi = 0$.

Lemma 7.4.1 *Let $\psi \in H^{\frac{1}{2}}(\mathbf{R}^d)$. There exists $\underline{\psi} \in H^1(\Omega)$ such that*

1. $\underline{\psi}|_\Sigma = \psi$,
2. $\underline{\psi}(x, y) = 0$ *if* $y \leq \eta(x) - \frac{h}{3}$,
3. *there exists* $C > 0$ *depending only on* d, h *such that*

$$\|\underline{\psi}\|_{H^1(\Omega)} \leq C(1 + \|\nabla_x \eta\|_{L^\infty(\mathbf{R}^d)})\|\psi\|_{H^{\frac{1}{2}}(\mathbf{R}^d)}.$$

Proof Let us denote $\mathbf{R}^{d+1}_- = \{(x, z) : x \in \mathbf{R}^d, z < 0\}$ and consider the function

$$\underline{\widetilde{\psi}}(x, z) = \chi(z)\mathcal{F}^{-1}_{\xi \to x}\big(e^{z\langle \xi \rangle}\widehat{\psi}\big)(x), \quad x \in \mathbf{R}^d, z < 0$$

where $\langle \xi \rangle = (1 + |\xi|^2)^{\frac{1}{2}}$ and $\chi \in C^\infty(\mathbf{R})$ is such that $\chi(z) = 1$ if $z \geq -\frac{h}{4}$, $\chi(z) = 0$ if $z < -\frac{h}{3}$.

Of course, we have

$$\underline{\widetilde{\psi}}(x, z)|_{z=0} = \psi(x) \quad \text{and} \quad \underline{\widetilde{\psi}} = 0 \quad \text{if } z \leq -\frac{h}{3}. \tag{7.4.1}$$

We will show that there exists $C > 0$ such that

$$\|\underline{\widetilde{\psi}}\|_{H^1(\mathbf{R}^{d+1}_-)} \leq C\|\psi\|_{H^{\frac{1}{2}}(\mathbf{R}^d)}.$$

Indeed, since $e^{z\langle \xi \rangle} \leq 1$ for $z < 0$, it is easy to see that $\|\underline{\widetilde{\psi}}\|_{L^2(\mathbf{R}^{d+1}_-)} \leq C\|\psi\|_{L^2(\mathbf{R}^d)}$. Next, since $\|v\|^2_{L^2} = (2\pi)^{-d}\|\widehat{v}\|^2_{L^2}$ and $\widehat{\partial_{x_j} v} = i\xi_j \widehat{v}$, we have

$$\begin{aligned} \|\nabla_x \underline{\widetilde{\psi}}\|^2_{L^2(\mathbf{R}^{d+1}_-)} &\leq C \int_{\mathbf{R}^d} \Big(\int_{-\infty}^0 e^{2z\langle \xi \rangle}\, dz\Big)|\xi|^2|\widehat{\psi}(\xi)|^2\, d\xi, \\ &\leq C \int_{\mathbf{R}^d} \frac{1}{2\langle \xi \rangle|}|\xi|^2|\widehat{\psi}(\xi)|^2\, d\xi \leq C'\|\psi\|^2_{H^{\frac{1}{2}}(\mathbf{R}^d)}. \end{aligned} \tag{7.4.2}$$

Finally, since $\partial_z \underline{\widetilde{\psi}} = \chi'(z)e^{z\langle D_x\rangle}\psi + \chi(z)\langle D_x\rangle e^{z\langle D_x\rangle}\psi$, the same calculation as above shows that

$$\|\partial_z \underline{\widetilde{\psi}}\|^2_{L^2(\mathbf{R}^{d+1}_-)} \leq C\|\psi\|^2_{H^{\frac{1}{2}}(\mathbf{R}^d)}. \tag{7.4.3}$$

Let us then set

$$\underline{\psi}(x,y) = \underline{\widetilde{\psi}}(x, y-\eta(x)). \tag{7.4.4}$$

According to (7.4.1) we have

$$\underline{\psi}(x,y)|_\Sigma = \psi \quad \text{and} \quad \underline{\psi}(x,y) = 0 \quad \text{if } y - \eta(x) \leq -\frac{h}{3}.$$

Next, setting $z = y - \eta(x)$ in the integral over y, we obtain

$$\iint_\Omega |\underline{\psi}(x,y)|^2\,dx\,dy = \int_{\mathbf{R}^d}\int_{-h}^{\eta(x)} |\underline{\widetilde{\psi}}(x, y-\eta(x))|^2\,dx\,dy \leq \iint_{\mathbf{R}^{d+1}_-} |\underline{\widetilde{\psi}}(x,z)|^2\,dx\,dz,$$
$$\leq C\|\psi\|^2_{L^2(\mathbf{R}^d)}.$$

Since

$$\nabla_x \underline{\psi}(x,y) = (\nabla_x \underline{\widetilde{\psi}} - \nabla_x\eta(x)\partial_y\underline{\widetilde{\psi}})(x, y-\eta(x)), \qquad \partial_y\underline{\psi}(x,y) = (\partial_z\underline{\widetilde{\psi}})(x, y-\eta(x)),$$

using the calculation above as well as (7.4.2), (7.4.3) and bounding $|\nabla_x\eta(x)|$ by $\|\nabla_x\eta\|_{L^\infty(\mathbf{R}^d)}$, we obtain point 3. of the lemma. □

Remark 7.4.2 We have $\eta(x) \geq -\frac{1}{2}h$, so $\eta(x) - \frac{h}{3} \geq -\frac{5}{6}h$. Since $\underline{\psi}$ is equal to zero for $y \leq \eta(x) - \frac{h}{3}$, it vanishes in the strip $B_h = \{(x,y) : -h < y < -\frac{5}{6}h\}$.

7.4.2 *Step 2. The Variational Theory*

Let us set

$$\phi = u - \underline{\psi}. \tag{7.4.5}$$

Solving problem (7.3.2) is then equivalent to solving the problem

$$\Delta_{x,y}\phi = -\Delta_{x,y}\underline{\psi}, \quad \text{in } \Omega, \quad \phi|_\Sigma = 0, \quad \partial_y\phi|_{y=-h} = 0, \tag{7.4.6}$$

since $\underline{\psi}$ is zero near $y = -h$, according to point 2. of Lemma 7.4.1.

This is the problem we will solve. We begin by specifying the spaces in which we will work.

Definition 7.4.3 In what follows, we denote by $C^{\infty,0}(\Omega)$ the space of elements of $C^{\infty}(\overline{\Omega})$ with real values, which are identically zero near Σ.

On this space, the quantity,

$$\|u\|_1 = \Big(\iint_{\Omega} |\nabla_{x,y} u(x,y)|^2 \, dx\, dy\Big)^{\frac{1}{2}} = \Big(\int_{\mathbf{R}^d} \Big(\int_{-h}^{\eta(x)} |\nabla_{x,y} u(x,y)|^2 \, dy\Big)\, dx\Big)^{\frac{1}{2}} \tag{7.4.7}$$

is a norm. This follows from the following Poincaré's lemma.

Lemma 7.4.4 *For all $\eta \in L^{\infty}(\mathbf{R}^d)$ such that $\|\eta\|_{L^{\infty}(\mathbf{R}^d)} \leq \frac{1}{2}h$ and all $u \in C^{\infty,0}(\Omega)$ we have*

$$\|u\|_{L^2(\Omega)} \leq \frac{3h}{2} \|\partial_y u\|_{L^2(\Omega)}. \tag{7.4.8}$$

Proof Since u is zero near Σ, we can write for $(x,y) \in \Omega$,

$$u(x,y) = -\int_{y}^{\eta(x)} \partial_y u(x,t)\, dt.$$

Since $|y - \eta(x)| \leq \frac{3h}{2}$, the Cauchy-Schwarz inequality implies that

$$|u(x,y)|^2 \leq \frac{3h}{2} \int_{y}^{\eta(x)} |\partial_y u(x,t)|^2 \, dt \leq \frac{3h}{2} \int_{-h}^{\eta(x)} |\partial_y u(x,t)|^2 \, dt.$$

We deduce that

$$\int_{\mathbf{R}^d} \int_{-h}^{\eta(x)} |u(x,y)|^2 \, dx\, dy \leq \Big(\frac{3h}{2}\Big)^2 \int_{\mathbf{R}^d} \int_{-h}^{\eta(x)} |\partial_y u(x,t)|^2 \, dx\, dt.$$

□

We then introduce the following space:

$$\mathcal{H}^{1,0}(\Omega) = \text{ completion of the space } C^{\infty,0}(\Omega) \text{ for the norm (7.4.7)}.$$

By density, we then obtain the following result.

Corollary 7.4.5 *Inequality* (7.4.8) *holds for all elements of* $\mathcal{H}^{1,0}(\Omega)$.

The space $\mathcal{H}^{1,0}(\Omega)$ can be endowed with a Hilbert space structure.

Proposition 7.4.6 *1. The quantity $((u,v)) = \big(\nabla_{x,y} u, \nabla_{x,y} v\big)_{L^2(\Omega)}$ is an inner product on the space $\mathcal{H}^{1,0}(\Omega)$.*

2. Equipped with this inner product the space $\mathcal{H}^{1,0}(\Omega)$ is a Hilbert space.

3. *On $\mathcal{H}^{1,0}(\Omega)$, the norms $\mathcal{N}(u) := \|u\|_{L^2(\Omega)} + \|\nabla_{x,y} u\|_{L^2(\Omega)}$ and $((u,u))^{\frac{1}{2}}$ are equivalent.*
4. *Every $u \in \mathcal{H}^{1,0}(\Omega)$ has a trace on Σ which is zero.*

Proof 1. It suffices to show that if $((u,u)) = 0$ then $u = 0$. This follows from Corollary 7.4.5.

2. Let $(u_k) \subset \mathcal{H}^{1,0}(\Omega)$ be a Cauchy sequence for the norm $((u,u))$. Then $(\nabla_{x,y} u_k)$ is a Cauchy sequence in $(L^2(\Omega))^{d+1}$. There exists $v \in (L^2(\Omega))^{d+1}$ such that $\nabla_{x,y} u_k \to v$ in $(L^2(\Omega))^{d+1}$. On the other hand, according to Corollary 7.4.5, the sequence (u_k) is Cauchy in $L^2(\Omega)$. Therefore, there exists $u \in L^2(\Omega)$ such that $(u_k) \to u$ in $L^2(\Omega)$. Convergence in $L^2(\Omega)$ implies convergence in $\mathcal{D}'(\Omega)$, so we deduce that $(\nabla_{x,y} u_k) \to \nabla_{x,y} u$. By uniqueness of the limit, we have $\nabla_{x,y} u = v$. In summary, $\mathcal{N}(u_k - u) \to 0$ as $k \to +\infty$. It remains to show that $u \in \mathcal{H}^{1,0}(\Omega)$. For $\varepsilon = \frac{1}{2^j}$ there exists k_j such that $\mathcal{N}(u_{k_j} - u) \leq \varepsilon_j$. Since $(u_{k_j}) \in \mathcal{H}^{1,0}(\Omega)$, by definition, there exists $\varphi_j \in C^{\infty,0}$ such that $\mathcal{N}(u_{k_j} - \varphi_j) \leq \varepsilon_j$. Then $\mathcal{N}(u - \varphi_j) \leq 2\varepsilon_j$ and thus $u \in \mathcal{H}^{1,0}(\Omega)$.

3. This follows from Corollary 7.4.5.

4. This point will be proved in the appendix.

□

Now let $\psi \in H^{\frac{1}{2}}(\mathbf{R}^d)$ (real-valued) and $\underline{\psi} \in H^1(\Omega)$ constructed in Lemma 7.4.1. We define a linear form L on $\mathcal{H}^{1,0}(\Omega)$ by the formula

$$L(v) = -\iint_\Omega \nabla_{x,y}\underline{\psi}(x,y)\nabla_{x,y} v(x,y)\,dx\,dy.$$

This is a linear form which is continuous on $\mathcal{H}^{1,0}(\Omega)$ because the right-hand side is, by the Cauchy-Schwarz inequality, bounded above by $\|\nabla_{x,y}\underline{\psi}|_{L^2(\Omega)}((v,v))^{\frac{1}{2}}$. We deduce from the representation lemma for continuous linear forms on Hilbert spaces that there exists a unique $\phi \in \mathcal{H}^{1,0}(\Omega)$ such that

$$L(v) = ((v,\phi)), \quad \forall v \in \mathcal{H}^{1,0}(\Omega),$$

that is,

$$-\iint_\Omega \nabla_{x,y}\underline{\psi}(x,y)\nabla_{x,y} v(x,y)\,dx\,dy = \iint_\Omega \nabla_{x,y} v(x,y)\nabla_{x,y}\phi(x,y)\,dx\,dy. \tag{7.4.9}$$

We will show that $\phi \in \mathcal{H}^{1,0}(\Omega)$ thus defined is indeed the solution to problem (7.4.6) that we are seeking.

First, by taking $v \in C_0^\infty(\Omega) \subset \mathcal{H}^{1,0}(\Omega)$, the equality (7.4.9) implies that

$$\Delta_{x,y}\phi = -\Delta_{x,y}\underline{\psi}, \tag{7.4.10}$$

in the sense of $\mathcal{D}'(\Omega)$.

Next, $\phi|_\Sigma = 0$, according to point 4. in the proof of Proposition 7.4.6.

Finally, the condition $\partial_y \phi|_{y=-h} = 0$ is contained in (7.4.9) and in the fact that $\phi \in \mathcal{H}^{1,0}(\Omega)$. This is explained in the appendix.

By taking $v = \phi$ in (7.4.9) and using the Cauchy-Schwarz inequality, we obtain

$$\|\nabla_{x,y}\phi\|_{L^2(\Omega)} \leq \|\nabla_{x,y}\underline{\psi}\|_{L^2(\Omega)}.$$

Using Lemma 7.4.1 we obtain

$$\|\nabla_{x,y}\phi\|_{L^2(\Omega)} \leq C(1 + \|\eta\|_{W^{1,\infty}(\mathbf{R}^d)})\|\psi\|_{H^{\frac{1}{2}}(\mathbf{R}^d)}. \tag{7.4.11}$$

In summary, we have obtained the solution to problem (7.3.2) in the form

$$u = \phi + \underline{\psi}, \tag{7.4.12}$$

where $\underline{\psi} \in H^1(\Omega)$ was determined in Lemma 7.4.1.

Thanks to (7.4.11) and point 3. of Lemma 7.4.1 we have

$$\|\nabla_{x,y}u\|_{L^2(\Omega)} \leq C(1 + \|\eta\|_{W^{1,\infty}(\mathbf{R}^d)})\|\psi\|_{H^{\frac{1}{2}}(\mathbf{R}^d)}. \tag{7.4.13}$$

7.4.3 Step 3. The Change of Variables

Set

$$\begin{aligned} &I_h = (-h, 0), \qquad \widetilde{\Omega} = \{(x,z) : x \in \mathbf{R}^d, z \in I_h\} = \mathbf{R}^d \times I_h, \\ &\rho(x,z) = \frac{1}{h}(z+h)\eta(x) + z. \end{aligned} \tag{7.4.14}$$

Lemma 7.4.7 *The map $\widetilde{\Omega} \to \Omega$, $(x,z) \mapsto (x, \rho(x,z))$ is bijective, of class $W^{1,\infty}$ as well is its inverse.*

Proof We have $\rho(x,-h) = -h$, $\rho(x,0) = \eta(x)$ and $\partial_z\rho(x,z) = 1 + \frac{1}{h}\eta(x) \geq \frac{1}{2}$ since we have $\|\eta\|_{L^\infty(\mathbf{R}^d)} \leq \frac{1}{2}h$. Therefore $-h < \rho(x,z) < \eta(x)$ if $-h < z < 0$. The map is indeed from $\widetilde{\Omega}$ to Ω. It is invertible, with inverse $(x,y) \mapsto \left(x, \frac{y-\eta(x)}{1+\frac{1}{h}\eta(x)}\right)$. Its regularity is that of η, that is, $W^{1,\infty}$. □

How do ∂_{x_j} and ∂_y transform under the action of this $W^{1,\infty}$ diffeomorphism? Let us set

$$\widetilde{f}(x,z) = f(x, \rho(x,z)).$$

Then,

$$\partial_{x_j}\widetilde{f} = \partial_{x_j} f + (\partial_{x_j}\rho)\,\partial_y f, \quad \partial_z\widetilde{f} = (\partial_z\rho)\partial_y f.$$

We deduce that for $y = \rho(x, z)$ we have

$$\partial_y f(x, y) = \frac{1}{\partial_z \rho(x, z)} \partial_z \widetilde{f}(x, z), \quad \partial_{x_j} f(x, y) = \big(\partial_{x_j} \widetilde{f} - \frac{\partial_{x_j} \rho}{\partial_z \rho} \partial_z \widetilde{f}\big)(x, z).$$

Let us set

$$\Lambda_1 = \frac{1}{\partial_z \rho} \partial_z, \quad \Lambda_{2,j} = \partial_{x_j} - \frac{\partial_{x_j} \rho}{\partial_z \rho} \partial_z = \partial_{x_j} - (\partial_{x_j} \rho) \Lambda_1. \tag{7.4.15}$$

We thus see that ∂_y transforms into Λ_1 and ∂_{x_j} transforms into $\Lambda_{2,j}$.

If we set $\widetilde{u}(x, z) = u(x, \rho(x, z))$ where u is the solution of (7.3.2) found in (7.4.12), we thus see that $\widetilde{u}$ is a solution of the problem

$$\big(\Lambda_1^2 + \sum_{j=1}^d (\Lambda_{2,j})^2\big)\widetilde{u} = 0 \quad \text{in } \widetilde{\Omega}, \quad \partial_z \widetilde{u}|_{z=-h} = 0. \tag{7.4.16}$$

On the other hand, setting $x = x,\ y = \rho(x, z)$ in the integral, we can write

$$\iint_\Omega |\partial_y u(x, y)|^2\, dx\, dy = \iint_{\widetilde{\Omega}} |\Lambda_1 \widetilde{u}(x, z)|^2 (1 + \frac{1}{h}\eta(x))\, dx\, dz,$$

(similar calculation for $\partial_{x_j} u$). Since $1 + \frac{1}{h}\eta(x) \geq \frac{1}{2}$, (7.4.13) shows that $\widetilde{u}$ satisfies

$$\|\Lambda_1 \widetilde{u}\|_{L^2(\widetilde{\Omega})} + \sum_{j=1}^d \|\Lambda_{2,j} \widetilde{u}\|_{L^2(\widetilde{\Omega})} \leq C(1 + \|\eta\|_{W^{1,\infty}(\mathbf{R}^d)}) \|\psi\|_{H^{\frac{1}{2}}(\mathbf{R}^d)}, \tag{7.4.17}$$

where $\widetilde{\Omega} = \mathbf{R}^d \times I_h$ with $I_h = (-h, 0)$.

Now, using (7.3.4), the fact that Σ is transformed into $z = 0$ by the diffeomorphism, and that by (7.4.14) we have $\nabla_x \rho|_{z=0} = \nabla_x \eta$, it follows, denoting $a \cdot b = \sum_{j=1}^d a_j b_j$,

$$G(\eta)\psi = U|_{z=0}, \quad U = \Lambda_1 \widetilde{u} - \nabla_x \rho \cdot \Lambda_2 \widetilde{u}. \tag{7.4.18}$$

Lemma 7.4.8 *There exists $C > 0$ depending only on d, h such that*

$$\|U\|_{L^2(I_h, L^2(\mathbf{R}^d))} + \|\partial_z U\|_{L^2(I_h, H^{-1}(\mathbf{R}^d))} \leq C(1 + \|\eta\|^2_{W^{1,\infty}(\mathbf{R}^d)}) \|\psi\|_{H^{\frac{1}{2}}(\mathbf{R}^d)}.$$

Proof Let $A = (1 + \|\eta\|^2_{W^{1,\infty}(\mathbf{R}^d)})\|\psi\|_{H^{\frac{1}{2}}(\mathbf{R}^d)}$.

First, we have

$$\begin{aligned}\|U\|_{L^2(I_h,L^2(\mathbf{R}^d))} &\leq \|\Lambda_1\widetilde{u}\|_{L^2(I_h,L^2(\mathbf{R}^d))} + \|\nabla_x\rho\|_{L^\infty(\mathbf{R}^d\times I_h))}\|\Lambda_2\widetilde{u}\|_{L^2(I_h,L^2(\mathbf{R}^d))},\\ &\leq C\big(\|\Lambda_1\widetilde{u}\|_{L^2(I_h,L^2(\mathbf{R}^d))} + \|\eta\|_{W^{1,\infty}(\mathbf{R}^d)}\|\Lambda_2\widetilde{u}\|_{L^2(I_h,L^2(\mathbf{R}^d))}\big).\end{aligned}$$

Using (7.4.17) we obtain

$$\|U\|_{L^2(I_h,L^2(\mathbf{R}^d))} \leq CA. \tag{7.4.19}$$

To estimate the term $\partial_z U$ we begin with a remark. First, the quantity $\nabla_x\partial_z\rho$ is well defined because it is equal to $\frac{1}{h}\nabla_x\eta$. Next, we can write

$$\begin{aligned}\partial_z U &= \partial_z\Lambda_1\widetilde{u} - (\nabla_x\partial_z\rho)\cdot\Lambda_2\widetilde{u} - (\nabla_x\rho)\cdot\partial_z\Lambda_2\widetilde{u},\\ &= (\partial_z\rho)\Lambda_1^2\widetilde{u} - (\nabla_x\partial_z\rho)\cdot\Lambda_2\widetilde{u} - (\partial_z\rho)(\nabla_x - \Lambda_2)\cdot\Lambda_2\widetilde{u},\\ &= (\partial_z\rho)(\Lambda_1^2 + \sum_{j=1}^d(\Lambda_{2,j})^2))\widetilde{u} - \nabla_x((\partial_z\rho)\Lambda_2\widetilde{u}),\end{aligned}$$

whence, since $(\Lambda_1^2 + \sum_{j=1}^d(\Lambda_{2,j})^2)\widetilde{u} = 0$,

$$\partial_z U = -\nabla_x((\partial_z\rho)\Lambda_2\widetilde{u}). \tag{7.4.20}$$

We thus have

$$\begin{aligned}\|\partial_z U\|_{L^2(I_h,H^{-1}(\mathbf{R}^d))} &\leq \|(\partial_z\rho)\Lambda_2\widetilde{u}\|_{L^2(I_h,L^2(\mathbf{R}^d))}\\ &\leq \|\partial_z\rho\|_{L^\infty(\mathbf{R}^d\times I_h)}\|\Lambda_2\widetilde{u}\|_{L^2(I_h,L^2(\mathbf{R}^d))},\\ &\leq C(1 + \|\eta\|_{L^\infty(\mathbf{R}^d)})\|\Lambda_2\widetilde{u}\|_{L^2(I_h,L^2(\mathbf{R}^d))}.\end{aligned}$$

Using again (7.4.17) we obtain

$$\|\partial_z U\|_{L^2(I_h,H^{-1}(\mathbf{R}^d))} \leq CA. \tag{7.4.21}$$

□

7.4.4 Step 4. An Interpolation Lemma

The following result will allow us to complete the proof of the main theorem.

Lemma 7.4.9 *Let $I_h = (-h, 0)$ and $s \in \mathbf{R}$. Let $f \in L^2_z(I_h, H^{s+\frac{1}{2}}(\mathbf{R}^d))$ which is such that $\partial_z f \in L^2_z(I_h, H^{s-\frac{1}{2}}(\mathbf{R}^d))$. Then f extends to an element of*

$C^0([-h,0], H^s(\mathbf{R}^d))$ and there exists a constant $C > 0$ (independent of f) such that

$$\sup_{z\in[-h,0]} \|f(z,\cdot)\|_{H^s(\mathbf{R}^d)} \leq C\big(\|f\|_{L^2_z(I_h,H^{s+\frac{1}{2}}(\mathbf{R}^d))} + \|\partial_z f\|_{L^2_z(I_h,H^{s-\frac{1}{2}}(\mathbf{R}^d))}\big).$$

Proof By working with $(I-\Delta_x)^{\frac{s}{2}} f$ we reduce to the case $s=0$. Next, if J is an interval of $\mathbf{R}$ we set

$$W(J) = \{f \in L^2_z(J, H^{\frac{1}{2}}(\mathbf{R}^d)) : \partial_z f \in L^2_z(J, H^{-\frac{1}{2}}(\mathbf{R}^d))\},$$

which we endow with the natural norm.

Case 1: $J = \mathbf{R}$.

It is classical that $C_0^\infty(\mathbf{R}\times\mathbf{R}^d)$ is dense in $W(\mathbf{R})$. If $v \in C_0^\infty(\mathbf{R}\times\mathbf{R}^d)$ it is easy to see that the map $z \mapsto \|v(z,\cdot)\|^2_{L^2(\mathbf{R}^d)}$ is differentiable on $\mathbf{R}$ and that

$$\frac{d}{dz}\|v(z,\cdot)\|^2_{L^2(\mathbf{R}^d)} = 2\mathrm{Re}\Big(v(z,\cdot), \partial_z v(z,\cdot)\Big)_{L^2(\mathbf{R}^d)}.$$

We deduce that

$$\frac{d}{dz}\|v(z,\cdot)\|^2_{L^2(\mathbf{R}^d)} \leq C\|v(z,\cdot)\|_{H^{\frac{1}{2}}(\mathbf{R}^d)}\|\partial_z v(z,\cdot)\|_{H^{-\frac{1}{2}}(\mathbf{R}^d)}.$$

As v is compactly supported in z, there exists $A>0$ such that $v(z,\cdot)$ is zero for $z \leq -A$. Integrating the above inequality between $-A$ and $z\in\mathbf{R}$ and using the Cauchy-Schwarz inequality, we obtain

$$\sup_{z\in\mathbf{R}}\|v(z,\cdot)\|_{L^2(\mathbf{R})} \leq C\|v\|^{\frac{1}{2}}_{L^2_z(\mathbf{R},H^{\frac{1}{2}}(\mathbf{R}^d))}\|\partial_z v\|^{\frac{1}{2}}_{L^2_z(\mathbf{R},H^{-\frac{1}{2}}(\mathbf{R}^d))} \leq C\|v\|_{W(\mathbf{R})}. \quad (7.4.22)$$

Now let $u \in W(\mathbf{R})$ and (u_j) a sequence in $C_0^\infty(\mathbf{R}\times\mathbf{R}^d)$ that converges to u in $W(\mathbf{R})$. According to inequality (7.4.22) applied to $u_j - u_k$, we see that the sequence (u_j) is Cauchy in $L^\infty(\mathbf{R}, L^2(\mathbf{R}^d))$. Since this space is complete, it converges to an element $\widetilde{u} \in L^\infty(\mathbf{R}, L^2(\mathbf{R}^d))$. The u_j are continuous in z and the convergence is uniform on $\mathbf{R}$ with values in $L^2(\mathbf{R}^d)$, so $\widetilde{u} \in C^0(\mathbf{R}, L^2(\mathbf{R}^d))$. Finally, the convergences in $W(\mathbf{R})$ and $L^\infty(\mathbf{R}, L^2(\mathbf{R}^d))$ imply convergence in the sense of distributions, so $u = \widetilde{u}$ in $\mathcal{D}'(\mathbf{R}\times\mathbf{R}^d)$. We can then pass to the limit in (7.4.22) to obtain the same inequality for u.

Case 2: $J = (0,+\infty)$.

If $u \in W(J)$ there exists $\widetilde{u} \in W(\mathbf{R})$ which satisfies $u = \widetilde{u}$ for $z \in (0,+\infty)$ and $\|\widetilde{u}\|_{W(\mathbf{R})} \sim \|u\|_{W(0,+\infty)}$. If u is regular, this is the extension by parity that is,

$$\widetilde{u}(z,\cdot) = u(z,\cdot) \text{ if } z>0, \quad \widetilde{u}(z,\cdot) = u(-z,\cdot) \text{ if } z<0.$$

Otherwise, we proceed by density of regular functions.

Case 1 implies, $\widetilde{u} \in C^0(\mathbf{R}, L^2(\mathbf{R}^d))$. Then, by definition,

$$u = \widetilde{u}|_{(0,+\infty)} \in C^0([0, +\infty), L^2(\mathbf{R}^d))$$

and we have an analogous estimate to (7.4.22) for u. This is obviously valid for the intervals $J = (a, +\infty)$ or $J = (-\infty, b)$.

Case 3. $J = (-h, 0)$.

There exist $\chi_j \in C^\infty(\mathbf{R})$, $j = 1, 2$ such that

$$\operatorname{supp} \chi_1 \subset \big(-\infty, -\frac{h}{4}\big), \quad \operatorname{supp} \chi_2 \subset \big(-\frac{3h}{4}, +\infty,\big) \text{ and } \chi_1(z) + \chi_2(z) = 1 \text{ on } (-h, 0).$$

Let $u \in W(J)$. We write $u = u_1 + u_2$ where $u_j(z, \cdot) = \chi_j(z)u(z, \cdot)$, $j = 1, 2$. It is easy to see that $u_1 \in W((-h, +\infty))$ and $u_2 \in W((-\infty, 0))$. According to case 2, we have

$$\begin{aligned} u_1 &\in C^0([-h, +\infty)), L^2(\mathbf{R}^d)) \subset C^0([-h, 0], L^2(\mathbf{R}^d)), \\ u_2 &\in C^0((-\infty, 0], L^2(\mathbf{R}^d)) \subset C^0([-h, 0]L^2(\mathbf{R}^d)), \end{aligned}$$

and thus $u \in C^0([-h, 0], L^2(\mathbf{R}^d))$. Finally, for $z \in [-h, 0]$ we can write

$$\begin{aligned} \|u(z, \cdot)\|_{L^2(\mathbf{R}^d)} &\le \|u_1(z, \cdot)\|_{L^2(\mathbf{R}^d)} + \|u_2(z, \cdot)\|_{L^2(\mathbf{R}^d)}, \\ &\le C\big(\|u_1\|_{W((-h,+\infty))} + \|u_2\|_{W((-\infty,0))}\big) \le C'\|u\|_{W((-h,0))}, \end{aligned}$$

because u_1 is zero for $z \ge 0$, u_2 is zero for $z \le -h$ and $u_j = \chi_j u$. □

7.4.5 Step 5. End of the Proof of Theorem 7.3.1

We use Lemma 7.4.8 and Lemma 7.4.9 with $s = -\frac{1}{2}$. The function U defined in (7.4.18) satisfies the conditions of Lemma 7.4.9. We deduce that it extends to a continuous function on the closed interval $[-h, 0]$ with values in $H^{-\frac{1}{2}}(\mathbf{R}^d)$. In particular we have $G(\eta)\psi = U(0) \in H^{-\frac{1}{2}}(\mathbf{R}^d)$. The estimate then follows from the two lemmas cited above.

We will now show an estimate complementary to the one proved in Theorem 7.3.1.

Theorem 7.4.10 *There exists $C > 0$ such that for all $\eta \in W^{1,\infty}(\mathbf{R}^d)$, we have*

$$\int_{\mathbf{R}^d} (G(\eta)\psi)(x)\psi(x)\,dx \ge \frac{C}{1 + \|\eta\|^2_{W^{1,\infty}(\mathbf{R}^d)}} \|\psi\|^2_{H^{\frac{1}{2}}(\mathbf{R}^d)},$$

for all $\psi \subset C_0^\infty(\mathbf{R}^d)$.

Proof Using (7.4.18) we get $G(\eta)\psi = U|_{z=0}$ where $U = \Lambda_1\widetilde{u} - \nabla_x\rho \cdot \Lambda_2\widetilde{u}$. Since $\widetilde{u}|_{z=0} = \psi$ we have

$$\int_{\mathbf{R}}^{d} (G(\eta)\psi)(x)\psi(x)\,dx = \int_{\mathbf{R}^d} (U\widetilde{u})|_{z=0}\,dx.$$

On the other hand, it follows from (7.4.14) that $\nabla_x\rho|_{z=-h} = 0$ and from (7.4.16) that $\partial_z\widetilde{u}|_{z=-h} = 0$. Consequently, $U|_{z=-h} = 0$. We deduce that

$$(1) := \int_{\mathbf{R}}^{d} (G(\eta)\psi)(x)\psi(x)\,dx = \int_{\mathbf{R}^d}\int_{-h}^{0} \partial_z(U\widetilde{u})(x,z)\,dz\,dx.$$

Thus,

$$(1) = \int_{\mathbf{R}^d}\int_{-h}^{0} (\widetilde{u}\partial_z U)(x,z)\,dx\,dz + \int_{\mathbf{R}^d}\int_{-h}^{0} (U\partial_z\widetilde{u})(x,z)\,dz\,dx = (2) + (3).$$

Using (7.4.20) and integrating by parts, we obtain

$$\begin{aligned}(2) &= \int_{\mathbf{R}^d}\int_{-h}^{0} \partial_z\rho(x,z)(\Lambda_2\widetilde{u}\cdot\nabla_x\widetilde{u})(x,z)\,dz\,dx,\\ &= \int_{\mathbf{R}^d}\int_{-h}^{0} \partial_z\rho(x,z)(\Lambda_2\widetilde{u}\cdot(\Lambda_2\widetilde{u} + (\nabla_x\rho)\Lambda_1\widetilde{u}))(x,z)\,dz\,dx.\end{aligned}$$

It follows that

$$\begin{aligned}(2) = \int_{\mathbf{R}^d}\int_{-h}^{0} &\{\partial_z\rho|\Lambda_2\widetilde{u}|^2\}(x,z)\,dz\,dx\\ &+ \int_{\mathbf{R}^d}\int_{-h}^{0} \big\{(\partial_z\rho)\,(\Lambda_1\widetilde{u})\,(\nabla_x\rho\cdot\Lambda_2\widetilde{u})\big\}(x,z)\,dz\,dx.\end{aligned}$$

Next, given the form of U, we have

$$(3) = \int_{\mathbf{R}^d}\int_{-h}^{0} \big\{(\Lambda_1\widetilde{u} - \nabla_x\rho\cdot\Lambda_2\widetilde{u})(\partial_z\rho)\Lambda_1\widetilde{u}\big\}(x,z)\,dz\,dx.$$

Therefore,

$$\begin{aligned}(3) = \int_{\mathbf{R}^d}\int_{-h}^{0} &\big\{(\partial_z\rho)(\Lambda_1\widetilde{u})^2\big\}(x,z)\,dz\,dx\\ &- \int_{\mathbf{R}^d}\int_{-h}^{0} \big\{(\partial_z\rho)\,(\Lambda_1\widetilde{u})\,(\nabla_x\rho\cdot\Lambda_2\widetilde{u})\big\}(x,z)\,dz\,dx.\end{aligned}$$

We deduce that

$$\int_{\mathbf{R}}^{d} (G(\eta)\psi)(x)\psi(x)\,dx = \int_{\mathbf{R}^d}\int_{-h}^{0} \partial_z\rho(x,z)\big\{(\Lambda_1\widetilde{u})^2) + |\Lambda_2\widetilde{u}(x,z)|^2\big\}(x,z)\,dz\,dx.$$

Since

$$\big\{(\partial_z\widetilde{u})^2 + |\nabla_x\widetilde{u}|^2\big\}(x,z) \le C(1+\|\eta\|^2_{W^{1,\infty}(\mathbf{R}^d)})(\big\{(\Lambda_1\widetilde{u})^2) + |\Lambda_2\widetilde{u}|^2\big\}(x,z),$$

it finally follows that

$$\int_{\mathbf{R}}^{d} (G(\eta)\psi)(x)\psi(x)\,dx \ge \frac{C}{1+\|\eta\|^2_{W^{1,\infty}(\mathbf{R}^d)}}\int_{\mathbf{R}^d}\int_{-h}^{0}\big\{(\partial_z\widetilde{u})^2 + |\nabla_x\widetilde{u}|^2\big\}(x,z)\,dz\,dx.$$

Using Lemma 7.4.9 with $s=\frac{1}{2}$ and the fact that $\widetilde{u}|_{z=0}=\psi$, we obtain

$$\int_{\mathbf{R}}^{d} (G(\eta)\psi)(x)\psi(x)\,dx \ge \frac{C}{1+\|\eta\|^2_{W^{1,\infty}(\mathbf{R}^d)}}\|\psi\|^2_{H^{\frac{1}{2}}(\mathbf{R}^d)}.$$

□

7.5 Formulation of the Surface Wave Equations Using the Dirichlet-Neumann Operator

In this section, we will reduce the surface wave equations obtained in the previous chapter in the open set Ω to equations on the upper boundary Σ of this domain. This process has the benefit of "gaining" a dimension since the equations, instead of depending on the variables (t, x_1, x_2, x_3), will now depend only on the variables (t, x_1, x_2).

Recall the system in (η, ϕ) obtained in the previous chapter.:

$$\begin{cases} (i) \ \dfrac{\partial\phi}{\partial t} + \dfrac{1}{2}\displaystyle\sum_{k=1}^{3}\Big(\dfrac{\partial\phi}{\partial x_k}\Big)^2 = -P - gx_3 \quad \text{in } \Omega, \\ (ii) \ \dfrac{\partial\eta}{\partial t} - \dfrac{\partial\phi}{\partial x_3} + \displaystyle\sum_{j=1}^{2}\dfrac{\partial\eta}{\partial x_j}\dfrac{\partial\phi}{\partial x_j} \quad \text{on } \Sigma, \\ (iii) \ \Delta_x\phi = 0 \quad \text{in } \Omega, \\ (iv) \ P = 0 \quad \text{on } \Sigma, \\ (v) \ \dfrac{\partial\phi}{\partial\nu} = 0 \quad \text{on } \Gamma. \end{cases} \tag{7.5.1}$$

We will deduce an equation for ψ, denoting $x = (x_1, x_2)$ and $x_3 = y$. Since we have $\psi(t, x) = \phi(t, x, \eta(t, x))$, we obtain

$$\partial_t \psi = \big(\partial_t \phi + \partial_y \phi \, \partial_t \eta\big)|_\Sigma = \big(\partial_t \phi + \partial_y \phi \, G(\eta)\psi\big)|_\Sigma, \tag{7.5.2}$$

$$\nabla_x \psi = \big(\nabla_x \phi + \partial_y \phi \, \nabla_x \eta\big)|_\Sigma. \tag{7.5.3}$$

On the other hand, according to (7.3.4),

$$G(\eta)\psi = \big(\partial_y \phi - \nabla_x \phi \cdot \nabla_x \eta\big)|_\Sigma. \tag{7.5.4}$$

We deduce from (7.5.3) that

$$\nabla_x \psi \cdot \nabla_x \eta = \big(\nabla_x \phi \cdot \nabla_x \eta + \partial_y \phi |\nabla_x \eta|^2\big)|_\Sigma,$$

whence, using (7.5.4),

$$(1 + |\nabla_x \eta|^2)\, \partial_y \phi|_\Sigma = \nabla_x \psi \cdot \nabla_x \eta + G(\eta)\psi. \tag{7.5.5}$$

We deduce from (7.5.3) that

$$\nabla_x \phi|_\Sigma = \nabla_x \psi - \frac{\nabla_x \psi \cdot \nabla_x \eta + G(\eta)\psi}{1 + |\nabla_x \eta|^2} \nabla_x \eta. \tag{7.5.6}$$

It follows from (7.5.5) and (7.5.6) that

$$|\nabla_{x,y}\phi|^2|_\Sigma = \frac{(\nabla_x \psi \cdot \nabla_x \eta + G(\eta)\psi)^2}{1 + |\nabla_x \eta|^2} + |\nabla_x \psi|^2 - 2\frac{\nabla_x \psi \cdot \nabla_x \eta + G(\eta)\psi}{1 + |\nabla_x \eta|^2} (\nabla_x \psi \cdot \nabla_x \eta). \tag{7.5.7}$$

According to the condition on P in (7.5.1) and the fact that $x_3 = \eta$ on Σ, we have

$$0 = \big(\partial_t \phi + \frac{1}{2}|\nabla_{x,y}\phi|^2 + gy + P\big)|_\Sigma = \big(\partial_t \phi + \frac{1}{2}|\nabla_{x,y}\phi|^2 + gy\big)|_\Sigma.$$

We deduce from the fact that $\phi|_\Sigma = \psi$, and from (7.5.3), (7.5.5), (7.5.7) that

$$\begin{aligned} 0 = \partial_t \psi &- \frac{\nabla_x \psi \cdot \nabla_x \eta + G(\eta)\psi}{1 + |\nabla_x \eta|^2} G(\eta)\psi + \frac{1}{2}\frac{(\nabla_x \psi \cdot \nabla_x \eta + G(\eta)\psi)^2}{1 + |\nabla_x \eta|^2} \\ &+ \frac{1}{2}|\nabla_x \psi|^2 - \frac{\nabla_x \psi \cdot \nabla_x \eta + G(\eta)\psi}{1 + |\nabla_x \eta|^2} (\nabla_x \psi \cdot \nabla_x \eta) + g\eta, \end{aligned}$$

whence we obtain

$$\partial_t \psi + \frac{1}{2}|\nabla_x \psi|^2 - \frac{1}{2}\frac{(\nabla_x \psi \cdot \nabla_x \eta + G(\eta)\psi)^2}{1 + |\nabla_x \eta|^2} + g\eta = 0.$$

In summary, the system satisfied by (η, ψ) is as follows:

$$\begin{cases} \partial_t \eta = G(\eta)\psi, \\ \partial_t \psi = -\dfrac{1}{2}|\nabla_x \psi|^2 + \dfrac{1}{2}\dfrac{(\nabla_x \psi \cdot \nabla_x \eta + G(\eta)\psi)^2}{1+|\nabla_x \eta|^2} - g\eta. \end{cases} \tag{7.5.8}$$

7.6 Appendix

7.6.1 *Trace on* Σ *of Elements of* $\mathcal{H}^{1,0}(\Omega)$

We will prove point 4. of the proof of Proposition 7.4.6, namely that the elements of $\mathcal{H}^{1,0}(\Omega)$ have a trace on Σ and that this trace is zero. This will follow from Lemma 7.4.9.

Indeed, let $u \in \mathcal{H}^{1,0}(\Omega)$ and denote as before by $\widetilde{u}$ its image under the diffeomorphism given by Lemma 7.4.7. We have, with the notations of (7.4.15),

$$\Lambda_1 \widetilde{u} \in L^2(\widetilde{\Omega}), \quad \Lambda_{2,j} \widetilde{u} \in L^2(\widetilde{\Omega}), 1 \leq j \leq d.$$

Since $\partial_z = (\partial_z \rho)\Lambda_1$ and $\partial_z \rho \in L^\infty(\widetilde{\Omega})$, we have $\partial_z \widetilde{u} \in L^2(\widetilde{\Omega})$. Next, by definition we have $\nabla_x \widetilde{u} = \Lambda_2 \widetilde{u} + (\nabla_x \rho)\Lambda_1 \widetilde{u}$. Thus, we also have $\nabla_x \widetilde{u} \in L^2(\widetilde{\Omega})$ since $\nabla_x \rho \in L^\infty(\widetilde{\Omega})$. The Poincaré inequality (10.4.2) shows that $\widetilde{u} \in L^2(\widetilde{\Omega})$. It follows that ,

$$\widetilde{u} \in L^2(I_h, H^1(\mathbf{R}^d)) \quad \text{and} \quad \partial_z \widetilde{u} \in L^2(I_h, L^2(\mathbf{R}^d)).$$

Lemma 7.4.9 with $s = \frac{1}{2}$ then implies that $\widetilde{u}$ extends to a continuous function on the closed interval $[-h, 0]$ with values in $H^{\frac{1}{2}}(\mathbf{R}^d)$. It follows then that the function $\widetilde{u}(x, 0) = u(x, \eta(x))$ belongs to $H^{\frac{1}{2}}(\mathbf{R}^d)$. Therefore, the function u has a trace on Σ. On the other hand, by the definition of $\mathcal{H}^{1,0}(\Omega)$, u is the limit of a sequence (φ_j) in $C^{\infty,0}(\Omega)$. Since all the φ_j are identically zero in a neighborhood of Σ, the trace of u is also zero on Σ.

7.6.2 *The Condition* $\partial_y \phi|_{y=-h} = 0$

We will explain the reason why the solution ϕ satisfies this condition on $\Gamma = \{(x, y) : y = -h\}$ and specify its meaning.

By Remark 7.4.2, ψ is equal to zero in the strip $B_h = \{(x, y) : -h < y < -\frac{5}{6}h\}$. The function ϕ is then a solution of the equation $\Delta_{x,y}\phi = 0$ in $\mathcal{D}'(B_h)$ and is therefore a C^∞ function in B_h (by the hypoellipticity of the Laplacian).

Let $\lambda \in C_0^\infty(\mathbf{R}^d)$ and $\theta \in C^\infty(\mathbf{R}), \theta(y) = 1$ if $-h \le y \le -\frac{7}{8}h$ and $\theta = 0$ if $y \ge -\frac{5}{6}h$.

Set $v(x, y) = \lambda(x)\theta(y)$. Then $v \in \mathcal{H}^{1,0}(\Omega)$ and $\operatorname{supp} v \subset B_h$.

The equality (7.4.9) can be written as

$$\iint_{B_h} \nabla_x \phi(x, y) \cdot \nabla_x v(x, y)\, dx\, dy + \iint_{B_h} \partial_y \phi(x, y) \partial_y v(x, y)\, dx\, dy = 0. \tag{7.6.1}$$

Suppose for a moment that

$$(\star) \quad \phi \in H^2(\widetilde{B}_h) \quad \text{where} \quad \widetilde{B}_h = \big\{(x, y) : -h < y < -\frac{7}{8}h\big\} \subset B_h.$$

Then, $\partial_y \phi \in H^1(\widetilde{B}_h)$. By the same reasoning as in the previous paragraph, $\partial_y \phi$ admits a trace on the set $\{(x, y) : y = -h\}$. We will integrate by parts in the integrals appearing in (7.6.1). Let us denote them I_1, I_2. In I_1 there is no boundary term because $\phi \in C^\infty(B_h)$ and $\lambda \in C_0^\infty(\mathbf{R}^d)$. We obtain

$$I_1 = -\iint_{B_h} \Delta_x \phi(x, y) v(x, y)\, dx\, dy.$$

On the other hand, since $v(x, -h) = \lambda(x)$ and $v(x, -\frac{5}{6}h) = 0$, we obtain

$$\begin{aligned}\int_{\mathbf{R}^d} \int_{-h}^{-\frac{5}{6}h} \partial_y \phi(x, y) \partial_y v(x, y)\, dy = &- \iint_{B_h} \partial_y^2 \phi(x, y) v(x, y)\, dy \\ &- \int_{\mathbf{R}^d} \partial_y \phi(x, -h) \lambda(x)\, dx.\end{aligned}$$

Using (7.6.1) it follows that

$$\iint_{B_h} \Delta_{x,y} \phi(x, y) v(x, y)\, dx\, dy + \int_{\mathbf{R}^d} \partial_y \phi(x, -h) \lambda(x)\, dx = 0.$$

Since $\Delta_{x,y} \phi(x, y) = 0$ almost everywhere (because $\Delta_{x,y} \phi \in L^2(B_h)$ by hypothesis), we finally obtain

$$\int_{\mathbf{R}^d} \partial_y \phi(x, -h) \lambda(x)\, dx = 0, \quad \forall \lambda \in C_0^\infty(\mathbf{R}^d),$$

which implies that $\partial_y \phi(x, -h) = 0$ almost everywhere. The third condition of (7.4.6) is therefore satisfied.

To finish, it remains to prove ($\star$). We apply the equality (7.4.9) to the function $v(x, y) = \chi(y)w(x, y)$ where $w \in \mathcal{H}^{1,0}(\Omega)$, supp $w \subset B_h$, and

$$\chi \in C^\infty(\mathbf{R}), \quad \chi(y) = 1 \text{ if } -h \leq y \leq -\frac{7}{8}h, \quad \chi(y) = 0 \text{ if } y \geq -\frac{5}{6}h.$$

Then $v \in \mathcal{H}^{1,0}(\Omega)$.

Since the supports in y of χ and $\underline{\psi}$ are disjoint, the first term of (7.4.9) is equal to zero. For simplicity, let us denote $X = (x, y)$. We have

$$\iint_{B_h} \nabla_X \phi(X) \cdot \nabla_X\big(\chi(y)w(X)\big)\, dX = 0.$$

We move the χ from w to ϕ, then we obtain

$$\begin{aligned}\iint_{B_h} \nabla_X\big(\chi(y)\phi(X)\big) \cdot \nabla_X w(X)\, dX = &\iint_{B_h} (\partial_y\chi)(y)\phi(X)\partial_y w(X)\, dX \\ &- \iint_{B_h} \partial_y\phi(X)(\partial_y\chi)(y)w(X)\, dX.\end{aligned} \tag{7.6.2}$$

For $\varepsilon > 0$ and $i = 1, \dots, d$ we set $(\tau_\varepsilon^i w)(x, y) = w(x + \varepsilon e_i, y)$ where e_i is the ith vector of the canonical basis of $\mathbf{R}^d$. Fix $i \in \{1, \dots, d\}$.

Since $B_h = \mathbf{R}^d \times (-h, -\frac{5}{6}h)$ and the support of w is contained in B_h, we also have $\tau_{-\varepsilon}^i w \in \mathcal{H}^{1,0}(\Omega)$, supp $\tau_{-\varepsilon}^i w \subset B_h$. We can therefore apply (7.6.2) to $\tau_{-\varepsilon}^i w$. Then in the integrals set $x' = x - \varepsilon e_i$. Since χ depends only on y we have

$$\begin{aligned}\iint_{B_h} \nabla_X\big(\chi(y)\tau_\varepsilon^i\phi(X)\big) \cdot \nabla_X w(X)\, dX = &\iint_{B_h} (\partial_y\chi)(y)\tau_\varepsilon^i\phi(X)\partial_y w(X)\, dX \\ &- \iint_{B_h} \partial_y\tau_\varepsilon^i\phi(X)(\partial_y\chi)(y)w(X)\, dX.\end{aligned} \tag{7.6.3}$$

Set

$$\Phi_\varepsilon = \frac{\tau_\varepsilon^i\phi - \phi}{\varepsilon}.$$

By subtracting equality (7.6.2) from (7.6.3), we obtain

$$\begin{aligned}\iint_{B_h} \nabla_X\big(\chi(y)\Phi_\varepsilon(X)\big) \cdot \nabla_X w(X)\, dX = &\iint_{B_h} (\partial_y\chi)(y)\Phi_\varepsilon(X)\partial_y w(X)\, dX \\ &- \iint_{B_h} \partial_y\Phi_\varepsilon(X)(\partial_y\chi)(y)w(X)\, dX = I_1 - I_2.\end{aligned} \tag{7.6.4}$$

Since $\Phi_\varepsilon \in \mathcal{H}^{1,0}$ and $\operatorname{supp} \chi \subset B_h$, we can apply the above equality to $w = \chi \Phi_\varepsilon$. The left-hand side is equal to $\|\nabla_X(\chi \Phi_\varepsilon)\|^2_{L^2(B_h)}$. Thus, we have

$$\|\nabla_X(\chi \Phi_\varepsilon)\|^2_{L^2(B_h)} \leq |I_1| + |I_2|. \tag{7.6.5}$$

Using the Cauchy-Schwarz inequality, we obtain

$$|I_1| \leq C\|\Phi_\varepsilon\|_{L^2(B_h)}\|\partial_y(\chi \Phi_\varepsilon)\|_{L^2(B_h)}.$$

On the other hand,

$$\begin{aligned}
I_2 &= \iint_{B_h} \partial_y \Phi_\varepsilon(X)(\partial_y \chi)(y)\chi(y)\Phi_\varepsilon(X)\, dX \\
&= \iint_{B_h} \partial_y\big(\chi(y)\Phi_\varepsilon(X)\big)(\partial_y\chi)(y)\Phi_\varepsilon(X)\, dX - \iint_{B_h} (\partial_y \chi(y))^2(\Phi_\varepsilon(X))^2\, dX, \\
&= I_3 - I_4.
\end{aligned}$$

Again, by Cauchy-Schwarz we have

$$|I_3|C \leq \|\partial_y(\chi \Phi_\varepsilon)\|_{L^2(B_h)}\|\Phi_\varepsilon\|_{L^2(B_h)}.$$

Eventually,

$$|I_4| \leq C\|\Phi_\varepsilon\|^2_{L^2(B_h)}.$$

Using inequality (7.6.5), the estimates for I_1, I_2 and the inequality $ab \leq \delta a^2 + \frac{1}{4\delta}b^2$, for $a > 0$, $b > 0$ and $\delta > 0$ small enough, we obtain

$$\|\nabla_X(\chi \Phi_\varepsilon)\|_{L^2(B_h)} \leq C\|\Phi_\varepsilon\|_{L^2(B_h)}. \tag{7.6.6}$$

We have $\chi(-\frac{5}{6}h) = 0$ so that for $-h < y < -\frac{5}{6}h$ we can write

$$(\chi \Phi_\varepsilon)(x, y) = -\int_y^{-\frac{5}{6}h} \partial_y(\chi \Phi_\varepsilon)(x, t)\, dt.$$

By the Cauchy-Schwarz inequality (as in the proof of Lemma 10.4.2) we obtain

$$\|\chi \Phi_\varepsilon\|_{L^2(B_h)} \leq C\|\partial_y(\chi \Phi_\varepsilon)\|_{L^2(B_h)}. \tag{7.6.7}$$

Using (7.6.6), (7.6.7) and the fact that $\chi(y) = 1$ for $y \in (-h, -\frac{7}{8}h)$ we finally obtain

$$\|\Phi_\varepsilon\|_{H^1(\widetilde{B}_h)} \leq C\|\Phi_\varepsilon\|_{L^2(B_h)}. \tag{7.6.8}$$

We have

$$\|\Phi_\varepsilon\|_{L^2(B_h)} \leq \|\partial_{x_i}\phi\|_{L^2(B_h)} \leq \|\phi\|_{\mathcal{H}^{1,0}(\Omega)}.$$

Indeed, since ϕ is a C^∞ function in $\widetilde{B}_h$, the Taylor formula with integral remainder allows us to write

$$\Phi_\varepsilon(x, y) = \int_0^1 \partial_i\phi(x + t\varepsilon e_i, y)\, dt.$$

Using the Cauchy-Schwarz inequality, we deduce

$$\iint_{B_h} |\Phi_\varepsilon(x, y)|^2\, dx\, dy \leq \int_0^1 \Big(\int_{-h}^{-\frac{5}{6}h} \int_{\mathbf{R}^d} |\partial_i\phi(x + t\varepsilon e_i, y|^2\, dx\, dy \Big)\, dt.$$

Setting $x' = x + t\varepsilon e_i$ in the integral over x, we obtain the desired result.

Therefore the left-hand side of (7.6.8) is uniformly bounded for all $\varepsilon > 0$. Since $H^1(\widetilde{B}_h)$ is a Hilbert space, there exists a subsequence (Φ_{ε_k}) which converges weakly in $H^1(\widetilde{B}_h)$ (thus in $\mathcal{D}'(\widetilde{B}_h)$) to $U \in H^1(\widetilde{B}_h)$. But in $\mathcal{D}'(\widetilde{B}_h)$ the sequence (Φ_{ε_k}) converges to $\partial_i\phi$. We deduce that $\partial_i\phi \in H^1(\widetilde{B}_h)$, that is, $\partial_i\partial_k\phi \in L^2(\widetilde{B}_h)$ for $1 \leq k \leq d$ and $\partial_i\partial_y\phi \in L^2(\widetilde{B}_h)$, for all $i = 1, \ldots, d$. Since $\partial_y^2\phi = -\sum_{k=1}^d \partial_k^2\phi$ in $\mathcal{D}'(\widetilde{B}_h)$, we have $\partial_y^2\phi \in L^2(\widetilde{B}_h)$, which implies that $\phi \in H^2(\widetilde{B}_h)$ and proves ($\star$).

Chapter 8
Spectral Theory of the Laplacian on the Sphere and on the Torus

8.1 Prerequisites

The Green formula for the Laplacian, Hilbert spaces.

8.2 The Context

Spectral theory for the Laplacian on various domains is an important topic in mathematical physics and in partial differential equations. We shall study in this book several cases for which we can describe the properties of the eigenvalues as well as those of the eigenfunctions. In this chapter we look to the case of the sphere and the torus. This leads to the study in the first case of the spherical harmonics and in the second case to the theory of multidimensional Fourier series.

8.3 Spectral Theory of the Laplacian on the Sphere

8.3.1 The Laplacian on the Sphere

Let $d \in \mathbf{N}$, $d \geq 2$ and $\Delta_{\mathbf{R}^d} = \sum_{j=1}^{d} \partial_j^2$ the Laplacian in $\mathbf{R}^d$. We denote the spherical coordinates of $\mathbf{R}^d$ by $(r, \omega) \in (0, +\infty) \times \mathbf{S}^{d-1}$ and we will set in what follows, for $1 \leq i, j \leq d$,

$$X_{ij} = x_i \partial_j - x_j \partial_i, \quad \text{and} \quad \tilde{X}_{ij} = \omega_i \partial_{\omega_j} - \omega_j \partial_{\omega_i}.$$

C. Zuily, *Selected Topics in Partial Differential Equations*, Universitext,
https://doi.org/10.1007/978-3-032-24082-8_8

Proposition 8.3.1 *For any function f of class C^2 on $\mathbf{R}^d$ we have*

$$\big(\Delta_{\mathbf{R}^d} f\big)(r\omega) = \Big(\partial_r^2 + \frac{d-1}{r}\partial_r + \frac{1}{r^2}\sum_{i<j}\widetilde{X}_{ij}^2\Big)\big[f(r\omega)\big]. \tag{8.3.1}$$

The operator $\sum_{i<j}\widetilde{X}_{ij}^2$, denoted $\Delta_{\mathbf{S}^{d-1}}$, is called the Laplace-Beltrami operator on the sphere $\mathbf{S}^{d-1}$.

Proof Point 1. Since $x = r\omega$, we have

$$\widetilde{X}_{ij}[f(r\omega)] = (\omega_i r\partial_{x_j} f - \omega_j r\partial_{x_i} f)(r\omega) = (X_{ij} f)(r\omega). \tag{8.3.2}$$

Point 2. (i) For $i < j$, x_i and ∂_j commute. Consequently,

$$X_{ij}(X_{ij} f) = x_i^2\partial_j^2 f - x_i\partial_i f - x_i x_j\partial_i\partial_j f - x_j\partial_j f - x_j x_i\partial_i\partial_j f + x_j^2\partial_i^2 f = \sum_{k=1}^{6} A_{ij}^k.$$

Thus, we have

$$\sum_{i<j}\big(A_{ij}^1 + A_{ij}^6\big) = \sum_{i=1}^{d}\sum_{j=i+1}^{d} x_i^2\partial_j^2 f + \sum_{j=1}^{d}\sum_{i=1}^{j-1} x_j^2\partial_i^2 f = \sum_{i\neq j} x_i^2\partial_j^2 f,$$

$$\sum_{i<j} A_{ij}^2 = -\sum_{i=1}^{d}\sum_{j=i+1}^{d} x_i\partial_i f = -\sum_{i=1}^{d}(d-i)x_i\partial_i f,$$

$$\sum_{i<j} A_{ij}^4 = -\sum_{j=1}^{d}\sum_{i=1}^{j-1} x_j\partial_j f = \sum_{j=1}^{d}(j-1)x_j\partial_j f, \quad \text{therefore,}$$

$$\sum_{i<j}\big(A_{ij}^2 + A_{ij}^4\big) = (d-1)\sum_{i=1}^{d} x_i\partial_i f,$$

$$\sum_{i<j}\big(A_{ij}^3 + A_{ij}^5\big) = -\sum_{i=1}^{d}\sum_{j=i+1}^{d} x_i x_j\partial_i\partial_j f - \sum_{j=1}^{d}\sum_{i=1}^{j-1} x_i x_j\partial_i\partial_j f = -\sum_{i\neq j} x_i x_j\partial_i\partial_j f.$$

(ii) Set $\widetilde{f}(r,\omega) = f(r\omega)$. It is easy to see that $\sum_{i=1}^{d} x_i\partial_{x_i} f = r\partial_r\widetilde{f}$. On the other hand,

$$
\begin{aligned}
\sum_{i\neq j} x_i x_j \partial_i \partial_j f &= \sum_{i=1}^{d}\sum_{\substack{j=1\\ j\neq i}}^{d} = \sum_{j=1}^{d} x_i\partial_i\Big(\sum_{j\neq i} x_j\partial_j f\Big),\\
&= \sum_{i=1}^{d} x_i\partial_i\Big(\sum_{j=1}^{d} x_j\partial_j f\Big) - \sum_{i=1}^{d}(x_i\partial_i)^2 f,\\
&= (r\partial_r)^2\widetilde{f} - \sum_{i=1}^{d} x_i^2\partial_i^2 f - \sum_{i=1}^{d} x_i\partial_i f = (r\partial_r)^2\widetilde{f} - r\partial_r\widetilde{f} - \sum_{i=1}^{d} x_i^2\partial_i^2 f,\\
&= r^2\partial_r^2\widetilde{f} - \sum_{i=1}^{d} x_i^2\partial_i^2 f.
\end{aligned}
$$

(iii) We have

$$
\sum_{i\neq j} x_i^2\partial_j^2 f = \sum_{i=1}^{d} x_i^2\Big(\sum_{j=1}^{d}\partial_j^2 f - \partial_i^2 f\Big) = |x|^2\Delta_{\mathbf{R}^d} - \sum_{i=1}^{d} x_i^2\partial_i^2 f.
$$

We deduce from (8.3.2), (i), (ii), (iii) that

$$
\sum_{i<j}\widetilde{X}_{ij}^2\widetilde{f} = \sum_{i<j} X_{ij}^2 f = |x|^2\Delta_{\mathbf{R}^d} f - \sum_{i=1}^{d} x_i^2\partial_i^2 f - r^2\partial_r^2\widetilde{f} + \sum_{i=1}^{d} x_i^2\partial_i^2 f - (d-1)\sum_{i=1}^{d} x_i\partial_i f.
$$

As $\sum_{i=1}^{d} x_i\partial_i f = r\partial_r\widetilde{f}$ and $|x| = r$, the equality above provides the desired result. □

Example 8.3.2 If $d = 2$ we can write $\omega_1 = \cos\theta, \omega_2 = \sin\theta, \theta \in (0, 2\pi)$. Then, $\widetilde{X}_{12} = \frac{\partial}{\partial\theta}$. Indeed, we have

$$
\begin{aligned}
\partial_\theta[f(\cos\theta,\sin\theta)] &= \big(-(\sin\theta)\partial_{\omega_1} f + (\cos\theta)\partial_{\omega_2} f\big)(\cos\theta,\sin\theta),\\
&= \big(-\omega_2\partial_{\omega_1} f + \omega_1\partial_{\omega_2} f\big)(\cos\theta,\sin\theta) = (\widetilde{X}_{12} f)(\omega_1,\omega_2).
\end{aligned}
$$

Thus $\Delta_{\mathbf{S}^1} = \big(\partial_\theta\big)^2$.

Exercise 8.3.3 Let $d = 3$. Consider the parametrization of the sphere $\mathbf{S}^2$:

$$
\omega_1 = \sin\theta\cos\varphi, \quad \omega_2 = \sin\theta\sin\varphi, \quad \omega_3 = \cos\theta, \quad \theta \in (0,\pi), \varphi \in (0, 2\pi).
$$

Then, in these coordinates,

$$
\Delta_{\mathbf{S}^2} = \frac{1}{\sin\theta}\frac{\partial}{\partial\theta}\sin\theta\frac{\partial}{\partial\theta} + \frac{1}{(\sin\theta)^2}\frac{\partial^2}{\partial\varphi^2}.
$$

Hint: note that $\theta = \mathrm{Arccos}(\omega_3)$, $\varphi = \arctan\big(\frac{\omega_2}{\omega_1}\big)$.

Lemma 8.3.4 *For* $u, v \in C^2(\mathbf{R}^{d+1} \setminus \{0\})$ *we have*

$$\int_{\mathbf{S}^d} (\Delta_{\mathbf{S}^d} u)(\omega) v(\omega)\, d\omega = \int_{\mathbf{S}^d} u(\omega)(\Delta_{\mathbf{S}^d} v)(\omega)\, d\omega.$$

Proof Let u, v be two C^2 functions with compact support in $\mathbf{R}^{d+1} \setminus \{0\}$. Let us set $F(r, \omega) = u(r\omega)$ and $G(r, \omega) = v(r\omega)$.

We use Green's formula for the Laplacian on the open set $\{x \in \mathbf{R}^{d+1} : |x| > r\}$. The outward normal derivative to this open set is $-\frac{\partial}{\partial r}$. We obtain

$$\iint_{|x|>r} \Delta u(x) v(x)\, dx = \iint_{|x|>r} u(x)\Delta v(x)\, dx + \int_{\mathbf{S}^d} \big[(\partial_r G)F - (\partial_r F)G\big](r, \omega)\, r^d\, d\omega. \tag{8.3.3}$$

On the other hand, from formula (8.3.1) we have

$$\iint_{|x|>r} \Delta u(x) v(x)\, dx = \int_r^{+\infty} \int_{\mathbf{S}^d} \big(\partial_t^2 F + \frac{d}{t}\partial_t F + \frac{1}{t^2}\Delta_{S^d} F\big)(t, \omega) G(t, \omega) t^d\, dt\, d\omega.$$

By integrating by parts in the t integral, we can write

$$\int_r^{+\infty} \int_{\mathbf{S}^d} \big[(\partial_t^2 F)(t, \omega) G(t, \omega) t^d\, dt\, d\omega = -\int_r^{+\infty} \int_{\mathbf{S}^d} (\partial_t F)(t, \omega)(\partial_t G)(t, \omega) t^d\, dt\, d\omega$$
$$-\int_r^{+\infty} \int_{\mathbf{S}^d} \frac{d}{t}(\partial_t F)(t, \omega) G(t, \omega) t^d\, dt\, d\omega - r^d \int_{\mathbf{S}^d} (\partial_r F)(r, \omega) G(r, \omega)\, d\omega.$$

We deduce from the above formulas that

$$\iint_{|x|>r} \Delta u(x) v(x)\, dx = -\int_r^{+\infty} \int_{\mathbf{S}^d} (\partial_t F)(t, \omega)(\partial_t G)(t, \omega) t^d\, dt\, d\omega$$
$$+ \int_r^{+\infty} \int_{\mathbf{S}^d} \frac{1}{t^2}(\Delta_{S^d} F)(t, \omega)\big] G(t, \omega) t^d\, dt\, d\omega - r^d \int_{\mathbf{S}^d} [(\partial_r F)G](r, \omega)\, d\omega.$$

By exchanging the roles of u and v we obtain

$$\iint_{|x|>r} u(x)\Delta v(x)\, dx = -\int_r^{+\infty} \int_{\mathbf{S}^d} (\partial_t G)(t, \omega)(\partial_t F)(t, \omega) t^d\, dt\, d\omega$$
$$+ \int_r^{+\infty} \int_{\mathbf{S}^d} \frac{1}{t^2}(\Delta_{S^d} G)(t, \omega) F(t, \omega) t^d\, dt\, d\omega - r^d \int_{\mathbf{S}^d} [(\partial_r G)F](r, \omega)\, d\omega.$$

Taking the difference between these two quantities and using (8.3.3) we finally get

$$\int_r^{+\infty}\int_{\mathbf{S}^d}\frac{1}{t^2}(\Delta_{S^d}G)(t,\omega)F(t,\omega)t^d\,dt\,d\omega$$
$$=\int_r^{+\infty}\int_{\mathbf{S}^d}\frac{1}{t^2}(\Delta_{S^d}F)(t,\omega)\big]G(t,\omega)t^d\,dt\,d\omega.$$

Both sides being differentiable functions of the variable $r\in(0,+\infty)$, by differentiating them and then taking $r=1$, we obtain

$$\int_{\mathbf{S}^d}(\Delta_{S^d}G)(1,\omega)\big]F(1,\omega)\,d\omega=\int_{\mathbf{S}^d}(\Delta_{S^d}F)(1,\omega)\big]G(1,\omega)\,d\omega,$$

which is the conclusion of Lemma 8.3.4. □

We now consider the Laplacian on the sphere $\mathbf{S}^d$ whose spectrum we will determine and study a basis of eigenvectors.

8.3.2 Spherical Harmonics

Definition 8.3.5 Let $k\in\mathbf{N}$.

1. We denote by $\mathcal{P}_k$ the set of homogeneous polynomials of degree k on $\mathbf{R}^{d+1}$.

$$P\in\mathcal{P}_k\Longleftrightarrow P(x)=\sum_{|\alpha|=k}a_\alpha x^\alpha,\quad a_\alpha\in\mathbf{C}.$$

2. We denote by $\mathcal{A}_k$ the set of elements of $\mathcal{P}_k$ that are harmonic.

$$P\in\mathcal{A}_k\Longleftrightarrow P\in\mathcal{P}_k\quad\text{and}\quad\Delta_{\mathbf{R}^{d+1}}P=0.$$

3. We denote by $\mathcal{H}_k$ the set of restrictions to $\mathbf{S}^d$ of the elements of $\mathcal{A}_k$.

$$P\in\mathcal{H}_k\Longleftrightarrow P=Q|_{\mathbf{S}^d},\ Q\in\mathcal{A}_k\Longleftrightarrow P=Q|_{\mathbf{S}^d},\ Q\in\mathcal{P}_k,\ \text{and}\ \Delta_{\mathbf{R}^{d+1}}Q=0.$$

The elements of $\mathcal{H}_k$ are called the **spherical harmonics of degree** k.

Lemma 8.3.6 *For $k\geq 2$, $\mathcal{H}_k$ is a vector space of dimension*

$$d_k=\binom{d+k}{d}-\binom{d+k-2}{d}.$$

For example, if $d=2$ we have $\dim\mathcal{H}_k=2k+1$ and Stirling's formula shows that

$$\dim\mathcal{H}_k\sim\frac{2}{(d-1)!}k^{d-1},\quad k\to+\infty.$$

Lemma 8.3.7 *If $P \in \mathcal{H}_k$ we have*

$$-\Delta_{\mathbf{S}^d} P = k(k+d-1)P,$$

that is, P is an eigenvector of $-\Delta_{\mathbf{S}^d}$ corresponding to the eigenvalue $k(k+d-1)$.

Proof By definition $P = Q|_{\mathbf{S}^d}$, $Q \in \mathcal{P}_k$, and $\Delta_{\mathbf{R}^{d+1}} Q = 0$. Using (8.3.1) with $d+1$ instead of d we can write for $r > 0$,

$$\begin{aligned} 0 &= \Delta_{\mathbf{R}^{d+1}}\big[Q(r\omega)\big] = \Delta_{\mathbf{R}^{d+1}}\big[r^k P(\omega)\big], \\ &= k(k-1)r^{k-2}P(\omega) + dkr^{k-2}P(\omega) + r^{k-2}\Delta_{\mathbf{S}^d} P(\omega), \\ &= r^{k-2}\big[\Delta_{\mathbf{S}^d} P(\omega) + k(k+d-1)P(\omega)\big], \end{aligned}$$

which proves the lemma. □

Proposition 8.3.8 *The finite linear combinations of elements of $\cup_{k=0}^{+\infty}\mathcal{H}_k$ form a dense set in $L^2(\mathbf{S}^d)$.*

Proof It suffices to show that this set is dense in the set of continuous functions on $\mathbf{S}^d$ equipped with the L^∞ norm. Let $f \in C^0(\mathbf{S}^d)$. According to the Weierstrass approximation theorem, it can be uniformly approximated by a sequence of polynomials restricted to $\mathbf{S}^d$. The following lemma completes the proof. □

Lemma 8.3.9 *The restriction to $\mathbf{S}^d$ of a polynomial on $\mathbf{R}^{d+1}$ is a sum of restrictions to $\mathbf{S}^d$ of harmonic polynomials.*

Proof of the lemma We show that if $P \in \mathcal{P}_k$ it can be written in the form

$$P(x) = P_0(x) + |x|^2 P_1(x) + \cdots + |x|^{2\ell} P_\ell(x),$$

where P_j is a homogeneous harmonic polynomial of degree $k-2j$, $j = 0, \ldots, \ell$.

We introduce an inner product $\langle P, Q\rangle$ on $\mathcal{P}_k$ by setting

$$\langle P, Q\rangle = P(D)\overline{Q}, \quad P, Q \in \mathcal{P}_k$$

where $P(D) = \sum_{|\alpha|=k} a_\alpha \partial^\alpha$. Since P and Q are homogeneous of the same degree, $\langle P, Q\rangle$ is a constant. Moreover, this is indeed a Hermitian bilinear form. It is an inner product because $\langle P, P\rangle = \sum_{|\alpha|=k} |a_\alpha|^2\alpha! > 0$ if $P \not\equiv 0$.

Every polynomial of degree ≤ 1 being harmonic, we may assume $k \geq 2$. Consider the linear map $\varphi : \mathcal{P}_k \to \mathcal{P}_{k-2}$ defined by $\varphi(P) = \Delta_{\mathbf{R}^{d+1}} P$.

Let us denote $\mathcal{Q}_k = \{Q \in \mathcal{P}_k = Q(x) = |x|^2 R_{k-2}(x), R_{k-2} \in \mathcal{P}_{k-2}\}$. We will show that $\mathcal{H}_k = \mathcal{Q}_k^\perp$ where the orthogonal is taken with respect to the above inner product. Indeed, let $P \in \mathcal{Q}_k^\perp$. Then for any $Q_0 \in \mathcal{P}_{k-2}$ we have

$$0 = \langle |x|^2 Q_0, P\rangle = \Delta Q_0(D)\overline{P} = Q_0(D)\Delta\overline{P}. \tag{8.3.4}$$

In particular, if $Q_0 = \Delta P$ we obtain $0 = (\Delta P)(D)\Delta \overline{P} = \langle \Delta P, \Delta P\rangle$, which implies that $\Delta \overline{P} = 0$ by the property of the inner product. Thus $P \in \mathcal{H}_k$. Conversely, if $P \in \mathcal{H}_k$ the equality (8.3.4) shows that $P \in \mathcal{Q}_k^{\perp}$.

We deduce that $\mathcal{P}_k = \mathcal{H}_k \bigoplus \mathcal{Q}_k$, which shows that any $P \in \mathcal{P}_k$ can be written as $P = Q + |x|^2 Q_0$ where $Q \in \mathcal{H}_k$ and $Q_0 \in \mathcal{P}_{k-2}$. An induction allows us to conclude. □

Proposition 8.3.10 *Let Y^k, Y^ℓ be two spherical harmonics of degree k and ℓ with $k \neq \ell$. Then,*

$$\int_{\mathbf{S}^d} Y^k(\omega) Y^\ell(\omega)\, d\omega = 0.$$

Proof We know that $Y^k = P_k|_{\mathbf{S}^d}$, $\Delta P_k = 0$, $Y^\ell = Q_\ell|_{\mathbf{S}^d}$, $\Delta Q_\ell = 0$. The outward normal derivative to $\mathbf{S}^d$ is $\frac{\partial}{\partial r}$. We have $\frac{\partial}{\partial r} P_k|_{\mathbf{S}^d} = kY^k(\omega)$ and $\frac{\partial}{\partial r} Q_l|_{\mathbf{S}^d} = \ell Y^\ell(\omega)$. By Green's formula we can write

$$\begin{aligned} 0 = \int_{|x|\leq 1} (P_k \Delta Q_\ell - Q_\ell \Delta P_k)\, dx &= \int_{\mathbf{S}^d} \Big(P_k \frac{\partial Q_\ell}{\partial r} - Q_\ell \frac{\partial P_k}{\partial r} \Big)\, d\omega \\ &= \int_{\mathbf{S}^d} \big(\ell Y^\ell(\omega) Y^k(\omega) - kY^k(\omega) Y^\ell(\omega) \big)\, d\omega \\ &= (\ell - k) \int_{\mathbf{S}^d} Y^k(\omega) Y^\ell(\omega)\, d\omega \end{aligned}$$

whence the result. □

Let us consider $\mathcal{H}_k$ as a subspace of $L^2(\mathbf{S}^d)$ equipped with the usual inner product. If $\{Y_1^k, \ldots, Y_{a_k}^k\}$, where $a_k = \dim \mathcal{H}_k$, is an orthonormal basis of $\mathcal{H}_k$, the previous proposition shows that the family $\cup_{k=0}^{+\infty}\{Y_1^k, \ldots, Y_{a_k}^k\}$ is an orthonormal basis of $L^2(\mathbf{S}^d)$. Therefore, if $u \in L^2(\mathbf{S}^d)$ one can write in a unique way

$$u = \sum_{k=0}^{+\infty} Y^k, \quad Y^k \in \mathcal{H}_k, \tag{8.3.5}$$

where the series converges in $L^2(\mathbf{S}^d)$. Indeed we have

$$Y^k = \sum_{j=1}^{a_k} b_j^k Y_j^k, \quad \text{where} \quad b_j^k = \big(u, Y_j^k\big)_{L^2(\mathbf{S}^d)}.$$

We can then deduce the spectral theory of the Laplacian on $\mathbf{S}^d$. Let $u \in L^2(\mathbf{S}^d)$ be an eigenvector associated with the eigenvalue λ that is,

$$-\Delta_{\mathbf{S}^d} u = \lambda u.$$

Using (8.3.5) we have $u = \sum_{k=0}^{+\infty} Y^k$, so that, using Lemma 8.3.7, we can write

$$\sum_{k=0}^{+\infty} k(k+d-1)Y^k = \sum_{k=0}^{+\infty} \lambda Y^k.$$

By uniqueness of the expansion, since the numbers $k(k+d-1), k \in \mathbf{N}$ are all distinct and $u \not\equiv 0$, there exists a unique k_0 such that $\lambda = k_0(k_0+d-1)$ and $u = Y^{k_0}$. In summary, we have obtained the following result

Theorem 8.3.11 *The spectrum of* $-\Delta_{\mathbf{S}^d}$ *consists of the sequence* $\lambda_k = k(k+d-1)$ *where* $k \in \mathbf{N}$ *and the eigenspace corresponding to the eigenvalue* λ_k *is the space* $\mathcal{H}_k$ *of spherical harmonics of degree* k.

We will now study, in the case of the sphere $\mathbf{S}^2$, some particular spherical harmonics. On $\mathbf{S}^2$ we use the coordinates $(\theta, \varphi) \in (0, \pi) \times (0, 2\pi)$ in which

$$\omega_1 = \sin\theta\cos\varphi, \quad \omega_2 = \sin\theta\sin\varphi, \quad \omega_3 = \cos\theta. \tag{8.3.6}$$

Recall that the space $\mathcal{H}_k$ is, when $d = 2$, of dimension $2k+1$. A basis of this space is then given by

$$Y_m^k(\theta, \varphi) = c_{k,m} P_m^k(\theta) e^{im\varphi}, \quad -k \le m \le k, \tag{8.3.7}$$

where the c_{km} are normalization constants and P_m^k are related to the Legendre polynomials.

Spherical Harmonics of Highest Degree

Let $P_k(x) = (x_1 + ix_2)^k$ in $\mathbf{R}^3$. This is a harmonic polynomial. Indeed, the function $(x_1 + ix_2)^k$ is holomorphic, hence harmonic, and does not depend on x_3. Then we have $P_k|_{\mathbf{S}^2} = (\sin\theta)^k e^{ik\varphi}$. It is of the form (8.3.7) with $m = k$ (the highest degree).

The masses of these spherical harmonics tend, as $k \to +\infty$, to concentrate as shows the following result.

Proposition 8.3.12 *Let* $Y^k(\omega) = k^{\frac{1}{4}}(\omega_1 + i\omega_2)^k$, $k \in \mathbf{N}$ $\omega \in \mathbf{S}^2$. *Then,*
1. *the sequence* (Y^k) *is bounded in* $L^2(\mathbf{S}^2)$,
2. *the sequence* (Y^k) *is orthogonal,*
3. *for all* $a \in C^0(\mathbf{S}^2)$ *we have*

$$\lim_{k\to+\infty} \int_{\mathbf{S}^2} a(\omega)|Y^k(\omega)|^2\, d\omega = \sqrt{\pi}\langle \delta_{\{\omega_3=0\}}, a\rangle = \sqrt{\pi}\int_0^{2\pi} a(\cos\theta, \sin\theta, 0)\, d\theta.$$

Proof We prove 1. and 3. at the same time. Indeed, let $a \in C^0(\mathbf{S}^2)$. We have

$$\begin{aligned} I_k := \int_{\mathbf{S}^2} a(\omega)|Y^k(\omega)|^2\, d\omega &= k^{\frac{1}{2}} \int_{\mathbf{S}^2} a(\omega)(\omega_1^2 + \omega_2^2)^k\, d\omega \\ &= k^{\frac{1}{2}} \int_{\mathbf{S}^2} a(\omega)(1 - \omega_3^2)^k\, d\omega \end{aligned}$$

We will parametrize the sphere $\mathbf{S}^2$ by $|\omega_3| \leq 1$ and $\omega_1^2 + \omega_2^2 = 1 - \omega_3^2$, then set

$$\omega_1 = \sqrt{1-\omega_3^2}\cos\theta,\, \omega_2 = \sqrt{1-\omega_3^2}\sin\theta,\, 0 \leq \theta \leq 2\pi.$$

We obtain

$$I_k = k^{\frac{1}{2}} \int_{-1}^{+1} \int_0^{2\pi} a\big(\sqrt{1-\omega_3^2}\cos\theta, \sqrt{1-\omega_3^2}\sin\theta, \omega_3\big)(1-\omega_3^2)^{k+\frac{1}{2}}\, d\omega_3\, d\theta.$$

Set $\omega_3 = \frac{\rho}{\sqrt{j}}$. It follows that

$$I_k = \int_{-\sqrt{k}}^{\sqrt{k}} \int_0^{2\pi} a\Big(\sqrt{1-\frac{\rho^2}{k}}\cos\theta, \sqrt{1-\frac{\rho^2}{k}}\sin\theta, \frac{\rho}{\sqrt{k}}\Big)(1-\frac{\rho^2}{k})^{k+\frac{1}{2}}\, d\rho\, d\theta.$$

We have $\lim_{k\to+\infty}(1-\frac{\rho^2}{k})^{k+\frac{1}{2}} = e^{-\rho^2}$ and $(1-\frac{\rho^2}{k})^{k+\frac{1}{2}} \leq e^{-\rho^2}$. First of all, if $a = 1$, the dominated convergence theorem shows that I_k, which is the square of the L^2 norm of Y^k, tends to a limit as k tends to $+\infty$. Thus, (Y^k) is bounded in L^2. Next, this same theorem shows that for $a \in C^0(\mathbf{S}^2)$ we have

$$\lim_{k\to+\infty} I_k = \Big(\int_{\mathbf{R}} e^{-\rho^2}\, d\rho\Big) \int_0^{2\pi} a(\cos\theta, \sin\theta, 0)\, d\theta = \sqrt{\pi} \int_0^{2\pi} a(\cos\theta, \sin\theta, 0)\, d\theta.$$

2. The polar coordinates on $\mathbf{S}^2$ are written as

$$\omega_1 - \sin\theta\cos\varphi, \quad \omega_2 = \sin\theta\sin\varphi, \quad \omega_3 = \cos\theta, \quad 0 < \theta < \pi, \quad 0 < \varphi < 2\pi,$$

with $d\omega = \sin\theta\, d\theta\, d\varphi$. Then, if $k \neq \ell$ we have

$$\begin{aligned}(Y^k, Y^\ell)_{L^2(\mathbf{S}^2)} &= k^{\frac{1}{4}}\ell^{\frac{1}{4}} \int_{\mathbf{S}^2} (\omega_1 + i\omega_2)^k (\omega_1 - i\omega_2)^\ell\, d\omega \\ &= k^{\frac{1}{4}}\ell^{\frac{1}{4}} \Big(\int_0^{2\pi} e^{i(k-\ell)\varphi}\, d\varphi\Big)\Big(\int_0^{\pi} (\sin\theta)^{k+\ell+1}\, d\theta\Big) = 0.\end{aligned}$$

□

Zonal Spherical Harmonics

These are the spherical harmonics of lowest degree, i.e. $m = 0$ in (8.3.7). These are the ones that are invariant under rotations in φ, i.e. $\frac{\partial}{\partial\varphi} Y_0^k = 0$. As we will see, they are concentrated around their pole. Let us first give their definition.

Let us fix a point ω_0 (for example $\omega_0 = (0, 0, 1)$, the north pole). Consider the linear map

$$\mathcal{H}_k \to \mathbf{C}, \quad Y \mapsto Y(\omega_0). \tag{8.3.8}$$

Since $\mathcal{H}_k$, equipped with the scalar product of $L^2(\mathbf{S}^2)$, is a Hilbert space, there exists a unique element $Z^k_{\omega_0} \in \mathcal{H}_k$ such that

$$Y(\omega_0) = \int_{\mathbf{S}^2} Y(\omega) Z^k_{\omega_0}(\omega)\, d\omega, \quad \forall Y \in \mathcal{H}_k. \tag{8.3.9}$$

The function $Z^k_{\omega_0}$ is called the spherical harmonic of degree k with pole at ω_0.

Lemma 8.3.13 *1. If $\{Y^1, Y^2, \dots, Y^{d_k}\}$ is an orthonormal basis of $\mathcal{H}_k$ then*

$$Z^k_{\omega_0}(\omega) = \sum_{\ell=0}^{d_k} \overline{Y^\ell(\omega_0)} Y^\ell(\omega).$$

2. $Z^k_{\omega_0}$ is real-valued and $Z^k_{\omega_0}(\omega) = Z^k_{\omega}(\omega_0)$.
3. If R is a rotation then $Z^k_{R\,\omega_0}(R\,\omega) = Z^k_{\omega_0}(\omega)$.

Proof 1. Since $\{Y^1, Y^2, \dots, Y^{d_k}\}$ is an orthonormal basis, we have

$$Z^k_{\omega_0} = \sum_{\ell=0}^{d_k} \left(Z^k_{\omega_0}, Y^\ell\right)_{L^2(\mathbf{S}^2)} Y^\ell.$$

Using (8.3.9) we have

$$\left(Z^k_{\omega_0}, Y^\ell\right)_{L^2(\mathbf{S}^2)} = \int_{\mathbf{S}^2} Z^k_{\omega_0}(\omega) \overline{Y^\ell(\omega)}\, d\omega = \overline{Y^\ell(\omega_0)},$$

which proves the first point.
2. We can choose the orthonormal basis to be real (the real or complex dimension of $\mathcal{H}_k$ is the same). Then 2. is an immediate consequence of 1.
3. Setting $\omega' = R\,\omega$ and using (8.3.9) we get

$$\int_{\mathbf{S}^2} Z^k_{R\,\omega_0}(R\,\omega) Y(\omega)\, d\omega = \int_{\mathbf{S}^2} Z^k_{R\,\omega_0}(\omega') Y(R^{-1}\,\omega')\, d\omega' = Y(R^{-1}\,R\,\omega_0) = Y(\omega_0).$$

The uniqueness of the representation of continuous linear forms shows that we have $Z^k_{R\,\omega_0}(R\,\omega) = Z^k_{\omega_0}(\omega)$, which proves 3. □

Corollary 8.3.14 *1. $Z^k_{\omega_0}(\omega_0) = d_k\, |\mathbf{S}^2|^{-1}$ where d_k is the dimension of $\mathcal{H}_k$ and $|\mathbf{S}^2|$ is the measure of the sphere $\mathbf{S}^2$.*
2. $\sum_{\ell=1}^{d_k} |Y^\ell(\omega)|^2 = d_k\, |\mathbf{S}^2|^{-1}$ for all $\omega \in \mathbf{S}^2$, independently of the chosen orthonormal basis $\{Y^1, Y^2, \dots, Y^{d_k}\}$.
3. $|Z^k_{\omega_0}(\omega)| \le d_k\, |\mathbf{S}^2|^{-1}, \quad \forall \omega, \omega_0 \in \mathbf{S}^2$.

Proof Let $\omega_1, \omega_2 \in \mathbf{S}^2$. There exists a rotation R such that $R\,\omega_1 = \omega_2$. By the previous lemma we have $Z^k_{\omega_2}(\omega_2) = Z^k_{\omega_1}(\omega_1)$. Consequently, $Z^k_{\omega}(\omega)$ is constant on $\mathbf{S}^2$. According to point 1. of Lemma 8.3.13 this constant c must be equal to $\sum_{\ell=1}^{d_k} |Y^\ell(\omega)|^2$ where $(Y^1, \ldots, Y^{d_k})$ is an orthonormal basis of $\mathcal{H}_k$. But then,

$$d_k = \sum_{\ell=1}^{d_k} \int_{\mathbf{S}^2} |Y^\ell(\omega)|^2\, d\omega = \int_{\mathbf{S}^2} c\, d\omega = c\,|\mathbf{S}^2|,$$

which proves points 1. and 2. To prove 3. we note that it follows from the definition of zonal functions that

$$Z^k_{\omega_0}(\omega) = \int_{\mathbf{S}^2} Z^k_{\omega_0}(\omega') Z^k_{\omega}(\omega')\, d\omega'. \tag{8.3.10}$$

On the other hand, if $(Y^1, \ldots, Y^{d_k})$ is an orthonormal basis of $\mathcal{H}_k$, then according to Lemma 8.3.13,

$$\|Z^k_\theta\|^2_{L^2(\mathbf{S}^2)} = \int_{\mathbf{S}^2} |Z^k_\theta(\omega')|^2\, d\omega' = \sum_{\ell=1}^{d_k} |Y^\ell(\theta)|^2 = a_k |\mathbf{S}^2|^{-1}.$$

According to (8.3.10), the Cauchy-Schwarz inequality, and the last equality, we obtain

$$|Z^k_{\omega_0}(\omega)| \le \|Z^k_{\omega_0}\|^2_{L^2(\mathbf{S}^2)} \|Z^k_{\omega}\|^2_{L^2(\mathbf{S}^2)} = \sqrt{d_k \Omega_2^{-1}} \sqrt{d_k |\mathbf{S}^2|^{-1}} = d_k |\mathbf{S}^2|^{-1}.$$

□

8.4 Spectral Theory of the Laplacian on the d-Dimensional Torus

The d-dimensional torus is defined here as $\mathbf{T}^d = \mathbf{R}^d/(2\pi\mathbf{Z})^d$.

Functions f from $\mathbf{T}^d$ to $\mathbf{C}$ can be viewed as a functions from $\mathbf{R}^d$ to $\mathbf{C}$ which are periodic of period 2π with respect to each variable that is,

$$f(x_1 + 2\pi, \ldots, x_d + 2\pi) = f(x_1, \ldots, x_d).$$

The Laplacian acting on a C^2 function f on the torus is given by the usual formula: $\Delta f = \sum_{j=1}^{d} \frac{\partial^2 f}{\partial x_j^2}$.

The scalar product of $L^2(\mathbf{T}^d)$ is given by

$$\langle f, g\rangle = \frac{1}{(2\pi)^d} \int_0^{2\pi} \cdots \int_0^{2\pi} f(x)\overline{g(x)}\, dx.$$

8.4.1 The Spectral Theory

For $\ell \in \mathbf{Z}^d$ the function $\varphi_\ell(x) = e^{i\langle \ell, x\rangle}$ for $x \in \mathbf{T}^d$ is an eigenfunction of $-\Delta$, with eigenvalue $|\ell|^2$. Here $\langle \ell, y\rangle = \sum_{j=1}^d \ell_j x_j$ and $|\ell|^2 = \sum_{j=1}^d \ell_j^2$.

Lemma 8.4.1 *The set of functions $\{\varphi_\ell : \ell \in \mathbf{Z}^d\}$ is a Hilbert basis of $L^2(\mathbf{T}^d)$.*

Remark 8.4.2 1. Recall that this means that the vectors of the family are independant and that the set of finite linear combinations of these vectors is dense in $L^2(\mathbf{T}^d)$.
2. This results leads to the theory of multidimensional Fourier series.

Proof First of all let us check that these functions are linearly independent. We proceed by induction. Assume $\varphi_{\ell_1}, \dots, \varphi_{\ell_k}$ are linearly independent and suppose that $\sum_{j=1}^{k+1} c_j \varphi_{\ell_j} = 0$. Multiplying both members by $\varphi_{-\ell_{k+1}}(x) = e^{-i\langle \ell_{k+1}, x\rangle}$ we obtain

$$c_{k+1} + \sum_{j=1}^{k} c_j \varphi_{\ell_j - \ell_{k+1}} = 0. \tag{8.4.1}$$

Applying $-\Delta$ to both members we get

$$0 = \sum_{j=1}^{k} c_j |\ell_j - \ell_{k+1}|^2 \varphi_{\ell_j - \ell_{k+1}} = e^{-i\langle \ell_{k+1}, x\rangle} \sum_{j=1}^{k} c_j |\ell_j - \ell_{k+1}|^2 \varphi_{\ell_j}.$$

Since $\ell_j \neq \ell_{k+1}$ for $j = 1, \dots k$ the induction hypothesis implies that $c_j = 0$, for $j = 1, \dots k$, so using (8.4.1), $c_{k+1} = 0$.

Let us show now that the set E of finite linear combinations of the φ_ℓ for $\ell \in \mathbf{Z}^d$ is dense in $L^2(\mathbf{T}^d)$. We use the Stone-Weierstrass theorem. First of all E is a subalgebra of $L^2(\mathbf{T}^d)$ since $\varphi_\ell \varphi_{\ell'} = \varphi_{\ell+\ell'}$. Let us prove that E separates points. Let $x_1, x_2 \in \mathbf{T}^d$ and assume that $x_1 \neq x_2$. Then $x_1 - x_2 \notin (2\pi\mathbf{Z})^d$. Assume now that $\varphi_\ell(x_1) = \varphi_\ell(x_2)$ for all $\ell \in \mathbf{Z}^d$. Then $e^{i\langle \ell, x_1 - x_2\rangle} = 1$ for all $\ell \in \mathbf{Z}^d$ so $\langle \ell, x_1 - x_2\rangle \in 2\pi\mathbf{Z}$ for all $\ell \in \mathbf{Z}^d$. Taking $\ell = (1, 0, \dots, 0), \dots, (0, \dots, 0, 1)$ we get $x_1 - x_2 \in (2\pi\mathbf{Z})^d$ which is a contradiction. Thus E separates points and the Stone-Weierstrass theorem ensures that E is dense in $L^2(\mathbf{T}^d)$. □

8.4.2 A Zygmund Inequality

We work in this paragraph on the 2-dimensional torus $\mathbf{T}^2 = \mathbf{R}^2/(2\pi\mathbf{Z})^2$. An eigenfunction of the Laplacian $-\Delta$ corresponding to the eigenvalue $\lambda \geq 0$ can be written as

$$e_\lambda(x) = \sum_{k \in \mathbf{Z}^2 : |k|^2 = \lambda} c_k e^{i\langle k, x\rangle},$$

where $k = (k_1, k_2) \in \mathbf{Z}^2$, $\langle k, x\rangle = k_1x_1 + k_2x_2$ and $|k|^2 = k_1^2 + k_2^2$. The following result has been proved by Antoni Zygmund in 1974.

Theorem 8.4.3 *For any $\lambda \geq 1$ we have*

$$\|e_\lambda\|_{L^4(\mathbf{T}^2)} \leq 3^{\frac{1}{4}} \|e_\lambda\|_{L^2(\mathbf{T}^2)}. \tag{8.4.2}$$

The important point here is that the constant appearing in the right hand side of the above inequality is independent of λ. (See the Comments).

Proof First of all using Parseval equality one can write

$$\|e_\lambda\|^2_{L^2(\mathbf{T}^2)} = \sum_{k\in\mathbf{Z}^2:|k|^2=\lambda} |c_k|^2. \tag{8.4.3}$$

Now, setting $A_\lambda = \{(k, m) \in \mathbf{Z}^2 \times \mathbf{Z}^2 : |k|^2 = \lambda,\ |m|^2 = \lambda\}$ we have

$$|e_\lambda(x)|^2 = \sum_{(k,m)\in A_\lambda} c_m\overline{c_k}e^{i\langle m-k),x\rangle}.$$

For $j \in \mathbf{Z}^2$ we set

$$A_{\lambda,j} = \{(k, m) \in A_\lambda : m - k = j\}.$$

Notice that if $|j| > 2\sqrt{\lambda}$ we have $A_{\lambda,j} = \emptyset$ because we would have

$$\sqrt{\lambda} = |m| = |j + k| \geq |j| - |k| > 2\sqrt{\lambda} - \sqrt{\lambda} = \sqrt{\lambda}.$$

Therefore we may assume that

$$|j| \leq 2\sqrt{\lambda}. \tag{8.4.4}$$

On the other hand if $j \neq j'$ we have $A_{\lambda,j} \cap A_{\lambda,j'} = \emptyset$. Moreover, $A_\lambda = \cup_{j\in\mathbf{Z}^2} A_{\lambda,j}$. Then,

$$|e_\lambda(x)|^2 = \sum_{j\in\mathbf{Z}^2}\Big(\sum_{(k,m)\in A_{\lambda,j}} c_m\overline{c_k}\Big)e^{i\langle j,x\rangle},$$

so that

$$\|e_\lambda\|^4_{L^4(\mathbf{T}^2)} = \||e_\lambda|^2\|^2_{L^2(\mathbf{T}^2)} = \sum_{j\in\mathbf{Z}^2}\Bigg|\sum_{(k,m)\in A_{\lambda,j}} c_m\overline{c_k}\Bigg|^2.$$

Since in $A_{\lambda,0}$ we have $m = k$ we get

$$\|e_\lambda\|^4_{L^4(\mathbf{T}^2)} = \Big(\sum_{k\in\mathbf{Z}^2:|k|^2=\lambda} |c_k|^2\Big)^2 + \sum_{j\in\mathbf{Z}^2\setminus 0}\Bigg|\sum_{(k,m)\in A_{\lambda,j}} c_m\overline{c_k}\Bigg|^2. \tag{8.4.5}$$

The key point is that, for fixed $j \neq 0$, the number of elements of $A_{\lambda,j}$, denoted by $\text{card}(A_{\lambda,j})$, is less or equal to 2. Indeed let $(k, m) \in A_{\lambda,j}$. We have

$$|k|^2 = \lambda \text{ and } \lambda = |m|^2 = |k + j|^2 = |k|^2 + 2\langle k, j\rangle + |j|^2 = \lambda + 2\langle k, j\rangle + |j|^2.$$

It follows that $2\langle k, j\rangle + |j|^2 = 0$. Set $\omega_j = \frac{j}{|j|} \in \mathbf{S}^2$ then $2\langle k, \omega_j\rangle = -|j|$. Therefore by (8.4.4), $\cos(k, \omega_j) = -\frac{|j|}{2\sqrt{\lambda}} := \alpha_j \in [-1, 0[$. There are two angles (possibly equal) $\theta \in]\frac{\pi}{2}, \frac{3\pi}{2}[$ such that $\cos\theta = \alpha_j$. So k belongs to the intersection of the circle $|k|^2 = \lambda$ with the two half straight lines making an angle θ with ω_j. There are at most two such k therefore two pairs (k, m) since $m = k + j$.

Applying the Cauchy-Schwarz inequality we obtain for $j \neq 0$,

$$\Big|\sum_{(k,m)\in A_{\lambda,j}} c_m\overline{c_k}\Big|^2 \leq \text{card}(A_{\lambda,j}) \sum_{(k,m)\in A_{\lambda,j}} |c_k|^2|c_m|^2 \leq 2 \sum_{(k,m)\in A_{\lambda,j}} |c_k|^2|c_m|^2.$$

It follows that

$$\begin{aligned}\sum_{j\in\mathbf{Z}^2\setminus 0}\Big|\sum_{(k,m)\in A_{\lambda,j}} c_m\overline{c_k}\Big|^2 &\leq 2\sum_{j\in\mathbf{Z}^2\setminus 0}\sum_{(k,m)\in A_{\lambda,j}} |c_k|^2|c_m|^2,\\ &\leq 2\sum_{(k,m)\in A_\lambda} |c_k|^2|c_m|^2 \leq 2\Big(\sum_{k\in\mathbf{Z}^2:|k|^2=\lambda} |c_k|^2\Big)^2.\end{aligned}$$

We deduce from (8.4.3) and (8.4.5) that

$$\|e_\lambda\|^4_{L^4(\mathbf{T}^2)} \leq 3\|e_\lambda\|^4_{L^2(\mathbf{T}^2)},$$

which proves the desired inequality. □

Remark 8.4.4 The mutltiplicity $m(\lambda)$ of the eigenvalue λ is an unbounded function of λ. Indeed this multiplicity is, (on $\mathbf{T}^d$), the number of way to write λ as a sum of d squares of elements of $\mathbf{Z}$. As proves the Proposition below, for any $n \in \mathbf{N}$ there exists $r = r(n) \geq 1$ such that on the circle of radius r there are at least n distinct points with coordinates in $\mathbf{Z}$. This is the exact contrapositive of the fact that $\exists C > 0 : m(\lambda) \leq C, \forall\lambda \geq 1$.

Proposition 8.4.5 *For any $n \in \mathbf{N}^*$ there exists an infinite number of circles with integer containing at least n points with coordinates in $\mathbf{Z}$.*

Proof For $r > 0$ denotes by $C(0, r)$ the circle centered at the origin with radius r.

For $t \in \mathbf{R}$ if we set $x = \frac{1-t^2}{1+t^2}$, $y = \frac{2t}{1+t^2}$ then $(x, y) \in C(0, 1)$.

For $j = 1, \ldots, n$ let $t_j = \frac{p_j}{q_j}$, $p_j, q_j, \in \mathbf{Z}, q_j \neq 0$.

Set $m_j = \left(\frac{1-t_j^2}{1+t_j^2}, \frac{2t_j}{1+t_j^2}\right) \in C(0, 1)$. Then,

$$m_j = \left(\frac{q_j^2 - p_j^2}{q_j^2 + p_j^2}, \frac{2p_j q_j}{q_j^2 + p_j^2}\right) = \frac{1}{q_j^2 + p_j^2}(q_j^2 - p_j^2, 2p_j q_j).$$

Set $\lambda = \prod_{k=1}^{n}(q_k^2 + p_k^2) \in \mathbf{N}^*$ and $M_j = \lambda m_j$. Then $M_j = (x_j, y_j) \in C(0, \lambda)$ and $x_j, y_j \in \mathbf{Z}$, for $j = 1, \dots, n$. □

8.5 Comments

The analogue of Theorem 8.4.3 on the torus $\mathbf{T}^d$, for $d \geq 3$ is still an open problem. The best result avaiable at the moment is the following estimate proved by J. Bourgain in 2013.

For any $\varepsilon > 0$ there exists $C_\varepsilon > 0$ such that for any e_λ satisfying $-\Delta e_\lambda = \lambda e_\lambda$ on the torus $\mathbf{T}^d$, with $d \geq 3$, we have

$$\|e_\lambda\|_{L^p(\mathbf{T}^d)} \leq C_\varepsilon \lambda^\varepsilon \|e_\lambda\|_{L^2(\mathbf{T}^d)}, \quad 2 \leq p \leq \frac{2d}{d-1},$$

which is worse than (8.4.2) when $\lambda \to +\infty$.

Chapter 9
The Hydrogen Atom

9.1 Prerequisites

Distributions. Sobolev spaces. L^p spaces. Fundamental solutions. Differential equations. Fourier transform. Holomorphic functions, residues.

9.2 Context

The hydrogen atom is composed of an electron and a proton (the nucleus) which interact according to Coulomb's law. According to the Bohr model, this electron orbits the nucleus in specific shells or orbits, each shell corresponding to an energy level. The lowest energy level is called the "ground state" while the others are called "excited states".

The description of the physical properties of this atom and the explanation of the Bohr model were the subject of many studies and controversies at the time when quantum mechanics was beginning to be developed. It is, in particular, to Schrödinger that we owe the theory that we will describe in what follows.

The nucleus of this atom (assumed to be at the origin) is supposed to constitute the center of gravity of the system where it is immobile (thus without kinetic energy). The kinetic energy of the atom is thus reduced to that of the electron, to which is associated the operator $-\frac{h^2}{2m}\Delta$ where Δ is the Laplacian. The potential energy, of purely electrostatic nature, is written as a function of the distance r from the electron to the nucleus, $V = -\frac{1}{4\pi\varepsilon_0}\frac{e^2}{r}$ (Coulomb's law). Here $h, m, 4\pi\varepsilon_0, e$ are physical constants that we will henceforth assume all equal to 1.

C. Zuily, *Selected Topics in Partial Differential Equations*, Universitext,
https://doi.org/10.1007/978-3-032-24082-8_9

9.3 The Problem

The purpose of this chapter is to mathematically account for these notions as well as the theory underlying them. In particular, we will determine the spectrum of the Schrödinger operator $-\Delta - \frac{1}{|x|}$ (see Theorem 9.5.2).

On the other hand, in the course of the proof we shall use some results of Chap. 8 on the spectrum of the Laplacian on the sphere.

9.4 Study of the Ground State

Let $\psi \in H^1(\mathbf{R}^3)$. We set

$$F(\psi) = \|\nabla_x \psi\|^2_{L^2(\mathbf{R}^3)} - \left\| \frac{\psi}{|x|^{\frac{1}{2}}} \right\|^2_{L^2(\mathbf{R}^3)}, \tag{9.4.1}$$

$$E = \inf\left\{ F(\psi) : \psi \in H^1(\mathbf{R}^3), \|\psi\|_{L^2(\mathbf{R}^3)} = 1 \right\} \geq -\infty. \tag{9.4.2}$$

In this section we shall show that $E = -\frac{1}{4}$, that the infimum is attained with the function $\widetilde{\psi} = \frac{1}{\sqrt{8\pi}} e^{-\frac{1}{2}|x|}$ and that the operator $-\Delta - \frac{1}{|x|} - E$ from $H^2(\mathbf{R}^3)$ to $L^2(\mathbf{R}^3)$ has a one-dimensional kernel, generated by $\widetilde{\psi}$. Here, $\Delta = \sum_{j=1}^3 \frac{\partial^2}{\partial x_j^2}$.

Remark 9.4.1 1. The energy of an electron in orbit is defined relative to that of an electron infinitely far from the nucleus, which by definition is equal to zero. Since the electron in orbit is more stable than an electron infinitely far from the nucleus, its energy is lower, thus negative.

2. The function ψ is called the "wave function." The quantity $F(\psi)$ is called the energy of ψ and E the "ground state" of the hydrogen atom, that is its lowest energy. Higher energies are called "excited states."

3. The fact that E is finite (and not $-\infty$) is a purely quantum fact, which is interpreted as ensuring the stability of matter.

9.4.1 *Preliminaries*

We begin by proving the Hardy-Morawetz inequality in any dimension $d \geq 3$.

Proposition 9.4.2 *Let $d \geq 3$. Then for all $u \in H^1(\mathbf{R}^d)$ we have*

$$\left(\int_{\mathbf{R}^d} \frac{|u(x)|^2}{|x|^2}\, dx \right)^{\frac{1}{2}} \leq \frac{2}{d-2} \left(\sum_{i=1}^d \int_{\mathbf{R}^d} |\frac{\partial u}{\partial x_i}(x)|^2\, dx \right)^{\frac{1}{2}}. \tag{9.4.3}$$

Proof 1. We begin with the case where $u \in C_0^\infty(\mathbf{R}^d)$.

Let $\omega \in S^{d-1}$ be fixed. Since $r^{d-3} = \frac{1}{d-2}\frac{d}{dr}r^{d-2}$, integrating by parts in the left-hand integral below and using the fact that $u \in C_0^\infty(\mathbf{R}^d)$ we obtain

$$A := \int_0^{+\infty} \frac{|u(r\omega)|^2}{r^2} r^{d-1}\, dr = -\frac{2}{d-2}\mathrm{Re} \int_0^{+\infty} u(r\omega)\big(\omega \cdot \nabla_x u\big)(r\omega) r^{d-2}\, dr, \tag{9.4.4}$$

where $\nabla_x u = (\partial_{x_i} u)_{i=1,\dots,d}$ denotes the gradient of u.

By writing $r^{d-2} = \frac{1}{r} r^{\frac{d-1}{2}} r^{\frac{d-1}{2}}$ in the right-hand integral of (9.4.4) and then using the Cauchy-Schwarz inequality we obtain

$$A \leq \frac{2}{d-2}\Big(\int_0^{+\infty} \frac{|u(r\omega)|^2}{r^2} r^{d-1}\, dr\Big)^{\frac{1}{2}} \Big(\int_0^{+\infty} |\nabla_x u(r\omega)|^2 r^{d-1}\, dr\Big)^{\frac{1}{2}},$$

from which we deduce

$$\int_0^{+\infty} \frac{|u(r\omega)|^2}{r^2} r^{d-1}\, dr \leq \Big(\frac{2}{d-2}\Big)^2 \int_0^{+\infty} |\nabla_x u(r\omega)|^2 r^{d-1}\, dr.$$

Integrating both sides over the sphere $\mathbf{S}^{d-1}$ and using polar coordinates in $\mathbf{R}^d$ we obtain the desired inequality.

2. Now let $u \in H^1(\mathbf{R}^d)$. There exists a sequence $(u_j)_{j\in\mathbf{N}} \subset C_0^\infty(\mathbf{R}^d)$ that converges to u in $H^1(\mathbf{R}^d)$. Applying (9.4.3) to the function $u_j - u_k \in C_0^\infty(\mathbf{R}^d)$, we see that the sequence $(\frac{1}{|x|}u_j)_j$ is Cauchy in $L^2(\mathbf{R}^d)$ and thus converges to v in $L^2(\mathbf{R}^d)$, hence in $\mathcal{D}'(\mathbf{R}^d)$. On the other hand, this sequence converges to $\frac{1}{|x|}u$ in $\mathcal{D}'(\mathbf{R}^d)$. Indeed, let $\varphi \in C_0^\infty(\mathbf{R}^d)$. We have

$$\langle \frac{1}{|x|}u_j, \varphi\rangle = \big(u_j, \frac{1}{|x|}\overline{\varphi}\big)_{L^2} \to \big(u, \frac{1}{|x|}\overline{\varphi}\big)_{L^2} = \langle \frac{1}{|x|}u, \varphi\rangle,$$

since $u_j \to u$ in $L^2(\mathbf{R}^d)$ and $\frac{1}{|x|}\overline{\varphi} \in L^2(\mathbf{R}^d)$ when $d \geq 3$. We deduce that we have $v = \frac{1}{|x|}u \in L^2(\mathbf{R}^d)$. Applying the inequality (9.4.3) to the function u_j and passing to the limit as $j \to +\infty$ we obtain the inequality (9.4.3) for u. □

We easily obtain the following corollary.

Corollary 9.4.3 *If $\psi \in H^1(\mathbf{R}^d)$, then $\frac{\psi}{|x|} \in L^2$, $\frac{\psi}{|x|^{\frac{1}{2}}} \in L^2$, as well as the inequality*

$$\int_{\mathbf{R}^d} \frac{|\psi(x)|^2}{|x|}\, dx \leq \frac{2}{d-2}\|\psi\|_{L^2}\|\nabla_x \psi\|_{L^2}. \tag{9.4.5}$$

Proof The first assertion follows from inequality (9.4.3). For the second, we use the Cauchy-Schwarz inequality and inequality (9.4.3). We obtain

$$\begin{aligned}\int_{\mathbf{R}^d} \frac{|\psi(x)|^2}{|x|}\,dx &\le \left(\int_{\mathbf{R}^d} |\psi(x)|^2\,dx\right)^{\frac{1}{2}} \left(\int_{\mathbf{R}^d} \frac{|\psi(x)|^2}{|x|^2}\,dx\right)^{\frac{1}{2}}, \\ &\le \frac{2}{d-2}\|\psi\|_{L^2}\|\nabla_x\psi\|_{L^2}.\end{aligned} \tag{9.4.6}$$

□

Remark 9.4.4 Inequality (9.4.5) already shows that the infimum of the function F defined in (9.4.1) is finite. Indeed let $d = 3$ and set $I^2 = \int_{\mathbf{R}^3} |\nabla_x\psi(x)|^2\,dx$. If $\|\psi\|_{L^2} = 1$, inequality (9.4.5) shows that

$$\int_{\mathbf{R}^3} \frac{|\psi(x)|^2}{|x|}\,dx \le 2I.$$

We deduce that $F(\psi) = I^2 - \int_{\mathbf{R}^3} \frac{|\psi(x)|^2}{|x|}\,dx \ge I^2 - 2I$. The function on $(0, +\infty)$ defined by $g(t) = t^2 - 2t$ admits a minimum equal to -1. Thus $F(\psi) \ge -1$.

In what follows, we will assume $d = 3$ and we will denote $X = X(\mathbf{R}^3)$ if X denotes one of the spaces $L^2, H^k, C_0^\infty, \mathcal{D}'$.

9.4.2 *Existence of a Minimizer*

In this section, we show the existence of a minimizer for the function F introduced in (9.4.1). Let us first define this term.

Definition 9.4.5 A vector $u \in H^1$ will be called a minimizer for the function F if $\|u\|_{L^2} = 1$ and $F(u) = E$.

We begin with the following result.

Lemma 9.4.6 *A minimizer, if it exists, satisfies* $F(u) = E < 0$.

Proof Let $\chi \in C_0^\infty$ such that $\|\chi\|_{L^2} = 1$. Set $\psi(x) = \varepsilon^{\frac{3}{2}}\chi(\varepsilon x)$, $\varepsilon > 0$. Then $\|\psi\|_{L^2} = 1$. Setting, $\varepsilon x = y$ in the integrals defining $F(\psi)$ we obtain

$$F(\psi) = \varepsilon\left(\varepsilon \int_{\mathbf{R}^3} |\nabla_x\chi(y)|^2\,dy - \int_{\mathbf{R}^3} \frac{|\chi(y)|^2}{|y|}\,dy\right) < 0$$

if ε is small enough, since $\chi \not\equiv 0$. Thus E, which is the infimum, is strictly negative. □

Proposition 9.4.7 *The function F admits a minimizer.*

Proof It will require several steps.

Step 1.

$$\exists (\psi_n)_{n\in\mathbf{N}} \subset H^1,\ \|\psi_n\|_{L^2} = 1 \text{ such that } \lim_{n\to+\infty} F(\psi_n) = E.$$

This follows from the definition of an infimum. Indeed, in the case where $E > -\infty$, for all $n \geq 1$, $E + \frac{1}{n}$ is no longer an infimum, that is, there exists $\psi_n \in H^1$ with $\|\psi_n\|_{L^2} = 1$ such that $E \leq F(\psi_n) \leq E + \frac{1}{n}$. If $E = -\infty$ then for all $n > 0$ there exists ψ_n as above such that $F(\psi_n) \leq -n$. In all cases $F(\psi_n) \to E$.

Step 2. The sequence (ψ_n) is uniformly bounded in H^1.

Indeed, first we have $\|\psi_n\|_{L^2} = 1$. Next, from Step 1 and the fact that $E < 0$, we have $F(\psi_n) < 0$ for n large enough. Using Corollary 9.4.3 we obtain

$$\|\nabla_x \psi_n\|_{L^2}^2 \leq \left\| \frac{\psi_n}{|x|^{\frac{1}{2}}} \right\|^2 \leq C \|\psi_n\|_{L^2} \|\nabla_x \psi_n\|_{L^2} = C \|\nabla_x \psi_n\|_{L^2}.$$

Thus $\|\nabla_x \psi_n\|_{L^2} \leq C$.

Step 3. There then exists a subsequence $(\psi_{\sigma(n)})$ which converges weakly in L^2 to $\widetilde{\psi}$ and $(\nabla_x \psi_{\sigma(n)})$ converges weakly in L^2 to $\nabla_x \widetilde{\psi}$.

Indeed, since $\|\psi_n\|_{L^2} = 1$ there exists a subsequence $(\psi_{\sigma_1(n)})$ which converges weakly in L^2 to $\widetilde{\psi}$. Thus $(\nabla_x \psi_{\sigma_1(n)})$ converges to $\nabla_x \widetilde{\psi}$ in $\mathcal{D}'$. Since $\|\nabla_x \psi_{\sigma_1(n)}\|_{L^2}$ is uniformly bounded (by C) there exists a subsequence $(\psi_{\sigma(n)})$ such that $(\nabla_x \psi_{\sigma(n)})$ converges weakly in L^2 (thus in $\mathcal{D}'$) to v. Then $v = \nabla_x \widetilde{\psi}$.

We then have

$$\|\widetilde{\psi}\|_{L^2} \leq 1, \quad \|\nabla_x \widetilde{\psi}\|_{L^2}^2 \leq \liminf_{n\to+\infty} \|\nabla_x \psi_{\sigma(n)}\|_{L^2}^2. \tag{9.4.7}$$

Indeed, for $\varphi \in L^2$ we have $|\left(\psi_{\sigma(n)}, \varphi\right)_{L^2}| \leq \|\varphi\|_{L^2}$. The left-hand side converges to $|(\widetilde{\psi}, \varphi)_{L^2}|$, so we obtain $\|\widetilde{\psi}\|_{L^2} = \sup\left\{|(\widetilde{\psi}, \varphi)_{L^2}| : \|\varphi\|_{L^2} = 1\right\} \leq 1$. The same argument can be applied to $\nabla_x \psi_{\sigma(n)}$.

Step 4. We have $\lim_{n\to+\infty} \left\| \frac{\psi_{\sigma(n)} - \widetilde{\psi}}{|x|^{\frac{1}{2}}} \right\|_{L^2} = 0$.

Indeed, let us set $\theta_n = \psi_{\sigma(n)} - \widetilde{\psi}$. Let $\varepsilon > 0$. Fix $R > 0$ such that $\frac{4}{R} \leq \frac{\varepsilon}{2}$. Then,

$$I_n =: \int_{\{|x|>R\}} \frac{|\theta_n(x)|^2}{|x|}\, dx \leq \frac{2}{R}(\|\psi_{\sigma(n)}\|_{L^2}^2 + \|\widetilde{\psi}\|_{L^2}^2) \leq \frac{4}{R} \leq \frac{\varepsilon}{2}.$$

By the Hardy-Morawetz inequality and the fact that (ψ_n) is bounded in H^1 by K we have

$$J_n = \int_{\{|x|\le R\}} \frac{|\theta_n(x)|^2}{|x|}\,dx \le \|\theta_n\|_{L^2(|x|\le R)} \left(\int \frac{|\theta_n(x)|^2}{|x|^2}\,dx\right)^{\frac{1}{2}},$$
$$\le C\|\theta_n\|_{L^2(|x|\le R)}\|\nabla_x\theta_n\|_{L^2} \le C\|\theta_n\|_{L^2(|x|\le R)}(K+\|\widetilde{\psi}\|_{H^1}).$$

Let $\chi \in C_0^\infty$ such that $\chi(x) = 1$ if $|x| \le 1$, $\chi(x) = 0$ if $|x| \ge 2$ and $0 \le \chi \le 1$. Then,

$$\|\theta_n\|_{L^2(|x|\le R)} \le \left\|\chi(\frac{x}{R})\theta_n\right\|_{L^2}.$$

On the other hand, we know from Step 3. that (θ_n) converges weakly to zero in H^1, which implies that $(\chi(\frac{x}{R})\theta_n)$ also converges weakly to zero in H^1. Since the map $u \mapsto u$ from $\{u \in H^1 : \operatorname{supp} u \subset \{|x| \le 2R\}\}$ into L^2 is compact, we deduce that $(\chi(\frac{x}{R})\theta_n)$ converges strongly (that is, in norm) to zero in L^2. Consequently, there exists $N \in \mathbf{N}$ such that for $n \ge N$, $J_n \le C\|\theta_n\|_{L^2(|x|\le R)}(K+\|\widetilde{\psi}\|_{H^1}) \le \frac{\varepsilon}{2}$. Therefore, $I_n + J_n \le \varepsilon$ for $n \ge N$.

Step 5. We have $F(\widetilde{\psi}) \le E$. Indeed, using Steps 3. and 4. we can write

$$F(\widetilde{\psi}) = \|\nabla_x\widetilde{\psi}\|_{L^2}^2 - \int \frac{|\widetilde{\psi}(x)|^2}{|x|}\,dx$$
$$\le \liminf_{n\to+\infty} \|\nabla_x\psi_{\sigma(n)}\|_{L^2}^2 - \lim_{n\to+\infty}\int \frac{|\psi_{\sigma(n)}|^2}{|x|}\,dx$$
$$\le \liminf_{n\to+\infty}\left(\|\nabla_x\psi_{\sigma(n)}\|_{L^2}^2 - \int \frac{|\psi_{\sigma(n)}|^2}{|x|}\,dx\right) = \liminf_{n\to+\infty} F(\psi_{\sigma(n)}) = E.$$

Step 6. We have $F(\widetilde{\psi}) \ge E\,\|\widetilde{\psi}\|_{L^2}^2$. Indeed, since $\theta = \frac{\widetilde{\psi}}{\|\widetilde{\psi}\|_{L^2}}$ belongs to H^1 and has norm equal to one, we have $F(\theta) \ge E$.

Step 7. $\widetilde{\psi}$ is a minimizer.

We deduce from Steps 5. and 6. that $E\|\widetilde{\psi}\|_{L^2}^2 \le F(\widetilde{\psi}) \le E$, so that $E(1-\|\widetilde{\psi}\|_{L^2}^2) \ge 0$. Since $E < 0$ we have $\|\widetilde{\psi}\|_{L^2}^2 \ge 1$. Using (9.4.7) we deduce that $\|\widetilde{\psi}\|_{L^2}^2 = 1$ and from the above inequality that $F(\widetilde{\psi}) = E$.

□

Lemma 9.4.8 *For $j = 1, 2$ let $E_j = \inf\{F(\psi) : \psi \in H^j, \|\psi\|_{L^2} = 1\}$. Then, we have $E_1 = E_2$.*

Proof Since $H^2 \subset H^1$ we have $E_1 \le E_2$. Let us show the reverse inequality.

Point 1. Let $\psi \in H^1$ and $(\varphi_n) \subset H^2$ such that $(\varphi_n) \to \psi$ in H^1. Then, $\lim_{n\to+\infty} F(\varphi_n) = F(\psi)$.

Indeed, we have $\nabla_x\varphi_n \to \nabla_x\psi$ in L^2 and by Corollary 9.4.3, $\frac{\varphi_n}{|x|^{\frac{1}{2}}} \to \frac{\psi}{|x|^{\frac{1}{2}}}$ in L^2. Therefore, $F(\varphi_n) \to F(\psi)$.

Point 2. Let $\psi \in H^1$ with $\|\psi\|_{L^2} = 1$ and $(\varphi_n) \subset H^2$ such that $\varphi_n \to \psi$ in H^1. Let $\varepsilon \in (0, 1)$. There exists $n \in \mathbf{N}$ such that $\|\varphi_n\|_{L^2} \ge \|\psi\|_{L^2} - \varepsilon = 1 - \varepsilon > 0$ and from Point 1 above, $F(\varphi_n) \le F(\psi) + \varepsilon$. We then write

$$E_2 \le F\left(\frac{\varphi_n}{\|\varphi_n\|_{L^2}}\right) \le \frac{1}{(1-\varepsilon)^2}(F(\psi)+\varepsilon).$$

Taking the infimum over all $\psi \in H^1$ such that $\|\psi\|_{L^2} = 1$ we obtain

$$E_2 \le \frac{1}{(1-\varepsilon)^2}(E_1+\varepsilon).$$

Since this is true for all $\varepsilon > 0$ we obtain $E_2 \le E_1$. □

9.4.3 Properties of the Minimizers

Proposition 9.4.9 *A minimizer $\widetilde{\psi}$ is an eigenfunction of the operator $\mathcal{H} = -\Delta - \frac{1}{|x|}$ corresponding to the eigenvalue E, that is in the sense of distributions we have*

$$-\Delta\widetilde{\psi} - \frac{\widetilde{\psi}}{|x|} = E\widetilde{\psi}.$$

Here, we have denoted $\Delta = \sum_{j=1}^{3} \frac{\partial^2}{\partial x_j^2}$ as the Laplacian.

Proof Let $\widetilde{\psi}$ be a minimizer. Let $\varepsilon > 0$ and $\varphi \in C_0^\infty$. Set $f = \widetilde{\psi} \pm \varepsilon\varphi$. Then,

$$\mathrm{Re}\left\langle -\Delta\widetilde{\psi} - \frac{\widetilde{\psi}}{|x|} - E\widetilde{\psi}, \overline{\varphi}\right\rangle = 0,$$

where $\langle\cdot,\cdot\rangle$ denotes the pairing between $\mathcal{D}'$ and C_0^∞.

Indeed, set $G(f) = F(f) - E\|f\|_{L^2}^2 \ge 0$. Since $G(\widetilde{\psi}) = 0$, according to Step 7 in the proof of Proposition 9.4.7, we obtain

$$\pm 2\varepsilon \mathrm{Re}\left(\int \nabla_x\widetilde{\psi}(x)\cdot\overline{\nabla_x\varphi(x)}\,dx - \int \frac{\widetilde{\psi}(x)\overline{\varphi(x)}}{|x|}\,dx - E\int \widetilde{\psi}(x)\overline{\varphi(x)}\,dx\right) \ge -\varepsilon^2 G(\varphi).$$

Dividing both sides by $2\varepsilon > 0$ and letting ε tend to zero, we obtain

$$\mathrm{Re}\left(\int \nabla_x\widetilde{\psi}(x)\cdot\overline{\nabla_x\varphi(x)}\,dx - \int \frac{\widetilde{\psi}(x)\overline{\varphi(x)}}{|x|}\,dx - E\int \widetilde{\psi}(x)\overline{\varphi(x)}\,dx\right) = 0,$$

which implies that

$$\mathrm{Re}\left\langle -\Delta\widetilde{\psi} - \frac{\widetilde{\psi}}{|x|} - E\widetilde{\psi}, \overline{\varphi}\right\rangle = 0,$$

in the sense of the duality between $\mathcal{D}'$ and C_0^∞.

By replacing φ with $i\varphi$, we deduce that $\mathcal{H}\widetilde{\psi} = E\widetilde{\psi}$ in the sense of disributions. □

Remark 9.4.10 The operator $-\Delta - \frac{1}{|x|}$ is called the "Schrödinger operator".

Proposition 9.4.11 *1. Let $\widetilde{\psi}$ be a minimizer. Then $\widetilde{\psi} \in D = \left\{u \in H^2 : \frac{u}{|x|} \in L^2\right\}$.*

2. Let Ker $(\mathcal{H} - E) = \{u \in D : (\mathcal{H} - E)u = 0\}$. If $u \in$ Ker $(\mathcal{H} - E), u \not\equiv 0$ then $\frac{u}{\|u\|_{L^2}}$ is a minimizer.

Proof 1. Since $\widetilde{\psi} \in H^1$, Corollary 9.4.3 shows that $\frac{\widetilde{\psi}}{|x|} \in L^2$. We deduce from the equation satisfied by $\widetilde{\psi}$ that $\Delta\widetilde{\psi} \in L^2$, which implies that $\widetilde{\psi} \in H^2$. Therefore, $\widetilde{\psi} \in D$.

2. If $u \not\equiv 0$, there exists $x_0 \in \mathbf{R}^3$ such that $u(x_0) \neq 0$. Since $H^2(\mathbf{R}^3) \subset C^0$, u is continuous; so there exists a neighborhood of x_0 in which u does not vanish. Consequently, $\|u\|_{L^2} \neq 0$. Moreover, since $u \in D$ we have

$$E\|u\|_{L^2}^2 = \Big(-\Delta u - \frac{u}{|x|}, u\Big)_{L^2} = \|\nabla_x u\|_{L^2}^2 - \left\|\frac{u}{|x|^{\frac{1}{2}}}\right\|_{L^2}^2 = F(u). \tag{9.4.8}$$

Therefore, $\frac{u}{\|u\|_{L^2}}$ is a minimizer.

Notice that from Corollary 9.4.3 we know that $D = H^2$. □

Proposition 9.4.12 *Let $\widetilde{\psi}$ be a minimizer. Then, $\widetilde{\psi}(x) \neq 0$ for all $x \in \mathbf{R}^3$.*

Proof Point 1. We show that if $u \in H^1$ is real-valued then

$$|u| \in H^1 \quad \text{and} \quad \nabla_x|u| = 1_{\{x:u(x)\neq 0\}}\frac{u}{|u|}\nabla_x u.$$

Indeed, we clearly have $|u| \in L^2$. Next, for $\varepsilon > 0$ set $u_\varepsilon = \sqrt{u^2+\varepsilon^2} - \varepsilon$. It is easy to see that $\varepsilon \le \sqrt{u^2+\varepsilon^2} \le |u| + \varepsilon$, so that $0 \le u_\varepsilon \le |u|$ and for almost every $x \in \mathbf{R}^3$, $\lim_{\varepsilon\to 0} u_\varepsilon(x) = |u(x)|$. By the dominated convergence theorem, we deduce that

$$\lim_{\varepsilon\to 0^+}\int u_\varepsilon(x)\varphi(x)\,dx = \int |u(x)|\varphi(x)\,dx,$$

for all $\varphi \in C_0^\infty$. This shows that $u_\varepsilon \to |u|$ in $\mathcal{D}'$. So, $\partial_j u_\varepsilon \to \partial_j|u|$ in $\mathcal{D}'$ for $j = 1, 2, 3$. On the other hand, $\partial_j u_\varepsilon = \frac{u\partial_j u}{\sqrt{u^2+\varepsilon^2}}$. We deduce that $\partial_j u_\varepsilon(x)$ converges to zero if $u(x) = 0$ and to $\frac{u(x)}{|u(x)|}\partial_j u(x)$ if $u(x) \neq 0$. Moreover, for $\varphi \in C_0^\infty$ we have $|\partial_j u_\varepsilon(x)\varphi(x)| \le |\partial_j u(x)\varphi(x)| \in L^1$. It follows from the dominated convergence theorem that

$$\partial_j u_\varepsilon \to 1_{\{u(x)\neq 0\}}\frac{u(x)}{|u(x)|}\partial_j u \quad \text{in} \quad \mathcal{D}'.$$

Consequently, $\partial_j|u| = 1_{\{u(x)\neq 0\}}\frac{u(x)}{|u(x)|}\partial_j u \in L^2$ for $j = 1, 2, 3$.

Point 2. If $\widetilde{\psi}$ is a minimizer, then $|\widetilde{\psi}|$ is also a minimizer.

Indeed, Point 1 shows that $|\widetilde{\psi}| \in H^1$ and $|\nabla_x|\widetilde{\psi}|| \leq |\nabla_x \widetilde{\psi}|$. This implies that $F(|\widetilde{\psi}|) \leq F(\widetilde{\psi})$. Since $F(\widetilde{\psi})$ is the minimum, we must have equality. Thus, $|\widetilde{\psi}|$ is a minimizer.

It follows from Proposition 9.4.11 that $|\widetilde{\psi}| \in D$ and from Proposition 9.4.9 that

$$(-\Delta + \omega^2)|\widetilde{\psi}| = \frac{|\widetilde{\psi}|}{|x|}, \quad \text{where} \quad E = -\omega^2, \omega > 0.$$

Point 3. For all $x \in \mathbf{R}^3$ we have

$$|\widetilde{\psi}(x)| = \frac{1}{4\pi} \int_{\mathbf{R}^3} \frac{e^{-\omega|x-y|}}{|x-y|} \frac{|\widetilde{\psi}(y)|}{|y|} \, dy. \tag{9.4.9}$$

Indeed, $E = \frac{1}{4\pi} \frac{e^{-\omega|x|}}{|x|}$, is a fundamental solution of the operator $-\Delta + \omega^2$, that is, $(-\Delta + \omega^2)E = \delta_0$ (see Lemma 9.6.2 in the appendix). The function E belongs to $L^1 \cap L^2$. We can therefore write

$$\begin{aligned} |\widetilde{\psi}(x)| &= (\delta_0 \star |\widetilde{\psi}|)(x) = ((-\Delta + \omega^2)E \star |\widetilde{\psi}|)(x) = (E \star (-\Delta + \omega^2)|\widetilde{\psi}|)(x), \\ &= \left(E \star \frac{|\widetilde{\psi}|}{|x|}\right)(x) = \int_{\mathbf{R}^3} \frac{e^{-\omega|x-y|}}{4\pi|x-y|} \frac{|\widetilde{\psi}(y)|}{|y|} \, dy. \end{aligned}$$

If there exists x_0 such that $|\widetilde{\psi}(x_0)| = 0$, (9.4.9) shows that $\widetilde{\psi}(y) = 0$ almost everywhere, which is impossible since $\|\widetilde{\psi}\|_{L^2} = 1$. Therefore $\widetilde{\psi}(x) \neq 0$ for all x in $\mathbf{R}^3$. □

Proposition 9.4.13 *For all $k \in \mathbf{N}$ we have $|x|^k|\widetilde{\psi}| \in L^2 \cap L^\infty$.*

Proof The statement is true for $k = 0$ since $|\widetilde{\psi}| \in H^2 \subset L^2 \cap L^\infty$. Suppose it is true at order $k \geq 0$. Then, since $|x|^{k+1} \leq C_k(|x-y|^{k+1} + |y|^{k+1})$ we have

$$|x|^{k+1}|\widetilde{\psi}| \leq C_k' \left(\int |x-y|^k e^{-\omega|x-y|} \frac{|\widetilde{\psi}(y)|}{|y|} \, dy + \int \frac{e^{-\omega|x-y|}}{|x-y|} |y|^k |\widetilde{\psi}(y)| \, dy \right).$$

Next, on the one hand,

$$|x|^k e^{-\omega|x|} \in L^1 \cap L^2, \quad \frac{|\widetilde{\psi}(y)|}{|y|} \in L^2,$$

and on the other hand, by induction,

$$\frac{e^{-\omega|x|}}{|x|} \in L^1 \cap L^2, \quad |y|^k|\widetilde{\psi}| \in L^2.$$

By Young's inequality, $L^1 \star L^2 \subset L^2$ and $L^2 \star L^2 \subset L^\infty$. Consequently, $|x|^{k+1}|\widetilde{\psi}|$ belongs to $L^2 \cap L^\infty$. □

Proposition 9.4.14 *Set $\mathcal{H} = -\Delta - \frac{1}{|x|}$. Then we have dim Ker $(\mathcal{H} - E) = 1$.*

Proof Point 1. If $u \in \text{Ker}(\mathcal{H} - E)$ then $u \equiv 0$ or $u \neq 0$ everywhere.

Indeed, let $u \in \text{Ker}(\mathcal{H} - E)$. If $u \not\equiv 0$ it follows from Proposition 9.4.11 that $\frac{u}{\|u\|_{L^2}}$ is a minimizer and from Proposition 9.4.12 that $|u(x)| > 0$ for all $x \in \mathbf{R}^3$.

Point 2. Let $u_1, u_2 \in \text{Ker}(\mathcal{H} - E)$ with $u_2 \not\equiv 0$. Let x_0 be such that $u_2(x_0) \neq 0$. Let us set $u = u_1 - \frac{u_1(x_0)}{u_2(x_0)} u_2 \in \text{Ker}(\mathcal{H} - E)$. We have $u(x_0) = 0$. It follows from Point 1. that $u \equiv 0$. So, $u_1 = \lambda u_2$, $\lambda \in \mathbf{C}$ which proves that dim $\text{Ker}(\mathcal{H} - E) = 1$. □

Proposition 9.4.15 *If $\widetilde{\psi} > 0$ is a minimizer then $\widetilde{\psi}$ is radial, that is, $\widetilde{\psi}(x) = U(|x|)$.*

Proof Recall that $\widetilde{\psi}$ is radial if and only if for any orthogonal matrix A we have $\widetilde{\psi}(Ax) = \widetilde{\psi}(x)$. Set $\widetilde{\psi}_A(x) = \widetilde{\psi}(Ax)$.

Let A be an orthogonal matrix. Since $|Ax| = |x|$ we have (see Lemma 9.6.1 in the appendix),

$$\begin{aligned} -\Delta\widetilde{\psi}_A(x) - \frac{\widetilde{\psi}_A(x)}{|x|} - E\widetilde{\psi}_A(x) &= (-\Delta\widetilde{\psi})(Ax) - \frac{\widetilde{\psi}(Ax)}{|Ax|} - E\widetilde{\psi}(Ax) \\ &= ((\mathcal{H} - E)\widetilde{\psi})(Ax) = 0. \end{aligned}$$

Moreover, $\|\widetilde{\psi}_A\|_{L^2} = \|\widetilde{\psi}\|_{L^2} = 1$. Consequently, according to Proposition 9.4.11, $\widetilde{\psi}_A$ is a minimizer. We deduce from Proposition 9.4.14 that there exists $c \neq 0$ such that $\widetilde{\psi}_A = c\,\widetilde{\psi}$. Since their L^2 norms are both equal to 1 and $\widetilde{\psi} > 0$ we have $c = 1$. Thus $\widetilde{\psi}$ is radial. □

9.4.4 Determination of E and $\widetilde{\psi}$

Proposition 9.4.16 *Let $\theta \in D = \left\{u \in H^2 : \frac{u}{|x|} \in L^2\right\}$ be a solution of $\mathcal{H}\theta = \lambda\theta$, with $\lambda = -\mu^2$, $\mu > 0$ such that*

$$\theta > 0, \quad |x|^k\theta \in L^2 \cap L^\infty \quad \forall k \in \mathbf{N}, \quad \theta \text{ is radial.}$$

Then, $F(\theta) = \lambda\|\theta\|_{L^2}^2$ where F is defined in (9.4.1).

Proof This follows from (9.4.8). □

Proposition 9.4.17 *Set $|x| = r$ and $\theta(x) = u(r)$.*

1. The function u is continuous on $[0, +\infty)$ and satisfies the equation

$$-u''(r) - \frac{2}{r}u' - \frac{1}{r}u(r) = \lambda u(r) \text{ in } \mathcal{D}'((0, +\infty)).$$

2. Set $v(r) = ru(r)$. Then, $r^k v \in L^\infty((0,+\infty))$ for all $k \in \mathbf{N}$, and thus v is in $L^1((0,+\infty))$. Moreover, it satisfies the equation $-v''(r) - \frac{1}{r}v(r) = \lambda v(r)$ in $\mathcal{D}'((0,+\infty))$.

3. Set $w(r) = 1_{\{r>0\}}v(r) \in L^1(\mathbf{R})$. Then w satisfies the differential equation $-rw'' - w - \lambda r w = 0$ in $\mathcal{S}'((0,+\infty))$.

Proof 1. Since $\theta \in H^2 \subset C^0$ the function u is continuous on $[0,+\infty)$. Next, since $\theta(x) = u(r)$ is radial, we have, in dimension $d = 3$, $\Delta\theta = u'' + \frac{2}{r}u'$.

2. By assumption, $r^k u \in L^\infty((0,+\infty))$ for all $k \in \mathbf{N}$. This implies that v satisfies the same property and that $v \in L^1((0,+\infty))$. Moreover, we have

$$v'' = ru'' + 2u' = r(u'' + \frac{2}{r}u') = r(-\frac{1}{r}u - \lambda u) = -\frac{v}{r} - \lambda v.$$

3. Set $T = -rw'' - w - \lambda r w$. Our goal is to show that $T = 0$. By the equation satisfied by v we see that $T = 0$ on $(0,+\infty)$ and, by the definition of w, also on $(-\infty, 0)$. Consequently, $\operatorname{supp} T \subset \{0\}$. We deduce that $T = \sum_{k=0}^N c_k \delta_0^{(k)}$, $c_k \in \mathbf{C}$. We will show that all the c_j are zero.

Let $\varphi \in C_0^\infty(\mathbf{R})$ such that $\varphi^k(0) \neq 0$ for $0 \le k \le N$ (take $\varphi(x) = e^x\theta(x)$ where $\theta \in C_0^\infty(\mathbf{R})$, $\theta(x) = 1$ for $|x| \le 1$). For $\varepsilon > 0$ set $\varphi_\varepsilon(r) = \varphi(\frac{r}{\varepsilon})$. We have

$$\begin{aligned}\langle T, \varphi_\varepsilon\rangle &= \big\langle w, -(r\varphi_\varepsilon)'' - \varphi_\varepsilon - \lambda r \varphi_\varepsilon\big\rangle,\\ &= -\int_0^{+\infty} v(r)(r\varphi_\varepsilon(r))''\,dr - \int_0^{+\infty} v(r)\left[\varphi\left(\frac{r}{\varepsilon}\right) + \lambda r\varphi\left(\frac{r}{\varepsilon}\right)\right] dr,\\ &= I_\varepsilon + J_\varepsilon.\end{aligned}$$

Setting $r = \varepsilon s$ in the integral we see that $|J_\varepsilon| \le \varepsilon\|v\|_{L^\infty}\int_0^{+\infty}(1+\varepsilon|\lambda|s)|\varphi(s)|\,ds$. Consequently $J_\varepsilon \to 0$ as $\varepsilon \to 0$. On the other hand, setting $r = \varepsilon s$ in the integral we obtain

$$\begin{aligned}I_\varepsilon &= -\int_0^{+\infty} v(r)\left[\frac{r}{\varepsilon^2}\varphi''\left(\frac{r}{\varepsilon}\right) + \frac{2}{\varepsilon}\varphi'\left(\frac{r}{\varepsilon}\right)\right] dr,\\ &= -\int_0^{+\infty} v(\varepsilon s)\left(s\varphi''(s) + 2\varphi'(s)\right) ds.\end{aligned}$$

Since u is continuous on $[0,+\infty)$ and $v = ru$ we have $v(0) = 0$. Using the dominated convergence theorem, we see that $\lim_{\varepsilon\to 0} I_\varepsilon = 0$. In summary, $\lim_{\varepsilon\to 0}\langle T, \varphi_\varepsilon\rangle = 0$.

On the other hand, we have $\langle T, \varphi_\varepsilon\rangle = \sum_{k=0}^N (-1)^k c_k \varepsilon^{-k}\varphi^{(k)}(0)$ and $\langle T, \varphi_\varepsilon\rangle \to 0$ as $\varepsilon \to 0$. This implies that $T = 0$. Indeed, we show that $c_N = \cdots = c_0 = 0$ as follows. We have $\varepsilon^N\langle T, \varphi_\varepsilon\rangle \to 0$ so $c_N = 0$. Then, $\varepsilon^{N-1}\langle T, \varphi_\varepsilon\rangle \to 0$ which implies $c_{N-1} = 0$ and we continue down to the term c_0. □

Proposition 9.4.18 *Let w be the function defined in Proposition 9.4.17.*

1. Set $f(\xi) = \widehat{w}(\xi) = \int_0^{+\infty} e^{-ir\xi} v(r)\,dr$. Then $f \in C^\infty(\mathbf{R})$ and can be extended to a holomorphic function (which we still denote by f) in the set $\{z \in \mathbf{C} : Im\, z < 0\}$.

2. The function f is a solution, in $\mathcal{S}'(\mathbf{R})$, of the differential equation

$$(\xi^2 + \mu^2) f'(\xi) + (2\xi + i) f(\xi) = 0, \quad \xi \in \mathbf{R}, \quad \lambda = -\mu^2, \mu > 0.$$

Proof 1. The fact that the function f is C^∞ on $\mathbf{R}$ follows from the fact that, according to point 2. of Proposition 9.4.17, we have $r^k v \in L^1(\mathbf{R})$ for all $k \in \mathbf{N}$ and from Lebesgue's theorem on the differentiability of integrals depending on a parameter. For Im z < 0 and $r > 0$, we have $|e^{izr} v(r)| = e^{-\text{Im}\, zr} |v(r)| \le |v(r)| \in L^1$. Consequently, we can set $f(z) = \int_0^{+\infty} e^{izr} v(r)\,dr$. Applying again this same Lebesgue theorem, we see that f is holomorphic in the set Im $z < 0$.

2. Recall that w is a solution of the equation $-rw'' - w - \lambda r w = 0$ in $\mathcal{S}'(\mathbf{R})$ and that $\widehat{w} = f$. Taking the Fourier transform of both sides and using the fact that

$$\mathcal{F}(rw'') = \frac{1}{i} \frac{d}{d\xi} (\xi^2 f), \quad \mathcal{F}(\lambda r w) = -\frac{\lambda}{i} \frac{d}{d\xi} f,$$

we deduce that $(\xi^2 + \mu^2) f' + (2\xi + i) f = 0$ in $\mathcal{S}'(\mathbf{R})$. □

We will use the following lemma.

Lemma 9.4.19 *Let G be a holomorphic function in the open set $\{z \in \mathbf{C} : Im\, z < 0\}$ and continuous in the set $\{z \in \mathbf{C} : Im\, z \le 0\}$. Suppose that $G|_{Im\, z=0} = 0$, then $G \equiv 0$.*

Proof We will use the following result due to Morera (see [2]).

Lemma 9.4.20 *Let $u : \mathbf{C} \to \mathbf{C}$ be a function which is holomorphic in the sets $Im\, z > 0$ and $Im\, z < 0$. If moreover u is continuous on $\mathbf{C}$, then u is holomorphic in $\mathbf{C}$.*

Let us go back to the proof of Lemma 9.4.19. We apply this result to the function defined by $u(z) = G(z)$ if Im $z \le 0$ and $u(z) = G(-z)$ if Im z > 0. This function u satisfies the above conditions (since $G = 0$ for Im $z = 0$), consequently, u is holomorphic in $\mathbf{C}$. As $u = 0$ for Im $z = 0$, the principle of isolated zeros proves that $u = 0$. □

We can now state the main result of this paragraph.

Proposition 9.4.21 *The minimal energy E of the function F is equal to $-\frac{1}{4}$.*

Proof Point 1. The function

$$G(z) = (z^2 + \mu^2) f'(z) + (2z + i) f(z)$$

is holomorphic in Im$z < 0$. By Proposition 9.4.18, G vanishes when Im $z = 0$. We deduce from Lemma 9.4.19 that $G = 0$ in Im $z < 0$.

Point 2. Let $\varepsilon > 0$ be small and $B = \{z \in \mathbf{C} : |z + i\mu| \leq \varepsilon\}$. By general theory, we know that the integral $\frac{1}{2i\pi}\int_B \frac{f'(z)}{f(z)}\,dz$ is equal to $n - p$ where n is the number of zeros of f and p the number of its poles. Since f is holomorphic in $\operatorname{Im} z < 0$ it has no poles. Consequently, the integral is an integer $n \in \mathbf{N}$.

Point 3. By Proposition 9.4.18 we have $\frac{f'(z)}{f(z)} = -\frac{2z+i}{z^2+\mu^2}$. Inside B, the function $\frac{f'}{f}$ has a simple pole at $z = -i\mu$. The residue at this pole is equal to,

$$\lim_{z\to -i\mu}(z + i\mu)\frac{f'(z)}{f(z)} = -\lim_{z\to -i\mu}\frac{2z+i}{z - i\mu} = \frac{1}{2\mu} - 1.$$

Consequently, by the residue theorem we have $\frac{1}{2i\pi}\int_B \frac{f'(z)}{f(z)}\,dz = \frac{1}{2\mu} - 1$. We deduce from Point 2 that $\frac{1}{2\mu} - 1 = n$ so $\mu = \frac{1}{2(n+1)}$.

Point 4. We deduce from the previous points that for any $\theta \in D, \theta > 0$, satisfying $(\mathcal{H} - \lambda)\theta = 0$ we must have $\lambda = -\frac{1}{4(n+1)^2}$. The minimum value of λ is then reached for $n = 0$. This minimum energy is precisely E. Consequently, we have $E = -\frac{1}{4}$. □

It remains only to compute w, which will provide a basis for the eigenspace associated with the eigenvalue $E = -\frac{1}{4}$.

Proposition 9.4.22 *We have* $w(r) = C1_{r>0}re^{-\frac{r}{2}}$ *and the space of minimizers is generated by the function* $\widetilde{\psi}(x) = \frac{1}{\sqrt{8\pi}}e^{-\frac{1}{2}|x|}$.

Proof Point 1. We have $\frac{f'(z)}{f(z)} = -\frac{2z+i}{z^2+\frac{1}{4}} = -2(z - \frac{i}{2})^{-1}$. This implies that we have $f(z) = C(z - \frac{i}{2})^{-2}$, where $C \in \mathbf{C}$. Next, setting $\varphi(r) = 1_{r>0}e^{-\frac{r}{2}} \in L^1(\mathbf{R})$ we have

$$\int_{\mathbf{R}} e^{-ir\xi}\varphi(r)\,dr = \int_0^{+\infty} e^{-r(\frac{1}{2}+i\xi)}\,dr = \frac{1}{\frac{1}{2} + i\xi}.$$

We deduce that $\mathcal{F}(r\varphi)(\xi) = -\frac{1}{i}(\frac{d}{d\xi}\widehat{\varphi})(\xi) = -(\xi - \frac{i}{2})^{-2}$.

Point 2. Point 1 shows that $r\varphi(r) = -\mathcal{F}^{-1}\left((\xi - \frac{i}{2})^{-2})\right)$. This proves that we have

$$w(r) = Cr\varphi(r) = C1_{r>0}re^{-\frac{r}{2}}$$

and implies that $v(r) = Cre^{-\frac{r}{2}}$ so $u(r) = Ce^{-\frac{r}{2}}$. Since the set of minimizers is a vector space of dimension 1, we deduce that it is generated by the function $\widetilde{\psi}(x) = Ce^{-\frac{1}{2}|x|}$ where $C = \frac{1}{\sqrt{8\pi}}$ to have $\|\widetilde{\psi}\|_{L^2} = 1$. □

9.5 Excited States

The goal is to determine the eigenvalues and eigenfunctions of the operator $H = -\Delta - \frac{1}{|r|}$ with domain $D(H) = H^2(\mathbf{R}^3)$, i.e.

$$-\Delta\psi - \frac{\psi}{|x|} = E\psi, \quad \psi \in H^2(\mathbf{R}^3), \quad \psi \not\equiv 0. \tag{9.5.1}$$

We shall use some results of the previous chapter.

Remark 9.5.1 *(i)* First of all, since the function $\frac{1}{|x|}$ is C^∞ in $\mathbf{R}^3 \setminus \{0\}$ and the operator $-\Delta$ is elliptic, any solution of Eq. (9.5.1) belongs to $C^\infty(\mathbf{R}^3 \setminus \{0\})$. On the other hand, since $H^2(\mathbf{R}^3) \subset C^0(\mathbf{R}^3) \cap L^\infty(\mathbf{R}^3)$, we have

$$\psi \in C^0(\mathbf{R}^3) \cap L^\infty(\mathbf{R}^3) \cap C^\infty(\mathbf{R}^3 \setminus \{0\}). \tag{9.5.2}$$

(ii) The eigenvalues are real. Indeed, multiplying Eq. (9.5.1) by $\overline{\psi}$ and integrating both sides over $\mathbf{R}^3$ we obtain

$$-\int_{\mathbf{R}^3} \Delta\psi(x)\overline{\psi}(x)\,dx - \int_{\mathbf{R}^3} \frac{|\psi(x)|^2}{|x|}\,dx = E\int_{\mathbf{R}^3} |\psi(x)|^2\,dx.$$

The second integral on the left-hand side is finite by Corollary 9.4.3. As for the first, it is equal to $\int_{\mathbf{R}^3} |\nabla\psi(x)|^2\,dx$, since $\psi \in H^2(\mathbf{R}^3)$.

The main result of this paragraph is the following, which studies the spectral theory of the Schrödinger operator.

Recall that $\mathcal{H}_\ell$ denotes the space of spherical harmonics of degree ℓ (see Definition 8.3.5).

Theorem 9.5.2 *Let $H = -\Delta - \frac{1}{|x|}$ with domain $D(H) = H^2(\mathbf{R}^3)$.*

(i) The eigenvalues of H are of the form $E = -\frac{1}{4(n+\ell+1)^2}$ where $n, \ell \in \mathbf{N}$.

(ii) They are associated with the eigenfunctions

$$\psi_{n,\ell}(x) = |x|^\ell Q_{n,\ell}(|x|)e^{-\frac{|x|}{2(n+\ell+1)}} P_\ell\Big(\frac{x}{|x|}\Big),$$

where $Q_{n,\ell}$ is a polynomial of degree $\leq n$ and $P_\ell \in \mathcal{H}_\ell$.

(iii) The dimension of the eigenspace associated with $E_k = -\frac{1}{(k+1)^2}$ is equal to $(k+1)^2$.

The rest of this paragraph is devoted to the proof of this result.

Lemma 9.5.3 *Let ψ be a solution of* (9.5.1). *There exist $\ell \in \mathbf{N}$, $P \in \mathcal{H}_\ell$ such that the function*

$$r \mapsto u_P(r) = \int_{\mathbf{S}^2} \psi(r\omega)P(\omega)\,d\omega$$

is not identically zero.

Proof Suppose that for all $\ell \in \mathbf{N}$, all $P \in \mathcal{H}_\ell$ we have

$$\int_{\mathbf{S}^2} \psi(r\omega)P(\omega)\,d\omega = 0, \quad \forall r > 0. \tag{9.5.3}$$

Then we still have (9.5.3) if P is a finite linear combination of elements of $\cup_{\ell=0}^{+\infty}\mathcal{H}_\ell$. But by Proposition 8.3.8, this set is dense in $L^2(\mathbf{S}^2)$. We easily deduce that $\psi(r\omega) = 0$ for all $r > 0$ and all $\omega \in \mathbf{S}^2$, in other words, that ψ is identically zero, which is absurd. □

Lemma 9.5.4 (i) *The function u_P belongs to $C^0([0,+\infty)) \cap L^\infty(]0,+\infty[)$ and satisfies the differential equation*

$$u_P'' + \frac{2}{r}u_P' + \Big(\frac{1}{r} - \frac{\ell(\ell+1)}{r^2} + E\Big)u_P = 0. \tag{9.5.4}$$

(ii) *We have $ru_P \in L^2((0,+\infty))$.*

Proof (i) The fact that u_P has the stated regularity follows from (9.5.2) and Lebesgue's theorem on differentiation under the integral sign. Next, according to Proposition 8.3.1 and (9.5.1) we have

$$E\psi(r\omega) = -\frac{\partial^2}{\partial r^2}\big[\psi(r\omega)\big] - \frac{2}{r}\frac{\partial}{\partial r}\big[\psi(r\omega)\big] - \frac{1}{r^2}\Delta_{\mathbf{S}^2}\big[\psi(r\omega)\big] - \frac{1}{r}\psi(r\omega). \tag{9.5.5}$$

By the Lebesgue's theorem mentioned above, we have

$$u_P''(r) = \int_{\mathbf{S}^2}\frac{\partial^2}{\partial r^2}\big[\psi(r\omega)\big]P(\omega)\,d\omega,\quad \frac{2}{r}u_P'(r) = \frac{2}{r}\int_{\mathbf{S}^2}\frac{\partial}{\partial r}\big[\psi(r\omega)\big]P(\omega)\,d\omega.$$

Using (9.5.5) we obtain

$$\begin{aligned} u_P''(r) + \frac{2}{r}u_P'(r) &= -\int_{\mathbf{S}^2} E\psi(r\omega)P(\omega)\,d\omega - \frac{1}{r^2}\int_{\mathbf{S}^2}\Delta_{\mathbf{S}^2}\big[\psi(r\omega)\big]P(\omega)\,d\omega \\ &\quad - \frac{1}{r}\int_{\mathbf{S}^2}\psi(r\omega)P(\omega)\,d\omega. \end{aligned}$$

Using Lemmas 8.3.4 and 8.3.7 we obtain

$$\begin{aligned} -\int_{\mathbf{S}^2}\Delta_{\mathbf{S}^2}\big[\psi(r\omega)\big]P(\omega)\,d\omega &= -\int_{\mathbf{S}^2}\psi(r\omega)(\Delta_{\mathbf{S}^2}P)(\omega)\,d\omega \\ &= \ell(\ell+1)\int_{\mathbf{S}^2}\psi(r\omega)P(\omega)\,d\omega, \end{aligned}$$

so that

$$\begin{aligned} u_P''(r) + \frac{2}{r}u_P'(r) &= -\int_{\mathbf{S}^2}\big[E - \frac{\ell(\ell+1)}{r^2} + \frac{1}{r}\big]\psi(r\omega)P(\omega)\,d\omega \\ &= -\Big(\frac{1}{r} - \frac{\ell(\ell+1)}{r^2} + E\Big)u_P(r) \end{aligned}$$

(ii) According to the Cauchy-Schwarz inequality, we can write

$$|ru_P(r)|^2 \leq \Big(\int_{\mathbf{S}^2} |\psi(r\omega)|^2 r^2\, d\omega\Big)\Big(\int_{\mathbf{S}^2} |P(\omega)|^2\, d\omega\Big) \leq C\int_{\mathbf{S}^2} |\psi(r\omega)|^2 r^2\, d\omega.$$

Integrating both sides over $(0, +\infty)$, we obtain

$$\int_0^{+\infty} |ru_P(r)|^2\, dr \leq C\int_0^{+\infty}\int_{\mathbf{S}^2} |\psi(r\omega)|^2 r^2\, d\omega\, dr = C\int_{\mathbf{R}^3} |\psi(x)|^2\, dx.$$

□

Lemma 9.5.5 *Let* $v = r^{\ell+1}u_P$. *Then* $v \in C^0([0, +\infty)) \cap C^\infty(]0, +\infty[)$, $v(0) = 0$. *Moreover, it satisfies the differential equation on* $]0, +\infty[$,

$$(rv)'' - (2\ell + 2)v' + (1 + rE)v = 0.$$

Proof The regularity of v follows from that of u_P. Next, $u_P = r^{-(\ell+1)}v$ so that

$$\begin{aligned} u_P' &= -(\ell + 1)r^{-(\ell+2)}v + r^{-(\ell+1)}v', \\ u_P'' &= (\ell + 1)(\ell + 2)r^{-(\ell+3)}v - 2(\ell + 1)r^{-(\ell+2)}v' + r^{-(\ell+1)}v''. \end{aligned}$$

Using (9.5.4) we obtain

$$r^{-(\ell+1)}\Big[v'' - \big(\frac{2(\ell+1)}{r} - \frac{2}{r}\big)v' + \big(\frac{(\ell+1)(\ell+2)}{r^2} - \frac{2(\ell+1)}{r^2} + \frac{1}{r} - \frac{\ell(\ell+1)}{r^2} + E\big)v\Big] = 0.$$

Consequently,

$$v'' - \frac{2\ell}{r}v' + \big(\frac{1}{r} + E\big) = 0 \Longleftrightarrow rv'' - 2\ell v' + (1 + rE))v = 0.$$

The lemma then follows immediately. □

Lemma 9.5.6 *Let* $w = 1_{r>0}v$. *Then* $w \in \mathcal{S}'(\mathbf{R})$ *and satisfies the differential equation in* $\mathcal{S}'(\mathbf{R})$

$$(rw)'' - (2\ell + 2)w' + (1 + rE)w = 0. \tag{9.5.6}$$

Proof We have $w = r^{\ell+1}1_{r>0}u_P$. Since $u_P \in L^\infty((0, +\infty))$ the function $1_{r>0}u_P$ belongs to $L^\infty(\mathbf{R}) \subset \mathcal{S}'(\mathbf{R})$. Therefore $r^{\ell+1}1_{r>0}u_P \in \mathcal{S}'(\mathbf{R})$.

Next, we have $(1_{r>0})' = \delta_{r=0}$. Since $v(0) = 0$ it follows that

$$w' = \delta_{r=0}v + 1_{r>0}v' = v(0)\delta_{r=0} + 1_{r>0}v' = 1_{r>0}v'.$$

Next, since $rw = 1_{r>0}(rv)$, $(rv)' = v + rv'$ and $r\delta_{r=0} = 0$ it follows that

$$\begin{aligned} (rw)' &= \delta_{r=0}(rv) + 1_{r>0}(rv)' = 1_{r>0}(rv)', \\ (rw)'' &= \delta_{r=0}(v + rv') + 1_{r>0}(rv)'' = 1_{r>0}(rv)''. \end{aligned}$$

The equation for w then follows from that for v. □

Lemma 9.5.7 *Let $f = \widehat{w} \in \mathcal{S}'(\mathbf{R})$. Then,*

$$\begin{aligned}
&(i)\quad f \text{ can be extended holomorphically in } \Omega_- = \{z \in \mathbf{C} : Im\, z < 0\},\\
&(ii)\quad f \text{ satifies in } \mathcal{S}'(\mathbf{R}),\ (\xi^2 - E)f' + \big((2\ell + 2)\xi + i\big)f = 0,\\
&(iii)\quad \text{there exists } g \in L^2(\mathbf{R}) \text{ such that } f = \Big(\frac{d}{d\xi}\Big)^{\ell} g.
\end{aligned}$$

Proof (i) Recall that $w = 1_{r>0} r^{\ell+1} u_P$. For $a > 0$ the function $e^{-ar} w(r)$ belongs to $L^1((0, +\infty))$. Set for $z \in \Omega_-$,

$$g(z) = \int_0^{+\infty} e^{-irz} r^{\ell+1} u_P(r)\, dr = \int_0^{+\infty} e^{-ir\mathrm{Re}\, z} e^{r\mathrm{Im}\, z} r^{\ell+1} u_P(r)\, dr.$$

Using Lebesgue's theorem on differentiation under the integral sign, we see that the function g is holomorphic in Ω_-. On the other hand, we can write $g(\xi + i\eta) = \mathcal{F}(e^{r\eta} w)(\xi)$. We thus see that the function g is a holomorphic extension of $\widehat{w}$ to the set Ω_-.

(ii) In $\mathcal{S}'(\mathbf{R})$ we have $\widehat{T'} = i\xi \widehat{T}$, $\widehat{T''} = -\xi^2 \widehat{T}$, $\widehat{rT} = i\frac{d}{d\xi}\widehat{T}$. Taking the Fourier transform of both sides of (9.5.6) we obtain

$$-\xi^2 i\frac{d}{d\xi} f - (2\ell + 2)i\xi f + f + iE\frac{d}{d\xi} f = 0.$$

(iii) According to Lemma 9.5.4 (ii), we have $ru_P \in L^2((0, +\infty))$. It follows that $r1_{r>0}u_P$ belongs to $L^2(\mathbf{R})$, so $w = r^\ell \varphi$ where $\varphi \in L^2(\mathbf{R})$, so that $f = \big(i\frac{d}{d\xi}\big)^\ell \widehat{\varphi}$. □

Lemma 9.5.8 *We cannot have $E = 0$.*

Proof If $E = 0$, Lemma 9.5.7 shows that $\xi^2 f' + ((2\ell + 2)\xi + i)f = 0$. Set $g = \xi^{2\ell+2} e^{-\frac{i}{\xi}} f$. Then,

$$g' = \big[\big((2\ell + 2)\xi^{2\ell+1} + i\xi^{2\ell}\big) f + \xi^{2\ell+2} f'\big] e^{-\frac{i}{\xi}} = \big[\xi^2 f' + ((2\ell + 2)\xi + i) f\big] \xi^{2\ell} e^{-\frac{i}{\xi}} = 0.$$

Therefore g is a constant and

$$f = \frac{C}{\xi^{2\ell+2}} e^{\frac{i}{\xi}}. \tag{9.5.7}$$

We will show that for all $\ell \in \mathbf{N}$ we have

$$\frac{e^{\frac{i}{\xi}}}{\xi^{2\ell+2}} = \Big(\frac{d}{d\xi}\Big)^{\ell} \Big(\frac{e^{\frac{i}{\xi}}}{\xi^2} P_\ell(\xi) + R_\ell(\xi)\Big), \tag{9.5.8}$$

where P_ℓ is a polynomial such that $P_\ell(0) \neq 0$ and $R_\ell \in L^\infty((0, 1))$.

To do this, we proceed by induction on ℓ. This equality is trivially true for $\ell = 0$ with $P_0 = 1$ and $R_0 = 0$. Suppose it is true at order ℓ and let us show it at order $\ell + 1$. We have

$$\frac{e^{\frac{i}{\xi}}}{\xi^{2\ell+4}} = \frac{e^{\frac{i}{\xi}}}{\xi^2}\frac{1}{\xi^{2\ell+2}} = \frac{d}{d\xi}\big(ie^{\frac{i}{\xi}}\big)\frac{1}{\xi^{2\ell+2}} = \frac{d}{d\xi}\frac{ie^{\frac{i}{\xi}}}{\xi^{2\ell+2}} + i(2\ell+2)\frac{e^{\frac{i}{\xi}}}{\xi^{2\ell+3}}.$$

Using the induction, we can write

$$\frac{e^{\frac{i}{\xi}}}{\xi^{2\ell+4}} = \Big(\frac{d}{d\xi}\Big)^{\ell+1}\Big(\frac{e^{\frac{i}{\xi}}}{\xi^2}iP_\ell(\xi) + iR_\ell(\xi)\Big) + i(2\ell+2)\frac{e^{\frac{i}{\xi}}}{\xi^{2\ell+3}}.$$

Next, we can write

$$\frac{e^{\frac{i}{\xi}}}{\xi^{2\ell+3}} = \frac{e^{\frac{i}{\xi}}}{\xi^2}\frac{1}{\xi^{2\ell+1}} = \frac{d}{d\xi}\big(ie^{\frac{i}{\xi}}\big)\frac{1}{\xi^{2\ell+1}} = \frac{d}{d\xi}\Big(\frac{ie^{\frac{i}{\xi}}}{\xi^{2\ell+1}}\Big) + i(2\ell+1)\frac{e^{\frac{i}{\xi}}}{\xi^{2\ell+2}}.$$

Using the induction, we deduce that

$$\frac{e^{\frac{i}{\xi}}}{\xi^{2\ell+4}} = \Big(\frac{d}{d\xi}\Big)^{\ell+1}\Big(\frac{e^{\frac{i}{\xi}}}{\xi^2}iP_\ell(\xi) + iR_\ell(\xi)\Big) + \alpha_\ell\frac{d}{d\xi}\Big(\frac{e^{\frac{i}{\xi}}}{\xi^{2\ell+1}}\Big) + \beta_\ell\Big(\frac{d}{d\xi}\Big)^{\ell}\Big(\frac{e^{\frac{i}{\xi}}}{\xi^2}P_\ell(\xi) + R_\ell(\xi)\Big).$$

Then,

$$\begin{aligned}\frac{d}{d\xi}\Big(\frac{e^{\frac{i}{\xi}}}{\xi^{2\ell+1}}\Big) &= \frac{d}{d\xi}\Big(\xi\frac{e^{\frac{i}{\xi}}}{\xi^{2\ell+2}}\Big) = \frac{e^{\frac{i}{\xi}}}{\xi^{2\ell+2}} + \xi\frac{d}{d\xi}\Big(\frac{e^{\frac{i}{\xi}}}{\xi^{2\ell+2}}\Big)\\ &= \Big(\frac{d}{d\xi}\Big)^{\ell}\Big(\frac{e^{\frac{i}{\xi}}}{\xi^2}P_\ell(\xi) + R_\ell(\xi)\Big) + \xi\Big(\frac{d}{d\xi}\Big)^{\ell+1}\Big(\frac{e^{\frac{i}{\xi}}}{\xi^2}P_\ell(\xi) + R_\ell(\xi)\Big)\\ &= \Big(\frac{d}{d\xi}\Big)^{\ell+1}\Big(\frac{e^{\frac{i}{\xi}}}{\xi^2}\xi P_\ell(\xi) + \xi R_\ell(\xi)\Big) + \gamma_\ell\Big(\frac{d}{d\xi}\Big)^{\ell}\Big(\frac{e^{\frac{i}{\xi}}}{\xi^2}P_\ell(\xi) + R_\ell(\xi)\Big).\end{aligned}$$

We deduce from the previous equalities that

$$\frac{e^{\frac{i}{\xi}}}{\xi^{2\ell+4}} = \Big(\frac{d}{d\xi}\Big)^{\ell+1}\Big(\frac{e^{\frac{i}{\xi}}}{\xi^2}(i+\alpha_\ell\xi)P_\ell(\xi) + (i+\alpha_\ell)\xi R_\ell(\xi)\Big) + \delta_\ell\Big(\frac{d}{d\xi}\Big)^{\ell}\Big(\frac{e^{\frac{i}{\xi}}}{\xi^2}P_\ell(\xi) + R_\ell(\xi)\Big).$$

On the other hand, we have

$$\begin{aligned}\frac{e^{\frac{i}{\xi}}}{\xi^2}P_\ell(\xi) + R_\ell(\xi) &= \frac{d}{d\xi}\big(ie^{\frac{i}{\xi}}\big)P_\ell(\xi) + R_\ell(\xi) = \frac{d}{d\xi}\big(ie^{\frac{i}{\xi}}P_\ell(\xi)\big) - ie^{\frac{i}{\xi}}P_\ell'(\xi) + R_\ell(\xi)),\\ &= \frac{d}{d\xi}\Big(\frac{e^{\frac{i}{\xi}}}{\xi^2}(i\xi^2P_\ell(\xi)) + \int_\xi^1 (R_\ell(t) - ie^{\frac{i}{t}}P_\ell(t))\,dt\Big),\end{aligned}$$

finally,

$$\frac{e^{\frac{i}{\xi}}}{\xi^{2\ell+4}} = \left(\frac{d}{d\xi}\right)^{\ell+1}\left(\frac{e^{\frac{i}{\xi}}}{\xi^2}P_{\ell+1}(\xi) + R_{\ell+1}(\xi)\right),$$

where $P_{\ell+1}(\xi) = (i + \alpha_\ell \xi + i\delta_\ell \xi^2)P_\ell(\xi)$ and $R_{\ell+1} \in L^\infty((0, 1))$.

Since $P_0(\xi) = 1$ and $P_{\ell+1}(0) = iP_\ell(0)$, (9.5.8) is proved at order $\ell + 1$.

Let us continue the proof of Lemma 9.5.8. It follows from (9.5.8) and (9.5.7) that the function f can be written as

$$f = \left(\frac{d}{d\xi}\right)^\ell\left(C\frac{e^{\frac{i}{\xi}}}{\xi^2}P_\ell(\xi) + CR_\ell(\xi)\right).$$

But this contradicts point (ii) of Lemma 9.5.7. Indeed, according to this point, we have, in particular, $f = \left(\frac{d}{d\xi}\right)^\ell g$ where $g \in L^2((0, 1))$. Comparing with the expression above, we deduce that for $\xi \in (0, 1)$,

$$g = C\frac{e^{\frac{i}{\xi}}}{\xi^2}P_\ell(\xi) + CR_\ell(\xi) + Q_\ell(\xi)$$

where Q_ℓ is a polynomial of degree less than or equal to $\ell - 1$. It follows that $\frac{e^{\frac{i}{\xi}}}{\xi^2}P_\ell(\xi)$ belongs to $L^2((0, 1))$, which is absurd since $P_\ell(0) \neq 0$.

Lemma 9.5.9 *We cannot have* $E > 0$.

Proof Suppose $E > 0$, i.e., $E = \omega^2$, $\omega \in \mathbf{R} \setminus \{0\}$. It follows from Lemma 9.5.7 that

$$f' = -\frac{2(\ell+1)\xi + i}{\xi^2 - \omega^2}f. \tag{9.5.9}$$

Let us place ourselves near the point $\xi = \omega$ for $\xi > \omega$. We write

$$\begin{aligned} -\frac{2(\ell+1)\xi + i}{\xi^2 - \omega^2} &= -(\ell+1)\frac{2\xi}{\xi^2 - \omega^2} - \frac{i}{2\omega}\left(\frac{1}{\xi - \omega} - \frac{1}{\xi + \omega}\right), \\ &= \frac{d}{d\xi}\left(\ln[(\xi^2 - \omega^2)^{-(\ell+1)}] - \frac{i}{\omega}\ln\left(\frac{\xi - \omega}{\xi + \omega}\right)\right) = \frac{d}{d\xi}A(\xi), \end{aligned}$$

so that Eq. (9.5.9) becomes $f'(\xi) = A'(\xi)f(\xi)$. Let us set $g(\xi) = e^{-A(\xi)}f(\xi)$. Then we have $g'(\xi) = e^{-A(\xi)}(f' - A'f) = 0$. Consequently, g is constant and we have $f(\xi) - Ce^{A(\xi)}$. We can thus write, for $\xi > \omega$,

$$f(\xi) = \frac{C}{(\xi - \omega)^{\ell+1+\frac{i}{2\omega}}}\frac{1}{(\xi + \omega)^{\ell+1-\frac{i}{2\omega}}}.$$

Set $\xi - \omega = \eta$. Then for ξ close to ω and $\xi > \omega$ we have $\eta \in (0, \varepsilon)$. We can write $f(\eta) = \frac{1}{\eta^{\ell+1+\frac{i}{2\omega}}} g(\eta)$, where $g(\eta) = \frac{1}{(\eta+2\omega)^{\ell+1-\frac{i}{2\omega}}}$. The function g is analytic in a neighborhood of zero so that we can write for small η, $g(\eta) = \sum_{n\geq 0} a_n(\omega, \ell)\eta^n$ with $a_0 = g(0) \neq 0$. Then,

$$f(\xi) = \sum_{n\geq 0} a_n(\omega, \ell)\eta^{n-(\ell+1+\frac{i}{2\omega})}.$$

Now, for $\alpha \in \mathbf{C} \setminus \mathbf{N}$ we have $\eta^{\alpha-\ell} = b_{\alpha,\ell}\big(\frac{d}{d\eta}\big)^{\ell}\eta^{\alpha}$. Taking $\alpha = n - 1 - \frac{i}{2\omega}$ we obtain

$$f(\xi) = \sum_{n\geq 0} c_n(\omega, \ell)\Big(\frac{d}{d\eta}\Big)^{\ell}\eta^{n-1-\frac{i}{2\omega}}, \quad c_0 \neq 0.$$

For ξ close to ω, $\xi > \omega$ we finally obtain

$$f(\xi) = \Big(\frac{d}{d\xi}\Big)^{\ell}\Big(\frac{1}{(\xi-\omega)^{1+\frac{i}{2\omega}}}\sum_{n\geq 0} c_n(\omega, \ell)(\xi-\omega)^n\Big) = \Big(\frac{d}{d\xi}\Big)^{\ell} h(\xi).$$

In a small neighborhood of ω, the function h is equivalent to $\frac{c_0}{(\xi-\omega)^{1+\frac{i}{2\omega}}}$ so that $|h(\xi)|$ is equivalent to $\frac{c_0}{\xi-\omega}$ which is not square integrable near ω. This again contradicts point (ii) of Lemma 9.5.7 and completes the proof of Lemma 9.5.8. □

It follows from Lemmas 9.5.8 and 9.5.9 that we have $E < 0$. We set $E = -\omega^2$. Taking into account Lemma 9.5.7 we have

$$f' = -\frac{2(\ell+1)\xi + i}{\xi^2 + \omega^2} f. \tag{9.5.10}$$

We have

$$-\frac{2(\ell+1)\xi + i}{\xi^2 + \omega^2} = \frac{d}{d\xi}A(\xi), \quad A(\xi) = \ln\frac{1}{(\xi^2+\omega^2)^{\ell+1}} - \frac{i}{\omega}\arctan\frac{\xi}{\omega}.$$

Setting $g = e^{-A} f$ we see that g satisfies $g' = e^{-A}(f' - A'f) = 0$. Thus g is constant and

$$f(\xi) = \frac{C}{(\xi^2+\omega^2)^{\ell+1}} e^{-\frac{i}{\omega}\arctan\frac{\xi}{\omega}}. \tag{9.5.11}$$

We deduce from point (i) of Lemma 9.5.7 that f extends holomorphically in $\Omega_- = \{z \in \mathbf{C} : \operatorname{Im} z < 0\}$. Let, for $\varepsilon > 0$ small, $D_\varepsilon = \{z \in \mathbf{C} : |z + i\omega| < \varepsilon\}$ and ∂D_ε the boundary of this disk. Since D_ε is entirely contained in Ω_- the function f has no pole there so that the integral $\frac{1}{2i\pi}\int_{\partial D_\varepsilon} \frac{f'(z)}{f(z)}\, dz$ is an integer, since it is equal to the number of zeros of f in D_ε. We deduce from (9.5.10) that

$$\frac{1}{2i\pi}\int_{\partial D_\varepsilon} g(z)\,dz = n \in \mathbf{N}, \quad g(z) = -\frac{2(\ell+1)z+i}{z^2+\omega^2}.$$

In D_ε the function g has a simple pole at $z=-i\omega$; since the index of ∂D_ε with respect to the point $z=-i\omega$ is equal to 1, the residue theorem shows that

$$\frac{1}{2i\pi}\int_{\partial D_\varepsilon} g(z)\,dz = \text{Res}(g,-i\omega) = \lim_{z\to -i\omega}(z+i\omega)g(z) = -(\ell+1)+\frac{1}{2\omega}.$$

We thus have $\frac{1}{2\omega}-(\ell+1)=n$, i.e.

$$\omega = \frac{1}{2(n+\ell+1)}. \tag{9.5.12}$$

Since $E=-\omega^2$, we see that the bound energies of problem (9.5.1) are of the form

$$E = -\frac{1}{4(n+\ell+1)^2}, \quad n,\ell \in \mathbf{N}.$$

It follows from (9.5.11) that

$$f(\xi) = \frac{C}{(\xi^2+\omega^2)^{\ell+1}}\left[e^{-2i\arctan\frac{\xi}{\omega}}\right]^{n+\ell+1}. \tag{9.5.13}$$

Lemma 9.5.10 *For all $x \in \mathbf{R}$ we have*

$$e^{-2i\arctan(x)} = \frac{1-ix}{1+ix}.$$

Proof Let $\theta = \arctan(x)$. Clearly, $\tan(\theta)=x$. Next,

$$\cos(2\theta) = 2\cos^2(\theta)-1 = \frac{2}{1+\tan^2(\theta)}-1 = \frac{2}{1+x^2}-1 = \frac{1-x^2}{1+x^2},$$
$$\sin(2\theta) = 2\sin(\theta)\cos(\theta) = 2\cos^2(\theta)\tan(\theta) = 2\frac{\tan(\theta)}{1+\tan^2(\theta)} = \frac{2x}{1+x^2}.$$

We deduce that

$$e^{-2i\arctan(x)} = \cos(2\theta)-i\sin(2\theta) = \frac{1-x^2}{1+x^2}-\frac{2ix}{1+x^2} = \frac{(1-ix)^2}{1+x^2} = \frac{1-ix}{1+ix},$$

□

We deduce from (9.5.13) and Lemma 9.5.10 that

$$f(\xi) = \frac{C}{\xi^2+\omega^2}\Big(\frac{1-i\frac{\xi}{\omega}}{1+i\frac{\xi}{\omega}}\Big)^{n+\ell+1} = \frac{C(\xi+i\omega)^n}{(\xi-i\omega)^{n+2\ell+2}},$$

so that

$$|f(\xi)| = \frac{C}{(\xi^2+\omega^2)^{\ell+1}}. \tag{9.5.14}$$

Recall that $f = \widehat{w}$ where $w(r) = 1_{r<0}\, v(r)$. As from (9.5.14) we have $f \in L^1(\mathbf{R})$, we can write, by inverse Fourier transform,

$$w(r) = \frac{1}{2\pi}\int_{\mathbf{R}} e^{ir\xi} f(\xi)\, d\xi = C' \int_{\mathbf{R}} e^{ir\xi} \frac{(\xi+i\omega)^n}{(\xi-i\omega)^{n+2\ell+2}}\, d\xi. \tag{9.5.15}$$

We will compute the integral above by the residue method. For this, we consider the meromorphic function in $\mathbf{C}$

$$F(z) = \frac{e^{irz}(z+i\omega)^n}{(z-i\omega)^{n+2\ell+2}},$$

which we will integrate over the oriented path, $\Gamma_R = \cup_{j=1}^4 \Gamma_j$, where

$$\Gamma_1 = \{z \in \mathbf{C} : \operatorname{Re} z \in (-R, R), \operatorname{Im} z = 0\},\quad \Gamma_2 = \{z \in \mathbf{C} : \operatorname{Re} z = R, \operatorname{Im} z \in (0, R)\},$$
$$\Gamma_3 = \{z \in \mathbf{C} : \operatorname{Re} z \in (R, -R), \operatorname{Im} z = R\},\quad \Gamma_4 = \{z \in \mathbf{C} : \operatorname{Re} z = -R, \operatorname{Im} z \in (R, 0)\}.$$

We also denote by D the domain bounded by Γ_R. The function F has inside D a single multiple pole at the point $z = i\omega$. As the path Γ_R has index equal to 1, the residue theorem implies that

$$\int_{\Gamma_R} F(z)\, dz = \operatorname{Res}(F, i\omega). \tag{9.5.16}$$

On the other hand,

$$\begin{aligned}\operatorname{Res}(F, i\omega) &= \frac{1}{(n+2\ell+1)!}\lim_{z\to i\omega}\Big(\frac{d}{dz}\Big)^{n+2\ell+1}\big[(z-i\omega)^{n+2\ell+2}F(z)\big]\\ &= \frac{1}{(n+2\ell+1)!}\lim_{z\to i\omega}\Big(\frac{d}{dz}\Big)^{n+2\ell+1}\big[e^{irz}(z+i\omega)^n\big].\end{aligned}$$

Using Leibniz's formula and (9.5.12) we obtain

$$\operatorname{Res}(F, i\omega) = Q_{n+2\ell+1}(r)e^{-\frac{r}{2(n+\ell+1)}}, \tag{9.5.17}$$

where $Q_{n+2\ell+1}(r)$ is a polynomial of degree less than or equal to $n+2\ell+1$ which satisfies $Q_{n+2\ell+1}(r) = r^{2\ell+1}\widetilde{Q}_{n,\ell}(r)$.

To finish the computation of the integral appearing in (9.5.15) it remains to let R tend to $+\infty$.

First, using (9.5.14) as well as the dominated convergence theorem, we have

$$\lim_{R\to+\infty}\int_{\Gamma_1} F(z)\,dz = \int_{\mathbf{R}} e^{ir\xi}\frac{(\xi+i\omega)^n}{(\xi-i\omega)^{n+2\ell+2}}\,d\xi. \tag{9.5.18}$$

Next, on Γ_2 we have $z = R+it$, $t\in(0,R)$, so that $|F(z)|\le C\frac{e^{-rt}}{R^{2\ell+2}}$. As $r>0$ we obtain

$$\lim_{R\to+\infty}\int_{\Gamma_2} F(z)\,dz = 0. \tag{9.5.19}$$

A completely similar reasoning shows that

$$\lim_{R\to+\infty}\int_{\Gamma_4} F(z)\,dz = 0. \tag{9.5.20}$$

Finally, on Γ_3 we have $z=t+iR$, $t\in(R,-R)$. Thus $|F(z)|\le C\frac{e^{-rR}}{R^{2\ell+2}}$ so that

$$\lim_{R\to+\infty}\int_{\Gamma_3} F(z)\,dz = 0. \tag{9.5.21}$$

It follows from (9.5.15) to (9.5.21) that

$$v(r) = r^{2\ell+1}\widetilde{Q}_{n,\ell}(r)e^{-\frac{r}{2(n+\ell+1)}},$$

and thus,

$$u_P(r) = r^{\ell}\widetilde{Q}_{n,\ell}(r)e^{-\frac{r}{2(n+\ell+1)}},$$

where $\widetilde{Q}_{n,\ell}(r)$ is a polynomial of degree less than or equal to n.

9.5.1 The Eigenvectors

To find the eigenvectors associated with the eigenvalue $E=-\frac{1}{4(n+\ell+1)^2}$, we note that we could have taken in the previous calculations, ψ in the form $\psi(r\omega)=U_\ell(r)P_\ell(\omega)$. Then $u_P=\alpha_\ell U$ where $\alpha_\ell=\|P_\ell\|^2_{L^2(\mathbf{S}^2)}\neq 0$. We thus obtain

$$\psi_{n,\ell}(x) = |x|^{\ell}Q_{n,\ell}(|x|)e^{-\frac{|x|}{2(n+\ell+1)}}P_\ell\Big(\frac{x}{|x|}\Big),$$

where $Q_{n,\ell}$ is a polynomial of degree less than or equal to n and P_ℓ is a spherical harmonic of degree ℓ.

The dimension of the eigenspace associated with the eigenvalue $E_k = -\frac{1}{(k+1)^2}$ where $k \in \mathbf{N}$ can be calculated. It is equal to $\dim E_k = \sum_{\ell=0}^{k} \dim \mathcal{H}_\ell$. Taking into account Lemma 8.3.6 we have

$$\dim E_k = \sum_{\ell=0}^{k} (2\ell + 1) = 2\frac{k(k+1)}{2} + k + 1 = (k+1)^2.$$

□

9.6 Appendix

9.6.1 Invariance of the Laplacian Under Rotation

We aim to show the following lemma.

Lemma 9.6.1 *Let $A \in O(d, \mathbf{R})$ be an orthogonal matrix and $u \in C^2(\mathbf{R}^d)$. If we set $v(x) = u(Ax)$ we have*

$$\Delta_x v(x) = (\Delta_y u)(Ax), \quad \forall x \in \mathbf{R}^d. \tag{9.6.1}$$

Proof This follows from the second-order Taylor formula.

Let us denote $\text{Hess}_x f = \left(\partial_{x_j} \partial_{x_k} f\right)_{1 \le j,k \le d}$ the Hessian matrix of a function f belonging to $C^2(\mathbf{R}^d)$ and note that $\Delta_x f = \text{Tr}(\text{Hess}_x f)$ where Tr denotes the trace of the matrix.

We can write for $x, x_0, y, y_0 \in \mathbf{R}^d$,

$$v(x) = v(x_0) + \nabla_x v(x_0) \cdot (x - x_0) + \frac{1}{2}\langle \text{Hess}_x v(x_0)(x - x_0), x - x_0 \rangle + o(|x - x_0|^2), \tag{9.6.2}$$

$$u(y) = u(y_0) + \nabla_y u(y_0) \cdot (y - y_0) + \frac{1}{2}\langle \text{Hess}_y u(y_0)(y - y_0), y - y_0 \rangle + o(|y - y_0|^2). \tag{9.6.3}$$

Let us apply the equality (9.6.3) to $y = Ax$, $y_0 = Ax_0$. We obtain

$$\begin{aligned} u(Ax) = u(Ax_0) &+ (\nabla_y u)(Ax_0) \cdot A(x - x_0) \\ &+ \frac{1}{2}\langle \text{Hess}_y u(Ax_0) A(x - x_0), A(x - x_0) \rangle + o(|A(x - x_0)|^2). \end{aligned} \tag{9.6.4}$$

Since $v(x) = u(Ax)$ and $|A(x - x_0)|^2 = |x - x_0|^2$ the equalities (9.6.2) and (9.6.4) allow us to deduce that

$$\nabla_x v(x_0) = (\nabla_y u)(Ax_0) \circ A, \quad \text{Hess}_x v(x_0) = {}^t\!A \, \text{Hess}_y u(Ax_0) A, \tag{9.6.5}$$

where ${}^t\!A$ denotes the transpose matrix of A. Since $\text{Tr}(BM) = \text{Tr}(MB)$ and ${}^t\!A\,A$ is the identity matrix, we deduce from (9.6.5) that

$$\begin{aligned}\Delta_x v(x_0) &= \mathrm{Tr}(\mathrm{Hess}_x\, v)(x_0) = \mathrm{Tr}({}^t\!A\, \mathrm{Hess}_y\, u(Ax_0)A) = \mathrm{Tr}(\mathrm{Hess}_y\, u(Ax_0){}^t\!A\, A),\\ &= \mathrm{Tr}(\mathrm{Hess}_y\, u(Ax_0)) = (\Delta_y u)(Ax_0),\end{aligned}$$

which proves (9.6.1). □

9.6.2 Tempered Fundamental Solution of the Operator $\Delta + \omega^2$ in $\mathbf{R}^3$

Lemma 9.6.2 *Let $\omega > 0$. The unique tempered fundamental solution of the operator $P = -\Delta + \omega^2$ in $\mathbf{R}^3$ is given by the function $E(x) = \frac{1}{4\pi}\frac{e^{-\omega|x|}}{|x|}$.*

Proof First, there is only one tempered fundamental solution. Indeed, let E_1, E_2 be in $\mathcal{S}'(\mathbf{R}^3)$ and satisfy $(-\Delta + \omega^2)E_j = \delta_0$ then $E = E_1 - E_2$ is a solution of $(-\Delta + \omega^2)E = 0$ so, by Fourier transform, $(|\xi|^2 + \omega^2)\widehat{E} = 0$ hence $\widehat{E} = 0$ since $|\xi|^2 + \omega^2 \neq 0$ and finally $E = 0$.

So we look for $E \in \mathcal{S}'(\mathbf{R}^3)$ satisfying $(-\Delta + \omega^2)E = \delta_0$, or, equivalently, $\widehat{E} = \frac{1}{|\xi|^2+\omega^2}$. Let us consider the function $u(x) = \frac{e^{-\omega|x|}}{|x|}$ which is in $L^1(\mathbf{R}^3)$. We will compute its Fourier transform. We have

$$\widehat{u}(\xi) = \int_{\mathbf{R}^3} e^{-ix\cdot\xi}\frac{e^{-\omega|x|}}{|x|}\,dx, \quad \forall \xi \in \mathbf{R}^d.$$

Since u is radial, so is $\widehat{u}$ and thus,

$$\widehat{u}(\xi) = \widehat{u}(0, 0, |\xi|) = \int_{\mathbf{R}^3} e^{-ix_3|\xi|}\frac{e^{-\omega|x|}}{|x|}\,dx.$$

We will switch to polar coordinates $(r, \theta, \varphi) \in (0, +\infty) \times (0, \pi) \times (0, 2\pi)$. In this case we have $x_3 = r\cos\theta$ and dx becomes $r^2 \sin\theta\, dr\, d\theta\, d\varphi$. We then obtain

$$\begin{aligned}\widehat{u}(\xi) &= \int_0^{+\infty}\int_0^{2\pi}\int_0^{\pi} e^{-ir|\xi|\cos\theta}\frac{e^{-\omega r}}{r} r^2 \sin\theta\, dr\, d\theta\, d\varphi,\\ &= 2\pi\int_0^{+\infty} re^{-\omega r}\Big(\int_0^{\pi} e^{-ir|\xi|\cos\theta}\sin\theta\, d\theta\Big)\,dr.\end{aligned}$$

Since $e^{-ir|\xi|\cos\theta}\sin\theta = \frac{d}{d\theta}\Big[\frac{1}{ir|\xi|}e^{-ir|\xi|\cos\theta}\Big]$, using the above equality we obtain

$$\begin{aligned}\widehat{u}(\xi) &= \frac{2\pi}{i|\xi|}\Big(\int_0^{+\infty} e^{-\omega r + ir|\xi}\,dr - \int_0^{+\infty} e^{-\omega r - ir|\xi|}\,dr\Big)\\ &= \frac{2\pi}{i|\xi|}\Big(\frac{1}{\omega - i|\xi|} - \frac{1}{\omega + i|\xi|}\Big) = \frac{4\pi}{|\xi|^2 + \omega^2}.\end{aligned}$$

With the above notations we thus have $\widehat{E} = \frac{1}{4\pi}\widehat{u}$ in $\mathcal{S}'(\mathbf{R}^3)$. Consequently we have $E = \frac{1}{4\pi}\frac{e^{-\omega|x|}}{|x|}$ as stated.

Note that there also exists a non-tempered fundamental solution given by

$$F(x) = \frac{1}{4\pi}\frac{e^{\omega|x|}}{|x|} \in L^1_{\text{loc}}(\mathbf{R}^3) \subset \mathcal{D}'(\mathbf{R}^3).$$

□

9.7 Comments

The reader interested in the questions raised in this chapter may profitably consult reference [13].

Chapter 10
Spectral Theory of the Dirichlet Problem for the Laplacian

10.1 Prerequisites

Hilbert spaces. Weak convergence. Operators on Hilbert spaces. Compact operators. Sobolev spaces. Compactness of the inclusion of $H_0^1(\Omega)$ into $L^2(\Omega)$, Ω open and bounded.[1]

10.2 The Problem

The purpose of this chapter is the study of the **spectrum** of the Dirichlet problem for the Laplacian in a **bounded** open set of $\mathbf{R}^d$.

This question has been the subject of very many works because, on the one hand, as we shall see, this theory plays, in the case of bounded open sets, the same role as that of the Fourier transform in $\mathbf{R}^d$, and on the other hand, it has been shown that there are very close links between this spectrum and the geometry of the open set: its volume, the surface area of its boundary, the number of holes, etc. To carry out this study, we will include this problem in a very general abstract framework, that of compact self-adjoint operators.

10.3 Spectral Theory of Compact Self-Adjoint Operators

In this section, H is a Hilbert space over $\mathbf{C}$, whose inner product will be denoted by $(\, , \,)$, and T is a continuous linear operator from H to H; we will write $T \in \mathcal{L}(H)$. We begin with some definitions and properties.

---

[1] This chapter is based on Chap. 13 of my book "Elements de distributions et d'équations aux dérivées partielles" Dunod (2002).

C. Zuily, *Selected Topics in Partial Differential Equations*, Universitext,
https://doi.org/10.1007/978-3-032-24082-8_10

10.3.1 Spectrum

The **spectrum** of T, denoted by $\sigma(T)$, is the subset of $\mathbf{C}$ defined by

$$\sigma(T) = \left\{\mu \in \mathbf{C} : T - \mu \operatorname{Id} \text{ is non invertible}\right\}, \tag{10.3.1}$$

where Id is the identity operator, $x \mapsto x$, from H to itself.

In particular, $\mu \in \mathbf{C}$ is called an **eigenvalue** of T if $T - \mu \operatorname{Id}$ is not injective, *i.e.* if there exists $x \in H, x \neq 0$ such that $T\,x = \mu\,x$. The set of eigenvalues is a subset of the spectrum.

Lemma 10.3.1 *The set $\sigma(T)$ is closed in $\mathbf{C}$.*

Proof We show that $(\sigma(T))^c$ is open. Let $\mu_0 \in \mathbf{C}$ such that $T - \mu_0 \operatorname{Id} = S_0$ is invertible. For $\mu \neq \mu_0$ we write $T - \mu \operatorname{Id} = (\mu_0 - \mu)\operatorname{Id} + S_0$ hence

$$S_0^{-1}(T - \mu \operatorname{Id}) = (T - \mu \operatorname{Id})\, S_0^{-1} = \operatorname{Id} + (\mu_0 - \mu)\, S_0^{-1}.$$

If $|\mu_0 - \mu|$ is small enough (such that $|\mu_0 - \mu|\,\|S_0^{-1}\| < 1$), the operator $\operatorname{Id} + (\mu_0 - \mu)\, S_0^{-1}$ is invertible (Neumann series) and thus $T - \mu \operatorname{Id}$ is invertible, that is, $\mu \in (\sigma(T))^c$. □

10.3.2 Adjoint

Proposition 10.3.2 *There exists a unique operator $T^* \in \mathcal{L}(H)$ such that*

$$(T\,x, y) = (x, T^* y), \quad \forall x, y \in H. \tag{10.3.2}$$

Proof *(a) Uniqueness*: if there exist T_1^*, T_2^* satisfying (10.3.2), we have

$$(x, (T_1^* - T_2^*)\, y) = 0 \quad \text{for all } x, y \in H.$$

Therefore, $(T_1^* - T_2^*)\, y = 0$, $\forall\, y \in H$, *i.e.* $T_1^* = T_2^*$.

(b) Existence: fix $y \in H$. The map $x \mapsto (T\,x, y)$ is a continuous linear form on H because $|(T\,x, y)| \leq \|T\,x\|\,\|y\| \leq \|T\|\,\|y\|\,\|x\|$. Therefore, there exists a unique element $z_y \in H$ such that $(T\,x, y) = (x, z_y)$, for all $x \in H$. The map $y \mapsto z_y$ is linear. Let us denote $z_y = T^* y$. Then T^* is continuous because

$$\|T^* y\| = \sup_{x \neq 0} \frac{|(x, T^* y)|}{\|x\|} = \sup_{x \neq 0} \frac{|(T\,x, y)|}{\|x\|} \leq \|T\|\,\|y\|.$$

□

The operator T^* is called **the adjoint** of T. We say that T is **self-adjoint** if $T = T^*$.

Proposition 10.3.3 *If $T \in \mathcal{L}(H)$ is self-adjoint, its eigenvalues are real.*

Proof Let x be an eigenvector corresponding to an eigenvalue $\mu \in \mathbf{C}$. We have

$$(T\,x, x) = \mu\,\|x\|^2 = (x, T\,x) = \overline{\mu}\,\|x\|^2,$$

hence $\mu = \overline{\mu}$. □

10.3.3 Positive Operators

$T \in \mathcal{L}(H)$ is said to be **positive** (and we write $T \geq 0$) if $(T\,x, x) \geq 0$, for all x in H.
If $T \geq 0$, its eigenvalues are positive or zero.

10.3.4 Compact Operators

$T \in \mathcal{L}(H)$ is said to be **compact** if the image by T of a ball in H is relatively compact in H.

Proposition 10.3.4 *The following conditions are equivalent.*
(i) T is compact.
(ii) The image by T of any bounded sequence contains a convergent subsequence.
(iii) T maps a weakly convergent sequence to a strongly convergent sequence (i.e. in the norm).

Proof The equivalence of (i) and (ii) is obvious.
$(iii) \Rightarrow (ii)$. Recall that $(x_n)_{n\in\mathbf{N}} \subset H$ is said to converge weakly to x (and we write $x_n \rightharpoonup x$) if $(x_n, y) \to (x, y)$ for all $y \in H$. Recall also that a closed ball in a Hilbert space is weakly compact. Let (x_n) be a bounded sequence; then there exists a subsequence (x_{n_k}) that converges weakly. According to (iii), the sequence $(T(x_{n_k}))$ is convergent, hence (ii).
$(ii) \Rightarrow (iii)$. We use the following result.

Lemma 10.3.5 *Every weakly convergent sequence is bounded.*

Proof This follows from the Banach-Steinhaus Theorem. If (x_n, y) tends to (x, y) for all y, there exists $M_y > 0$ such that $|(x_n, y)| \leq M_y$ for all $n \in \mathbf{N}$. Since the application $T_n : y \mapsto (y, x_n)$ is linear, there exists $M > 0$ such that $|(x_n, y)| \leq M\,\|y\|$, for all $y \in H$. It follows that we have $\|x_n\| = \sup_{y\neq 0} \frac{|(x_n, y)|}{\|y\|} \leq M$. □

Suppose $x_n \rightharpoonup x$; since $\|x_n\|$ is bounded, we deduce from (ii) that the sequence $(T\,x_n)_{n\in\mathbf{N}}$ has at least one cluster point. If this is unique, the sequence $(T\,x_n)$ will be convergent (since $(T\,x_n)_{n\in\mathbf{N}}$ is contained in a compact set). Let y be a cluster point.

There exists a subsequence (x_{n_k}) such that $T\,x_{n_k} \to y$. Next, $T\,x_n \rightharpoonup T\,x$, because $(T\,x_n - T\,x, y) = (x_n - x, T^* y) \to 0$ so $y = T\,x$. □

We will admit in the following the result below.

Proposition 10.3.6 *Let $T \in \mathcal{L}(H)$ be a compact self-adjoint operator. Then T admits as eigenvalues the numbers $\pm\|T\|$.*

We can now state the main result of this paragraph. Recall that $Ker(S)$ denotes the kernel of the operator S.

Theorem 10.3.7 *Let $T \in \mathcal{L}(H)$ be a compact self-adjoint operator.*

(1) *The nonzero eigenvalues of T form a finite set or a sequence $\{\mu_n\}$ of real numbers tending to zero.*

(2) *If μ_n is a nonzero eigenvalue, the space $E_{\mu_n} = Ker\,(T - \mu_n\, Id)$ is finite-dimensional.*

(3) $H = \overset{+\infty}{\underset{j=1}{\oplus}} E_{\mu_n} \oplus E_0$ *where* $E_0 = Ker\,T$.

(4) *If* $\dim H = +\infty$, $\sigma(T) = \{0\} \cup \big(\bigcup_{j=1}^{+\infty}\{\mu_j\}\big)$ *where the μ_j are the nonzero eigenvalues. If* $\dim H < +\infty$, $\sigma(T) = \{eigenvalues\}$.

Proof *First point*: let μ_i and μ_j be two distinct eigenvalues. Then E_{μ_i} and E_{μ_j} are orthogonal.

Indeed, let $x_i \in E_{\mu_i}$, $x_j \in E_{\mu_j}$, *i.e.* $T\,x_i = \mu_i\,x_i$, $T\,x_j = \mu_j\,x_j$. Then we have $(T\,x_i, x_j) = (x_i, T\,x_j)$ hence $\mu_i(x_i, x_j) = \mu_j(x_i, x_j)$ since the μ_i are real. Therefore, $(\mu_i - \mu_j)(x_i, x_j) = 0$, which implies $(x_i, x_j) = 0$.

Second point: $\mu_j \neq 0 \Rightarrow \dim E_{\mu_j} < +\infty$.

Let B_j be the unit ball of E_{μ_j}, then $T(B_j) = \mu_j\,B_j$ *i.e.* $B_j = \frac{1}{\mu_j}\,T(B_j)$. Since T is compact, $T(B_j)$ is relatively compact in H, hence in E_{μ_j} which is a closed subspace of H. It follows from the Riesz Theorem that $\dim E_{\mu_j} < +\infty$ which proves (2).

Third point: let us show (1). Let $\varepsilon > 0$; we will prove that there are at most finitely many eigenvalues μ such that $|\mu| \geq \varepsilon$. Suppose, on the contrary, that there are infinitely many; in particular, there exists a sequence (μ_{k_j}) such that $\mu_{k_j} \neq \mu_{k_\ell}$ if $j \neq \ell$ and $|\mu_{k_j}| \geq \varepsilon$. Let x_{k_j} be an eigenvector of norm 1 associated to μ_{k_j}. Then,

$$\|T\,x_{k_j} - T\,x_{k_\ell}\|^2 = \|T\,x_{k_j}\|^2 + \|T\,x_{k_\ell}\|^2 + 2\,\mathrm{Re}\,(T\,x_{k_j}, T\,x_{k_\ell}) = \mu_{k_j}^2 + \mu_{k_\ell}^2,$$

since $(T\,x_{k_j}, T\,x_{k_\ell}) = \mu_{k_j}\,\mu_{k_\ell}(x_{k_j}, x_{k_\ell}) = 0$, according to the first point. Therefore we have $\|T\,x_{k_j} - T\,x_{k_\ell}\|^2 \geq 2\,\varepsilon^2$. In summary, we have a sequence (x_{k_j}) in the unit ball of H, such that $(T\,x_{k_j})$ admits no convergent subsequence. This contradicts the fact that T is compact. Consequently, in each interval $\left[\frac{1}{n+1}, \frac{1}{n}\right]$ there is a finite number of eigenvalues: these therefore form a sequence which is either finite or infinite. If it is infinite, for every $n \in \mathbf{N}$, there are at most a finite number such that $|\mu| \geq \frac{1}{n}$. Therefore, the sequence tends to zero.

Fourth point: let us show (3). Let $V = \overset{+\infty}{\underset{j=1}{\oplus}} E_{\mu_j}$ be the Hilbert sum of the spaces E_{μ_j} such that $\mu_j \neq 0$. Let N be the orthogonal of V. Then N is stable under T. Indeed, note that $x \in V^{\perp}$ if and only if $x \in E_{\mu_j}^{\perp}$ for all j. Then $(T\,x, y_j) = (x, T\,y_j) = \mu_j(x, y_j) = 0$ if $y_j \in E_{\mu_j}$. Therefore $T\,x \in V^{\perp} = N$. The restriction T' of T to N is a compact and self-adjoint operator which has no nonzero eigenvalue; according to Proposition 10.3.6 we have $T' = 0$ *i.e.* $T|_N = 0$ and thus $N \subset \operatorname{Ker} T$. On the other hand, according to the first point, $\operatorname{Ker} T$ (if it is not reduced to zero) is orthogonal to all the E_{μ_j} *i.e.* $\operatorname{Ker} T \subset V^{\perp} = N$. Therefore $N = \operatorname{Ker} T$ and $H = V \oplus V^{\perp} = V \oplus N = \overset{+\infty}{\underset{j=1}{\oplus}} E_{\mu_j} \oplus \operatorname{Ker} T$.

Fifth point: let us show (4). Let $\Lambda = \{0\} \cup \big(\overset{+\infty}{\underset{1}{\cup}} \{\mu_j\}\big)$. Let us show that Λ^c is included in $(\sigma(T))^c$. Let $\mu \neq 0$ such that $\mu \neq \mu_j$ for all j. Then $T - \mu\,\mathrm{Id}$ is injective, since all the eigenvalues are in Λ. Let us show that $T - \mu\,\mathrm{Id}$ is surjective. Let $y \in H$. According to (3), y can be written as $y = y_0 + \sum_{n,j}(y, e_j^{(n)})\, e_j^{(n)}$ where $(e_j^{(n)})$ is an orthonormal basis of E_{μ_n} (which is finite-dimensional) and $y_0 \in \operatorname{Ker} T$. Set

$$x = -\frac{1}{\mu}\, y_0 + \sum_{n,j} (y, e_j^{(n)})\, \frac{e_j^{(n)}}{\mu_n - \mu}.$$

Since $(\mu_n) \to 0$, for n large enough we have $|\mu_n - \mu| \geq \frac{1}{2}\,|\mu| > 0$ so,

$$\frac{|(y, e_j^{(n)})|^2}{|\mu_n - \mu|^2} \leq \Big(\frac{2}{|\mathbf{R}^d \mu|}\Big)^2\, |(y, e_j^{(n)})|^2,$$

which shows that the series defining x converges in H. It is easy to see that we have $(T - \mu\,\mathrm{Id})\,x = y$; thus $T - \mu\,\mathrm{Id}$ is bijective and $\mu \in (\sigma(T))^c$. We have thus shown the inclusion $\sigma(T) \subset \Lambda$. Conversely, we always have $\overset{+\infty}{\underset{1}{\cup}} \{\mu_j\} \subset \sigma(T)$. If $\dim H = +\infty$, the sequence (μ_j) tends to 0; thus $0 \in \sigma(T)$, since $\sigma(T)$ is closed *i.e.* $\Lambda \subset \sigma(T)$. □

10.4 The Dirichlet Problem for the Laplacian and Its Spectral Theory

10.4.1 The Dirichlet Problem

Let Ω be a **bounded** subset of $\mathbf{R}^d$ and $\Delta = \sum_{j=1}^{d} \partial_j^2$ be the Laplacian in $\mathbf{R}^d$, where $\partial_j = \frac{\partial}{\partial x_j}$. We first study the Dirichlet problem and its properties. The main result is the following.

Theorem 10.4.1 *1. For any $f \in H^{-1}(\Omega)$ there exists a unique $u \in H_0^1(\Omega)$ such that $-\Delta u = f$ in $\mathcal{D}'(\Omega)$.*

2. *Consider the space $E = \{u \in H_0^1(\Omega) : \Delta u \in L^2(\Omega)\}$ equipped with the norm $\|u\|_E = \|u\|_{H_0^1} + \|\Delta u\|_{L^2}$. Then the map*

$$-\Delta : E \to L^2(\Omega)$$

is an isomorphism.

3. *The operator $T = (-\Delta)^{-1}$ from $L^2(\Omega)$ to $L^2(\Omega)$ is compact.*
4. *The operator T is self adjoint in $L^2(\Omega)$.*
5. *The operator T is positive.*

The proof of this theorem will require several steps. We begin by the following result.

Proposition 10.4.2 (i) *On $H_0^1(\Omega)$ the quantity*

$$((u, v)) = \sum_{j=1}^{d} \left(\partial_j u, \partial_j v\right)_{L^2(\Omega)}$$

is a scalar product.

(ii) *Equipped with this scalar product $H_0^1(\Omega)$ is a Hilbert space.*

Proof For (i) the only non trivial thing to verify is the following assertion:

$$u \in H_0^1(\Omega), \quad ((u, u)) = 0 \implies u = 0. \tag{10.4.1}$$

This will be a consequence of the following "Poincaré inequality".

Lemma 10.4.3 *There exists a positive constant C depending on the diameter of Ω such that for every $u \in H_0^1(\Omega)$ we have*

$$\|u\|_{L^2} \leq C \sum_{j=1}^{d} \|\partial_j u\|_{L^2}. \tag{10.4.2}$$

Proof of the Lemma. Consider first the case where $u \in C_0^\infty(\Omega)$ and extend u by zero outside its support. Denote by δ the diameter of Ω and let $x^0 \in \Omega$. Then $\Omega \subset B(x^0, \delta)$ (the open ball) and for every $x' = (x_2, \ldots, x_d) \in \mathbf{R}^{d-1}$ the point $X = (\delta + x_1^0, x')$ does not belong to $B(x^0, \delta)$. Therefore $u(\delta + x_1^0, x') = 0$. It follows that for any $x = (x_1, x') \in \Omega$ one can write

$$u(x) = \int_{\delta + x_1^0}^{x_1} \partial_1 u(t, x')\, dt.$$

Using the Cauchy-Schwarz inequality one deduce that

$$|u(x)|^2 \leq |\delta + x_1^0 - x_1| \int_{\delta+x_1^0}^{x_1} \left|\partial_1 u(t,x')\right|^2 dt \leq 2\delta \int_{\delta+x_1^0}^{x_1} |\partial_1 u(t,x')|^2 \, dt.$$

Now, since $|x_1 - x_1^0| < \delta$ if $x \in \Omega$ we can write

$$\begin{aligned}\int_\Omega |u(x)|^2 \, dx &\leq 2\delta \int_{|x_1-x_1^0|<\delta} \int_{x' \in \mathbf{R}^{d-1}} \int_{t\in\mathbf{R}} |\partial_1 u(t,x')|^2 \, dt \, dx' \, dx_1, \\ &\leq 4\delta^2 \int_\Omega |\partial_1 u(x)|^2 \, dx,\end{aligned}$$

since u has its support contained in Ω. This proves the desired inequality for $u \in C_0^\infty(\Omega)$.

Now, if $u \in H_0^1(\Omega)$ there exists a sequence $(u_k) \subset C_0^\infty(\Omega)$ which converges to u in $H^1(\Omega)$. We have just to apply the inequality (10.4.2) to u_k and pass to the limit when $k \to +\infty$. We obtain (10.4.2) for u. □

The assertion (10.4.1) follows immediately from (10.4.2) and proves (i) in Proposition 10.4.2. The point (ii) in this Proposition follows immediately from the fact that $L^2(\Omega)$ is a complete normed space. □

End of the proof of Theorem 10.4.1 Point 1. Since $H^{-1}(\Omega)$ is the dual of $H_0^1(\Omega)$ the representation theorem of linear forms on a Hilbert space implies that

$$\forall f \in H^{-1}(\Omega), \ \exists u \in H_0^1(\Omega) \text{ unique, such that } \langle f, v\rangle = ((u,v)), \quad \forall v \in H_0^1(\Omega).$$

In particular, taking $v \in C_0^\infty(\Omega)$ we get

$$\langle f, v\rangle = ((u,v)) = -\sum_{j=1}^{d} \langle \partial_j^2 u, v\rangle = -\langle \Delta u, v\rangle,$$

which shows that $-\Delta u = f$ in $\mathcal{D}'(\Omega)$.

Point 2. $-\Delta$ maps trivially E in $L^2(\Omega)$. Moreover by definition of E this map is continuous. Let us show that it is surjective. Indeed since $L^2(\Omega) \subset H^{-1}(\Omega)$, by point 1. in Theorem 10.4.1 for any $f \in L^2(\Omega)$ there exists $u \in H_0^1(\Omega)$ such that $-\Delta u = f$. So $u \in H_0^1(\Omega)$ and $\Delta u \in L^2(\Omega)$ thus $u \in E$. Point 1. in Theorem 10.4.1 shows also that $-\Delta$ is injective thus bijective. The continuity of the inverse map is a consequence of the Banach Theorem.

Point 3. Recall that the identity map from $H_0^1(\Omega)$ to $L^2(\Omega)$ is compact, since Ω is bounded. Moreover recall that the composition of a continuous map and a compact map is also compact. Now $(-\Delta)^{-1}$ is continuous from $L^2(\Omega)$ to E by point 2., the identity map, $u \mapsto u$, is continuous from E to $H_0^1(\Omega)$ and compact from $H_0^1(\Omega)$ to $L^2(\Omega)$. Thus the composition is compact.

Point 4. Let us show that T is self-adjoint, *i.e.* denoting by (,) the inner product of L^2,

$$(T f, g) = (f, T g), \quad \forall f, g \in L^2(\Omega). \tag{10.4.3}$$

The equality $u = T\,f \in E$ is equivalent to $-\Delta\,u = f$; similarly $v = T\,g$ is equivalent to $-\Delta\,v = g$. Thus, (10.4.3) can be written as

$$(u, -\Delta\,v) = (-\Delta\,u, v)\,, \quad u, v \in E. \tag{10.4.4}$$

We will show that

$$-\,(u, \Delta\,v) = \sum_{i=1}^{n} \left(\frac{\partial u}{\partial x_i}, \frac{\partial v}{\partial x_i}\right), \quad u, v \in E. \tag{10.4.5}$$

If $v \in E$ and $u \in C_0^\infty(\Omega)$, the equality (10.4.5) is obvious because

$$(u, \partial_i^2\,v) = \langle \partial_i^2\,\overline{v}, u\rangle = -\langle \partial_i\,\overline{v}, \partial_i\,u\rangle = -\int \partial_i\,\overline{v}\cdot\partial_i\,u\,dx = -(\partial_i\,u, \partial_i\,v)\,.$$

If $u \in H_0^1$, let $(u_j) \in C_0^\infty$ such that $u_j \to u$ in H_0^1. Then $(u_j, \Delta\,v) \to (u, \Delta\,v)$ since $|(u_j - u, \Delta\,v)| \le \|u_j - u\|_{L^2}\,\|\Delta\,v\|_{L^2} \to 0$ and $(\partial_i\,u_j, \partial_i\,v) \to (\partial_i\,u, \partial_i\,v)$ because we can write $|(\partial_i\,u_j - \partial_i\,u, \partial_i\,v)| \le \|u_j - u\|_{H^1}\,\|v\|_{H^1} \to 0$. Thus, (10.4.5) is proved and (10.4.4) follows by swapping u and v.

Point 5. T is positive because, according to (10.4.3), (10.4.4), and (10.4.5), we have

$$(T\,f, f) = (u, -\Delta\,u) = \sum_{i=1}^{n} \|\partial_i\,u\|_{L^2}^2\,.$$

Notice that, by Point 2., T is injective. Therefore, $\mathrm{Ker}\,T = \{0\}$. □

10.4.2 Spectral Theory of the Dirichlet Problem

Using Theorem 10.4.1 we shall apply Theorem 10.3.7 to $T = (-\Delta)^{-1}$ and $H = L^2(\Omega)$.

The spectrum of T consists of zero and a sequence of eigenvalues $\mu_n > 0$ which tend to zero, i.e., the problem $T\,f = \mu\,f$, $f \in L^2(\Omega)$ has a solution if and only if $\mu = \mu_n$. Now,

$$T\,f = \mu\,f\,, \quad f \in L^2 \Longleftrightarrow -\Delta\,f = \lambda\,f\,, \quad f \in H_0^1\,, \quad \mu = \frac{1}{\lambda}\,.$$

Notice that $\lambda = 0$ is not an eigenvalue since

$$u \in H_0^1(\Omega) \text{ and } \Delta u = 0 \Longrightarrow u = 0.$$

Consequently, the problem

$$-\Delta u = \lambda u, \quad u \in H_0^1(\Omega) \tag{10.4.6}$$

is solvable if and only if λ belongs to a sequence (λ_n) of strictly positive numbers tending to $+\infty$. We order them

$$0 < \lambda_1 \leq \lambda_2 \leq \cdots \leq \lambda_n \leq \cdots \to +\infty \tag{10.4.7}$$

repeating the multiple eigenvalues.

According to Theorem 10.3.7, (3), there exists an orthonormal basis (e_n) of L^2 such that $T\, e_n = \mu_n\, e_n$, or equivalently,

$$-\Delta e_n = \lambda_n e_n, \quad e_n \in H_0^1(\Omega)\,. \tag{10.4.8}$$

Note that, by hypoellipticity of the Laplacian, we have $e_n \in C^\infty(\Omega)$ and if Ω has a C^∞ boundary, $e_n \in C^\infty(\overline{\Omega})$.

We will show that (e_n) is an orthogonal basis of $H_0^1(\Omega)$ and that

$$\|e_n\|_{H_0^1} = \sqrt{\lambda_n}\,. \tag{10.4.9}$$

Indeed, $(T\, e_n, e_m)_{L^2} = (u_n, -\Delta\, v_m)$ where $-\Delta\, v_m = e_m$, i.e., $v_m = T\, e_m$. Then,

$$\mu_n(e_n, e_m)_{L^2} = \sum_{i=1}^{n} \Big(\frac{\partial u_n}{\partial x_i}, \frac{\partial v_m}{\partial x_i}\Big) = \sum_{i=1}^{n} \Big(\frac{\partial}{\partial x_i} T\, e_n, \frac{\partial}{\partial x_i} T\, e_m\Big) = \mu_n\, \mu_m (e_n, e_m)_{H_0^1},$$

so $\mu_m (e_n, e_m)_{H_0^1} = \delta_{nm}$. In summary,

$$\begin{cases} u \in L^2(\Omega) \Longleftrightarrow u = \displaystyle\sum_{n=1}^{+\infty} c_n\, e_n \quad \text{and} \quad \sum_{n=1}^{+\infty} |c_n|^2 < +\infty\,; \\ \|u\|_{L^2}^2 = \displaystyle\sum_{n=1}^{+\infty} |c_n|^2. \\ u \in H_0^1(\Omega) \Longleftrightarrow u = \displaystyle\sum_{n=1}^{+\infty} c_n\, e_n \quad \text{and} \quad \sum_{n=1}^{+\infty} \lambda_n\, |c_n|^2 < +\infty\,; \\ \|u\|_{H_0^1}^2 = \displaystyle\sum_{n=1}^{+\infty} \lambda_n\, |c_n|^2. \end{cases} \tag{10.4.10}$$

If Ω has C^∞ boundary, we can characterize the spaces $H_0^1(\Omega) \cap H^k(\Omega)$. Let us detail the case $k = 2$. We know that

$$u \in H_0^1(\Omega), \quad \Delta u \in L^2(\Omega) \Longleftrightarrow u \in H_0^1(\Omega) \cap H^2(\Omega).$$

Write $u = \sum_{n\geq 1} c_n e_n$ where the series is convergent in ℓ^2 (i.e., $\sum |c_n|^2 < +\infty$). Thus, the series converges in $\mathcal{D}'(\Omega)$. We then have, in $\mathcal{D}'(\Omega)$,

$$\Delta u = \sum_{n\geq 1} c_n \Delta e_n = -\sum_{n\geq 1} c_n \lambda_n e_n.$$

If $\Delta u \in L^2$, the series converges in ℓ^2, that is $\sum_{n\geq 1} \lambda_n^2 |c_n|^2 < +\infty$ and conversely. In the same way, one shows that for $k \geq 1$,

$$u \in H^k(\Omega) \cap H_0^1(\Omega) \Longleftrightarrow u = \sum_{n\geq 1} c_n e_n \quad \text{with} \quad \sum_{n\geq 1} \lambda_n^k |c_n|^2 < +\infty,$$

and $\|u\|_{H^k}^2 \sim \sum_{n\geq 1} \lambda_n^k |c_n|^2$.

10.5 Application to the Mixed Problem

We will see how the spectral theory of the Laplacian allows us to solve the Cauchy problem for the heat equation, when the spatial variable x belongs not to $\mathbf{R}^d$ (where the Fourier transform can be used) but to a bounded open subset of $\mathbf{R}^d$.

10.5.1 The Heat Equation

Theorem 10.5.1 *Let Ω be a bounded open subset of $\mathbf{R}^d$. Let $g \in L^2(\Omega)$. Then there exists a unique $u \in C^0([0, +\infty[, L^2(\Omega))$ such that*

$$\begin{cases} \frac{\partial u}{\partial t} - \Delta u = 0 \text{ in } \mathcal{D}'(]0, +\infty[\times\Omega) \\ u(0, \cdot) = g \\ u(t, \cdot) \in H_0^1(\Omega) \quad \forall t > 0, \end{cases} \tag{10.5.1}$$

and moreover, for all $\ell \in \mathbf{N}$, $k \in \mathbf{N}$, $\partial_t^\ell u \in C^0(]0, +\infty[, H^k(\Omega) \cap H_0^1(\Omega))$. This solution is given by the formula

$$u = \sum_{n=1}^{+\infty} e^{-\lambda_n t} (g, e_n)\, e_n, \tag{10.5.2}$$

where $(,)$ denotes the inner product of $L^2(\Omega)$, (λ_n) is the set of eigenvalues of the Laplacian, given by (10.4.6), (10.4.7) and (e_n) is an orthonormal basis of $L^2(\Omega)$ consisting of corresponding eigenvectors.

Proof Note that the hypoellipticity of the heat operator implies that any solution of (10.5.1) belongs to $C^\infty(]0,+\infty[\times\Omega)$.

(a) Existence: First, the formula (10.5.2) indeed defines, for $t \ge 0$, an element of $L^2(\Omega)$; indeed, $e^{-\lambda_n t}|(g, e_n)| \le |(g, e_n)|$ and $\Sigma\,|(g, e_n)|^2 < +\infty$ since $g \in L^2$. Obviously, $u_0 = g$. Next, for $t > 0$, $u(t,\cdot) \in H_0^1(\Omega)$. Indeed, we can write $\lambda_n\, e^{-\lambda_n t}\,|(g, e_n)| \le \frac{1}{t}\,|(g, e_n)|$ and it suffices to apply (10.4.10). Let us show that $\frac{\partial u}{\partial t} - \Delta u = 0$ in $\mathcal{D}'(]0,+\infty[\times\Omega)$. Recall that we have

$$\langle u, \psi\rangle = \int_0^{+\infty}\sum_{n=1}^{+\infty} e^{-\lambda_n t}\,(g, e_n)\,\langle e_n, \psi(t,\cdot)\rangle\, dt\,, \quad \psi \in C_0^\infty(]0,+\infty[\times\Omega)\,.$$

Lemma 10.5.2 *We have*

$$\langle u, \psi\rangle = \sum_{n=1}^{+\infty}\int_0^{+\infty} e^{-\lambda_n t}\,(g, e_n)\,\langle e_n, \psi(t,\cdot)\rangle\, dt\,, \quad \psi \in C_0^\infty(]0,+\infty[\times\Omega)\,.$$

Proof We will use Fubini's theorem. It suffices to check that the function $e^{-\lambda_n t}|(g, e_n)|\,|\langle e_n, \psi(t,\cdot)\rangle|$ is integrable with respect to the product measure. For $t > 0$, the function $x \mapsto \overline{\psi}(t, x)$ belongs to $L^2(\Omega)$ and we have $\|\overline{\psi}(t,\cdot)\|_{L^2}^2 = \sum_{n\ge1}|(e_n, \overline{\psi}(t,\cdot)|^2 = \sum_{n\ge1}|\langle e_n, \psi(t,\cdot)\rangle|^2 < +\infty$. Moreover, the function $\theta(t) = \|\overline{\psi}(t,\cdot)\|_{L^2}$ belongs to $C_0^0(]0,+\infty[)$. Then,

$$\begin{aligned}(1) = \sum_{n\ge1} e^{-\lambda_n t}\,|(g, e_n)|\,|\langle e_n, \psi(t,\cdot)\rangle| \le \quad &\Big(\sum_{n\ge1}|(g, e_n)|^2\Big)^{1/2}\Big(\sum_{n\ge1}|\langle e_n, \psi(t,\cdot)\rangle|^2\Big)^{1/2}\\ \le \|g\|_{L^2}\,\theta(t)\,.\end{aligned}$$

Thus, $\int_0^{+\infty}(1)\,dt \le \|g\|_{L^2}\int_0^{+\infty}\theta(t)\,dt < +\infty$ and the Fubini-Tonelli Theorem allows us to conclude. □

We then have

$$\langle \Delta u, \psi\rangle = \langle u, \Delta\psi\rangle = \sum_{n\ge1}\int_0^{+\infty} e^{-\lambda_n t}\,(g, e_n)\langle \Delta e_n, \psi(t,\cdot)\rangle\, dt\,.$$

But $-\Delta e_n = \lambda_n\, e_n$ so

$$\langle \Delta u, \psi\rangle = \sum_{n\ge1}\int_0^{+\infty}(g, e_n)(-\lambda_n)\, e^{-\lambda_n t}\,\langle e_n, \psi(t,\cdot)\rangle\, dt\,.$$

We use the fact that $-\lambda_n\, e^{-\lambda_n t} = \frac{d}{dt}\, e^{-\lambda_n t}$ and integrate by parts, which is possible since the function $t \mapsto \langle e_n, \psi(t,\cdot)\rangle$ belongs to $C_0^\infty(]0,+\infty[)$. Since we have $\psi(0,\cdot) = \psi(+\infty,\cdot) = 0$ we can write

$$\langle \Delta u, \psi \rangle = -\sum_{n\geq 1} \int_0^{+\infty} e^{-\lambda_n t}\, (g, e_n) \Big\langle e_n, \frac{\partial \psi}{\partial t}(t, \cdot) \Big\rangle\, dt = -\Big\langle u, \frac{\partial \psi}{\partial t} \Big\rangle = \Big\langle \frac{\partial u}{\partial t}, \psi \Big\rangle .$$

Let us show that $u \in C^0([0, +\infty[, L^2(\Omega))$. Let $t_0 \in [0, +\infty[$ and (t_j) be a sequence in $[0, +\infty[$ tending to t_0. Then,

$$\|u(t_0, \cdot) - u(t_j, \cdot)\|_{L^2}^2 = \sum_{n\geq 1} \big| e^{-\lambda_n t_0} - e^{-\lambda_n t_j} \big|^2\, |(g, e_n)|^2 .$$

Since $|e^{-\lambda_n t_0} - e^{-\lambda_n t_j}|^2\, |(g, e_n)|^2 \leq 4\, |(g, e_n)|^2$, it suffices to apply the dominated convergence theorem to conclude.

Next, it is easy to see that in $\mathcal{D}'(\mathbf{R}^d)$ we have

$$\partial_t^\ell u = \sum_{n\geq 1} e^{-\lambda_n t}\, (-\lambda_n)^\ell\, (g, e_n)\, e_n .$$

For $t > 0$, $\partial_t^\ell u(t, \cdot) \in H^k \cap H_0^1$ because

$$\lambda_n^k\, \lambda_n^{2\ell}\, e^{-2\lambda_n t}\, |(g, e_n)|^2 \leq \frac{(k + 2\ell)!}{(2t)^{k+2\ell}}\, |(g, e_n)|^2$$

and $\sum_{n\geq 1} |(g, e_n)|^2 < +\infty$. Finally, for $t_0 \in]0, +\infty[$ and $\frac{t_0}{2} \leq t_j \leq \frac{3}{2}\, t_0$, $t_j \to t_0$,

$$\|\partial_t^\ell u(t_0, \cdot) - \partial_t^\ell u(t_j, \cdot)\|_{H^k}^2 \leq C \sum_{n\geq 1} \lambda_n^{k+2\ell}\, |e^{-\lambda_n t_0} - e^{-\lambda_n t_j}|^2\, |(g, e_n)|^2,$$

and it suffices to apply the dominated convergence theorem to conclude that u belongs to the space $C^\ell(]0, +\infty[, H^k(\Omega) \cap H_0^1(\Omega))$.

(b) Uniqueness: let $u \in C^0([0, +\infty[, L^2(\Omega)) \cap C^0(]0, +\infty[, H^2(\Omega) \cap H_0^1(\Omega))$ such that

$$\frac{\partial u}{\partial t} - \Delta u = 0 \text{ in } \mathcal{D}'(]0, +\infty[\times \Omega), \quad u(0, \cdot) = 0 .$$

We will show that $u \equiv 0$.

First, note that it follows from the hypotheses that Δu belongs to $C^0(]0, +\infty[, L^2(\Omega))$ and thus, by the equation, that $\frac{\partial u}{\partial t} \in C^0(]0, +\infty[, L^2(\Omega))$. Denoting $(\,,\,)$ the inner product of $L^2(\Omega)$, we have for all $t > 0$,

$$2\text{Re}\Big(\Big(\frac{\partial u}{\partial t} - \Delta u \Big)(t, \cdot),\, u(t, \cdot) \Big) = 0 .$$

It is easy to see that, for $t > 0$,

$$2\text{Re}\Big(\frac{\partial u}{\partial t}(t, \cdot),\, u(t, \cdot) \Big) = \frac{d}{dt}\, \|u(t, \cdot)\|_{L^2}^2 ,$$

$$-(\Delta u(t,\cdot), u(t,\cdot)) = \sum_{i=1}^{n} \left\| \frac{\partial u}{\partial x_i}(t,\cdot) \right\|_{L^2}^2,$$

the second equality resulting from (10.4.5), since $u(t,\cdot) \in H_0^1(\Omega)$ and $\Delta u(t,\cdot)$ belongs to $L^2(\Omega)$ for $t > 0$. We deduce that

$$\frac{d}{dt} \|u(t,\cdot)\|^2 + \sum_{i=1}^{n} \left\| \frac{\partial u}{\partial x_i}(t,\cdot) \right\|_{L^2}^2 = 0.$$

Integrating this equality between $\varepsilon > 0$ and $T > 0$ we obtain

$$\|u(T,\cdot)\|_{L^2}^2 \le \|u(\varepsilon,\cdot)\|_{L^2}^2.$$

Since $u \in C^0([0,+\infty[, L^2(\Omega))$ and $u(0,\cdot) = 0$ we have $\lim_{\varepsilon\to 0^+} \|u(\varepsilon,\cdot)\|^2 = 0$, which proves that $u(T,\cdot) \equiv 0$, for all $T > 0$. □

Remark 10.5.3 Suppose that Ω is of class C^∞ and that $g \in C_0^\infty(\Omega)$. Then the solution u, given by the Theorem 10.5.1 belongs to $C^0([0,+\infty[, H^N(\Omega))$, for all $N \in \mathbf{N}$. In particular, by taking $N > \frac{d}{2}$ we see that u belongs to $C^0([0,+\infty[, C^0(\overline{\Omega}))$, thus to $C^0([0,+\infty[\times\overline{\Omega})$. Indeed, for all $N \in \mathbf{N}$ we have $C_0^\infty(\Omega) \subset H^N(\Omega) \cap H_0^1(\Omega)$, so that $\sum \lambda_n^N |(g, e_n)|^2 < +\infty$. It follows from formula (10.5.1) that for all $t \ge 0$ we have $u(t,\cdot) \in H^N(\Omega)$. On the other hand, if $t_0 \in [0,+\infty[$ and $(t_j) \subset [0,+\infty[$ is a sequence tending to t_0 we have

$$(1) = \|u(t_j,\cdot) - u(t_0,\cdot)\|_{H^N}^2 \le C \sum_{n=1}^{+\infty} \left| e^{-\lambda_n t_j} - e^{-\lambda_n t_0} \right|^2 \lambda_n^N |(g, e_n)|^2.$$

By using the dominated convergence theorem (by bounding the exponentials by 1), we see that (1) $\to 0$ as $j \to +\infty$; this proves that $u \in C^0([0,+\infty[, H^N(\Omega))$.

To conclude this paragraph, here is an interesting result that is useful for proving uniqueness results.

Maximum Principle for the Heat Equation

Let us denote $Q =]0,+\infty[\times\Omega$. Then we have $\overline{Q} = [0,+\infty[\times\overline{\Omega}$ and the boundary of Q is the set $\partial Q = [0,+\infty[\times\partial\Omega \cup \{0\} \times \overline{\Omega}$; it is a closed set. Let $T > 0$ and $K_T = [0,T] \times \overline{\Omega}$ the truncated "cylinder". Let us set

$$\Sigma_T = K_T \cap \partial Q = \{0 \le t \le T,\ x \in \partial\Omega\} \cup \{t = 0,\ x \in \overline{\Omega}\}.$$

This is the parabolic boundary of K_T. It is a compact subset of $[0,+\infty[\times\mathbf{R}^d$.

Theorem 10.5.4 (Maximum principle) *Let $u \in C^0(\overline{Q}) \cap C^2(Q)$ be a real-valued function.*
(1) *If $\frac{\partial u}{\partial t} - \Delta u \leq 0$ in Q, then $\sup_{K_T} u = \sup_{\Sigma_T} u$.*
(2) *If $\frac{\partial u}{\partial t} - \Delta u \geq 0$ in Q, then $\inf_{K_T} u = \inf_{\Sigma_T} u$.*
(3) *If $\frac{\partial u}{\partial t} - \Delta u = 0$ in Q, then $\sup_{K_T} u = \sup_{\Sigma_T} u$, $\inf_{K_T} u = \inf_{\Sigma_T} u$.*

Proof First, note that what statement (1) asserts, for example, is that the maximum on the compact K_T of such a function is attained on its parabolic boundary.

For $\varepsilon > 0$ and $(t, x) \in Q$ set $u_\varepsilon(t, x) = u(t, x) + \varepsilon\, |x|^2$. Clearly, $u(t, x)$ is bounded by $u_\varepsilon(t, x)$ in Q. Moreover, $u_\varepsilon \in C^0(\overline{Q}) \cap C^2(Q)$ and

$$\Big(\frac{\partial u_\varepsilon}{\partial t} - \Delta u_\varepsilon\Big)(t, x) \leq -2n\, \varepsilon\,, \quad (t, x) \in Q\,. \tag{10.5.3}$$

Since K_T is compact, let $m_\varepsilon = (t_\varepsilon, x_\varepsilon) \in K_T$ be a point where u_ε attains its maximum. Suppose $m_\varepsilon \notin \Sigma_T$. Then $x_\varepsilon \in \Omega$ and $0 < t_\varepsilon \leq T$. Since $u_\varepsilon(t_\varepsilon, x) \leq u_\varepsilon(t_\varepsilon, x_\varepsilon)$, the function from $\overline{\Omega}$ to $\mathbf{R}$, $x \mapsto u(t_\varepsilon, x)$, attains its maximum at x_ε. Therefore we have, $\Delta u_\varepsilon(t_\varepsilon, x_\varepsilon) \leq 0$. On the other hand,

$$\frac{\partial u_\varepsilon}{\partial t}\,(t_\varepsilon, x_\varepsilon) = \lim_{h \to 0^+} \frac{u_\varepsilon(t_\varepsilon - h, x_\varepsilon) - u(t_\varepsilon, x_\varepsilon)}{-h}\,.$$

For $h > 0$ small enough we have $0 < t_\varepsilon - h < T$ thus $(t_\varepsilon - h, x_\varepsilon) \in K_T$. It follows that $u_\varepsilon(t_\varepsilon - h, x_\varepsilon) - u_\varepsilon(t_\varepsilon, x_\varepsilon) \leq 0$. Consequently, $\frac{\partial u}{\partial t}\,(t_\varepsilon, x_\varepsilon) \geq 0$, so $\big(\frac{\partial u}{\partial t} - \Delta u\big)(t_\varepsilon, x_\varepsilon) \geq 0$, which contradicts (10.5.3). We deduce that $m_\varepsilon \in \Sigma_T$.

We can then write

$$\sup_{K_T} u \leq \sup_{K_T} u_\varepsilon = \sup_{\Sigma_T}\,(u + \varepsilon\, |x|^2) \leq \sup_{\Sigma_T} u + C\,\varepsilon\,,$$

since, as $\overline{\Omega}$ is compact, we have $|x|^2 \leq C$ for $x \in \overline{\Omega}$. Letting ε tend to zero we obtain

$$\sup_{K_T} u \leq \sup_{\Sigma_T} u \leq \sup_{K_T} u\,,$$

since $\Sigma_T \subset K_T$, which proves (1).

Assertion (2) follows from (1) applied to the function $-u$. Finally, (3) follows trivially from (1) and (2). □

Corollary 10.5.5 *Let $u \in C^0(\overline{Q}) \cap C^2(Q)$ be a real-valued function which satisfies $\frac{\partial u}{\partial t} - \Delta u \geq 0$ in Q. If $u|_{\partial Q} \geq 0$ then $u \geq 0$ in Q.*

Proof For any $T > 0$ we have $\inf_{\Sigma_T} u \geq 0$. According to Theorem 10.5.4, (2) we have $\inf_{K_T} u \geq 0$ so $u \geq 0$ on K_T for all $T > 0$, hence $u \geq 0$ in Q. Of course, there are analogous statements if $\frac{\partial u}{\partial t} - \Delta u \leq 0$; it suffices to replace u by $-u$. □

10.5.2 The Wave Equation

Theorem 10.5.6 (The homogeneous case) *Let Ω be a bounded open subset of $\mathbf{R}^d$, I an interval of $\mathbf{R}$, $t_0 \in I$, $f \in H_0^1(\Omega)$, $g \in L^2(\Omega)$. Then there exists a unique u in $C^0(I, H_0^1(\Omega))$ such that $\partial_t u \in C^0(I, L^2(\Omega))$ which solves the problem*

$$\begin{cases} \Box u = 0 \text{ in } \mathcal{D}'(\overset{\circ}{I} \times \Omega) \\ u(t_0, \cdot) = f \quad \partial_t u(t_0, \cdot) = g \\ u(t, \cdot) \in H_0^1(\Omega)\,, t \in I\,. \end{cases} \tag{10.5.4}$$

This solution is given by the formula

$$u(t, \cdot) = \sum_{n=1}^{+\infty} \Big[(f, e_n) \cos[(t - t_0)\sqrt{\lambda_n}] + \frac{(g, e_n)}{\sqrt{\lambda_n}} \sin[(t - t_0)\sqrt{\lambda_n}]\Big] e_n\,. \tag{10.5.5}$$

Proof *(a) Existence*: Let us first show that for $t \in I$, the series in the right-hand side of (10.5.5) converges in $H_0^1(\Omega)$. To prove this fact it suffices, denoting by $c_n(t)\, e_n$ the general term, to note that $\lambda_n\, |c_n|^2 \leq 2(\lambda_n |(f, e_n)|^2 + |(g, e_n)|^2)$ and then use the fact that $f \in H_0^1$ and $g \in L^2$. Next, we have $u \in C^0(I, H_0^1(\Omega))$. Indeed, let $\bar{t} \in I$ and $(t_j) \subset I$, $t_j \to \bar{t}$. Then,

$$\|u(t_j, \cdot) - u(\bar{t}, \cdot)\|^2_{H_0^1} \leq 2 \sum_{n \geq 1} \Big(\lambda_n |(f, e_n)|^2 \Big| \cos[(t_j - t_0)\sqrt{\lambda_n}] - \cos[(\bar{t} - t_0)\sqrt{\lambda_n}]\Big|^2$$
$$+ |(g, e_n)|^2 \Big| \sin[(t_j - t_0)\sqrt{\lambda_n}] - \sin[(\bar{t} - t_0)\sqrt{\lambda_n}]\Big|^2\Big).$$

The dominated convergence theorem implies $\lim_{j \to +\infty} \|u(t_j, \cdot) - u(\bar{t}, \cdot)\|^2_{H_0^1} = 0$. Next, it is easy to see that in $\mathcal{D}'(\Omega)$ we have

$$\partial_t u = \sum_{n \geq 1} \Big(- \sqrt{\lambda_n}\, (f, e_n) \sin[(t - t_0)\sqrt{\lambda_n}] + (g, e_n) \cos[(t - t_0)\sqrt{\lambda_n}]\Big)\, e_n$$

and a similar argument shows that $\partial_t u \in C^0(I, L^2(\Omega))$. It then easily follows that $u(t_0, \cdot) = f$, $\partial_t u(t_0, \cdot) = g$.

Finally, let us show that $\Box u = 0$ in $\mathcal{D}'(\overset{\circ}{I} \times \Omega)$. For $\psi \in C_0^\infty(\overset{\circ}{I} \times \Omega)$ we have

$$\langle \Delta u, \psi \rangle = \int_I \langle u(t, \cdot) \Delta \psi(t, \cdot)\rangle\, dt = \int_I \sum_{n \geq 1} c_n(t) \langle e_n, \Delta \psi(t, \cdot)\rangle\, dt,$$
$$= \int_I \sum_{n \geq 1} c_n(t) \langle \Delta e_n, \psi(t, \cdot)\rangle\, dt = \int_I \sum_{n \geq 1} c_n(t)(-\lambda_n) \langle e_n, \psi(t, \cdot)\rangle\, dt\,.$$

Lemma 10.5.7 *We have* $\langle \Delta u, \psi \rangle = \sum_{n \geq 1} \int_I (-\lambda_n)\, c_n(t) \langle e_n, \psi(t, \cdot)\rangle\, dt.$

Let us admit this lemma for a moment. We have $(-\lambda_n)\, c_n(t) = \frac{d^2}{dt^2}\, c_n(t)$. Integrate by parts twice in the integral over I. The boundary terms vanish because ψ belongs to $C_0^\infty(\overset{\circ}{I} \times \Omega)$. We deduce

$$\begin{aligned}\langle \Delta u, \psi \rangle &= \sum_{n \geq 1} \int_I c_n(t) \frac{d^2}{dt^2} \langle e_n, \psi(t, \cdot)\rangle\, dt = \sum_{n \geq 1} \int_I c_n(t) \langle e_n, \partial_t^2\, \psi(t, \cdot)\rangle\, dt,\\ &= \int_I \sum_{n \geq 1} c_n(t) \langle e_n, \partial_t^2\, \psi(t, \cdot)\rangle\, dt = \langle u, \partial_t^2\, \psi \rangle = \langle \partial_t^2\, u, \psi \rangle\,,\end{aligned}$$

hence $\Box u = 0$ in $\mathcal{D}'(\overset{\circ}{I} \times \Omega)$. □

Proof of Lemma 10.5.7 Since $\psi \in C_0^\infty(\overset{\circ}{I} \times \Omega)$, the function $\theta(t) = \|\psi(t, \cdot)\|_{H^1}$ belongs to $C_0^0(I)$. On the other hand,

$$\sum_{n \geq 1} \lambda_n |c_n(t)|^2 \leq 2(\|f\|_{H^1}^2 + \|g\|_{L^2}^2).$$

Then, using the Cauchy-Schwarz inequality we obtain

$$\sum_{n \geq 1} \lambda_n\, |c_n(t)|\, |\langle e_n, \psi(t, \cdot)\rangle| \leq 2(\|f\|_{H^1}^2 + \|g\|_{L^2}^2)^{1/2}\, \theta(t) \in L^1(I).$$

Lemma 10.5.7 follows then from the Fubini-Tonelli Theorem, since the function $(\lambda_n\, c_n(t) \langle e_n, \psi(t, \cdot)\rangle)$ is integrable with respect to the product measure.

(b) Uniqueness: the difference v of two solutions is an element of $C^0(I, H_0^1(\Omega))$ which is such that $\partial_t v \in C^0(I, L^2(\Omega))$ and satisfies

$$\Box v = 0 \text{ in } \mathcal{D}'(\overset{\circ}{I} \times \Omega), \quad v(t_0, \cdot) = \partial_t v(t_0, \cdot) = 0, \quad v(t, \cdot) \in H_0^1(\Omega)\,,\ t \in I. \tag{10.5.6}$$

Take $t_1 \in \overset{\circ}{I}$ and $\varphi \in C_0^\infty(\Omega)$. Let w be a solution of the problem

$$\Box w = 0 \text{ in } \mathcal{D}'\ \overset{\circ}{I} \times \Omega), \quad w(t_1, \cdot) = 0, \quad \partial_t w(t_1, \cdot) = \varphi, \quad w(t, \cdot) \in H_0^1(\Omega)\,,\ t \in I\,. \tag{10.5.7}$$

Using the equalities $\langle \Box v, \psi \rangle = 0$, $\langle \Box w, \psi \rangle = 0$, for $\psi \in C_0^\infty(\overset{\circ}{I} \times \Omega)$ of the form $\psi = \theta(t)\, \chi(x)$, where $\theta \in C_0^\infty(\overset{\circ}{I})$, $\chi \in C_0^\infty(\Omega)$, one easily shows that

$$\partial_t^2 v - \Delta v = 0\,, \quad \partial_t^2 w - \Delta w = 0\,. \tag{10.5.8}$$

Then, $\partial_t^2 v \in C^0(I, H^{-1}(\Omega))$; similarly we have $w \in C^0(I, H_0^1(\Omega))$, $\partial_t w \in C^0(I, L^2(\Omega))$ and $\partial_t^2 w \in C^0(I, H^{-1}(\Omega))$. Set

$$F(t) = \int_\Omega v(t,x)\,\partial_t w(t,x)\,dx - \int_\Omega \partial_t v(t,x)\,w(t,x)\,dx. \tag{10.5.9}$$

One shows that F is differentiable on I and

$$\begin{aligned} F'(t) = \int_\Omega \partial_t v(t,x)\,\partial_t w(t,x)\,dx + \langle v(t,\cdot), \partial_t^2 w(t,\cdot)\rangle \\ - \langle \partial_t^2 v(t,\cdot), w(t,\cdot)\rangle - \int_\Omega \partial_t v(t,x)\,\partial_t w(t,x)\,dx, \end{aligned} \tag{10.5.10}$$

where $\langle\,,\,\rangle$ denotes the duality between $H_0^1(\Omega)$ and $H^{-1}(\Omega)$.

We deduce from (10.5.8) that $F'(t) = \langle v(t,\cdot), \Delta w(t,\cdot)\rangle - \langle \Delta v(t,\cdot), w(t,\cdot)\rangle$. Since $v(t,\cdot)$ and $w(t,\cdot)$ are in H_0^1, it is easy to see, using the density of $C_0^\infty(\Omega)$ in $H_0^1(\Omega)$ that

$$\langle v(t,\cdot), \Delta w(t,\cdot)\rangle\rangle = -\sum_{j=1}^n \int_\Omega \frac{\partial v}{\partial x_j}\cdot\frac{\partial w}{\partial x_j}\,dx.$$

Similarly, $\langle \Delta v(t,\cdot), w(t,\cdot)\rangle = -\sum_{j=1}^n \int_\Omega \frac{\partial v}{\partial x_j}\cdot\frac{\partial w}{\partial x_j}\,dx$. We deduce that $F'(t) = 0$ for $t \in I$. It follows that $F(t) = F(t_0)$. Consequently,

$$F(t) = \int_\Omega v(t_0,x)\,\partial_t w(t_0,x)\,dx - \int_\Omega \partial_t v(t_0,x)\,w(t_0,x)\,dx = 0,$$

by (10.5.6). Therefore we have $F(t_1) = 0$ which, using (10.5.7), can be written as $\int_\Omega u(t_1,x)\,\varphi(x)\,dx = 0$. Since φ is arbitrary, we deduce that $u(t_1,x) = 0$ almost everywhere and since t_1 is arbitrary we have $u(t,\cdot) = 0$ for $t \in I$. □

Theorem 10.5.8 (The inhomogeneous problem) *Let Ω be a bounded open subset of $\mathbf{R}^d$, I an interval of $\mathbf{R}$, $t_0 \in I$. Let $F = (F_t) \in C^0(I, L^2(\Omega))$. There exists a unique function $u \in C^0(I, H_0^1(\Omega)) \cap C^1(\mathring{I}, L^2(\Omega))$ such that*

$$\Box u = F \text{ in } \mathcal{D}'(\mathring{I}\times\Omega), \quad u(t_0,\cdot) = 0, \quad \partial_t u(t_0,\cdot) = 0. \tag{10.5.11}$$

This solution is given by

$$u(t,\cdot) = \sum_{n\geq 1}\Big(\int_{t_0}^t \frac{\sin(\sqrt{\lambda_n}(t-s))}{\sqrt{\lambda_n}}\,F_n(s)\,ds\Big)\,e_n(\cdot), \tag{10.5.12}$$

where $F(t,\cdot) = \sum_{n\geq 1} F_n(t)\,e_n(\cdot)$.

Uniqueness follows from Theorem 10.5.6. The proof of existence is entirely analogous to that of Theorem 10.5.6 and is left to the reader.

10.6 Comments

Interested readers will benefit from reading the book [8].

Chapter 11
Weyl's Law

11.1 Prerequisites

Results[1] from Chap. 10. Distributions. Fundamental solutions. Beppo Levi and Fubini-Tonelli theorems.

11.2 The Problem

This chapter follows the previous one. We are interested here in the asymptotic behavior (i.e. as $n \to +\infty$) of the eigenvalues of the Dirichlet problem for the Laplacian. The main object of our study is what is called "the counting function $N(\lambda)$," that is, the number of eigenvalues less than λ, and we will be interested in its asymptotic behavior (as λ tends to $+\infty$). For reasons of difficulty, we will limit ourselves to describing the first term of this expansion. The result presented here is due to the German mathematician Hermann Weyl (1885–1955).

11.3 The Results

We denote by $(\lambda_n)_{n\geq 1}$ the sequence of eigenvalues of the Dirichlet problem for the Laplacian in a bounded open set Ω of $\mathbf{R}^d$ with C^∞ boundary.

Let us introduce the counting function. For $\lambda > 0$ we set

$$N(\lambda) = \text{ cardinal}\,\{n \geq 1 : \lambda_n \leq \lambda\}. \tag{11.3.1}$$

Recall that $f(\lambda) \sim g(\lambda)$, $\lambda \to +\infty$ means that $\lim_{\lambda\to+\infty} \frac{f(\lambda)}{g(\lambda)} = 1$.

[1] This chapter is based on Chap. 13 of my book "Elements de distributions et d'équations aux dérivées partielles" Dunod (2002).

C. Zuily, *Selected Topics in Partial Differential Equations*, Universitext,
https://doi.org/10.1007/978-3-032-24082-8_11

The main result of this section is the following.

Theorem 11.3.1 *We have*

$$N(\lambda) \sim (2\pi)^{-d}\, C_d\, |\Omega|\, \lambda^{d/2} \quad \lambda \to +\infty \tag{11.3.2}$$

where $|\Omega| = \int_\Omega dx$ *is the volume of* Ω *and* $C_d = \int_{|x|<1} dx$.

Corollary 11.3.2 *Under the hypotheses of Theorem 11.3.1, we have*

$$\lambda_n \sim \frac{(2\pi)^2}{[C_d\, |\Omega|]^{2/d}}\, n^{2/d}, \quad n \to +\infty. \tag{11.3.3}$$

Proof of Corollary 11.3.2 We know that $\lambda_n \to +\infty$. Take in (11.3.2), $\lambda = \lambda_{n_0}$ for n_0 large enough. Then $N(\lambda_{n_0}) = n_0$ hence $n_0 \sim (2\pi)^{-d}\, C_d\, |\Omega|\, \lambda_{n_0}^{d/2}$. □

Example 11.3.3 Suppose $\Omega =]0, 1[$, $d = 1$. The Dirichlet problem then reads

$$-u'' = \lambda u, \quad u(0) = u(1) = 0.$$

It is easy to see that $e_n(x) = \frac{1}{\sqrt{2}} \sin(n\pi x)$ and $\lambda_n = n^2\pi^2$. Moreover, $|\Omega| = 1$ and $C_d = |[-1, 1]| = 2$ so that the equivalence (11.3.3) is here an equality.

11.4 Proof of Theorem 11.3.1

We begin by showing the following result.

Lemma 11.4.1 *Let* $(\lambda_n)_{n\geq 1}$ *be a sequence of positive real numbers such that*

$$\text{the series } \sum_{n=1}^{+\infty} e^{-\lambda_n t} \text{ converges for all } t > 0, \tag{11.4.1}$$

$$\exists\, C_0 > 0, \quad \alpha > 0: \ \sum_{n=1}^{+\infty} e^{-\lambda_n t} \sim \frac{C_0}{t^\alpha}, \quad t \to 0^+. \tag{11.4.2}$$

Then,

$$card\{j \in \mathbf{N}^* : \lambda_j \leq \lambda\} \sim \frac{C_0\, \lambda^\alpha}{\alpha\, \Gamma(\alpha)}, \quad \lambda \to +\infty,$$

where $card(A)$ *denotes the cardinality of the set* A *and* Γ *is the Gamma function.*

Proof Note that we have

$$\text{card } \{j \in \mathbf{N}^* : \lambda_j \leq \lambda\} = \text{card}\,\{j \in \mathbf{N}^* : \frac{\lambda_j}{\lambda} \leq 1\} = \sum_{j \in \mathbf{N}^*} \mathbf{1}_{[0,1]}\Big(\frac{\lambda_j}{\lambda}\Big)$$

where

$$\mathbf{1}_{[0,1]}(x) = \begin{cases} 1, & \text{if } x \in [0, 1], \\ 0, & \text{if } x \notin [0, 1]. \end{cases}$$

We must therefore show that

$$\lim_{\lambda \to +\infty} \lambda^{-\alpha} \sum_{j \in \mathbf{N}^*} \mathbf{1}_{[0,1]}\Big(\frac{\lambda_j}{\lambda}\Big) = \frac{C_0}{\alpha\,\Gamma(\alpha)}. \tag{11.4.3}$$

We interpret the quantity $\lambda^{-\alpha} \sum_{j \in \mathbf{N}^*} \mathbf{1}_{[0,1]}\big(\frac{\lambda_j}{\lambda}\big)$ as being the value at the point $\mathbf{1}_{[0,1]}$ of the linear form given by

$$g \longrightarrow \lambda^{-\alpha} \sum_{j \in \mathbf{N}^*} g\Big(\frac{\lambda_j}{\lambda}\Big).$$

Let $\mathcal{E}$ be the vector space of functions $f : [0, +\infty[\to \mathbf{R}$, integrable (in the Riemann sense) on $[0, +\infty[$ such that $\sup_{s \geq 0} e^s\,|f(s)| < +\infty$, which we equip with the norm $\|f\| = \sup_{s \geq 0} e^s\,|f(s)|$. For $t > 0$, we consider the linear form on $\mathcal{E}$ defined by

$$\mathcal{L}_t(f) = t^\alpha \sum_{j \in \mathbf{N}^*} f(\lambda_j t). \tag{11.4.4}$$

We have

$$|\mathcal{L}_t(f)| \leq t^\alpha \sum_{j \in \mathbf{N}^*} e^{-\lambda_j t}\, e^{\lambda_j t}\, |f(\lambda_j t)| \leq t^\alpha \sum_{j \in \mathbf{N}^*} e^{-\lambda_j t}\, \|f\| \leq K(t)\, \|f\|,$$

according to (11.4.1). Thus $\mathcal{L}_t$ is continuous on $\mathcal{E}$.

Next, consider, for $k \in \mathbf{N}^*$, the functions $f_k(s) = e^{-ks} \in \mathcal{E}$. It follows from (11.4.2) that

$$\mathcal{L}_t(f_k) = t^\alpha \sum_{j \in \mathbf{N}^*} e^{-\lambda_j kt} \to \frac{C_0}{k^\alpha} = \frac{C_0}{\Gamma(\alpha)} \int_0^{+\infty} f_k(s)\, s^{\alpha-1}\, ds, \quad t \to 0^+,$$

because

$$\int_0^{+\infty} e^{-ks}\, s^{\alpha-1}\, ds = \frac{1}{k^\alpha} \int_0^{+\infty} e^{-x}\, x^{\alpha-1}\, dx = \frac{\Gamma(\alpha)}{k^\alpha}.$$

The linear form $\mathcal{L}_0$ on $\mathcal{E}$ defined by

$$\mathcal{L}_0(f) = \frac{C_0}{\Gamma(\alpha)} \int_0^{+\infty} f(s)\, s^{\alpha-1}\, ds \tag{11.4.5}$$

is such that

$$|\mathcal{L}_0(f)| \le \frac{C_0}{\Gamma(\alpha)} \Big(\int_0^{+\infty} e^{-s}\, s^{\alpha-1}\, ds \Big) \|f\| \le C_0\, \|f\|\,.$$

It is therefore continuous on $\mathcal{E}$ and $\|\mathcal{L}_0\| \le C_0$. Moreover, if we denote by $\mathcal{E}_0$ the vector space generated by the f_k, $k \in \mathbf{N}^*$, we have

$$\lim_{t\to 0^+} \mathcal{L}_t = \mathcal{L}_0 \text{ on } \mathcal{E}_0 \subset \mathcal{E}.$$

By continuity we have

$$\lim_{t\to 0^+} \mathcal{L}_t(f) = \mathcal{L}_0(f)\,, \quad \forall\, f \in \overline{\mathcal{E}_0} \text{ (the closure of } \mathcal{E}_0 \text{ in } \mathcal{E})\,. \tag{11.4.6}$$

Lemma 11.4.2 *Let* $f : [0, +\infty[\to \mathbf{R}$ *be a continuous function with compact support in* $[0, +\infty[$. *Then* $f \in \overline{\mathcal{E}_0}$.

Let us admit this result for a moment. Then, for all $f \in C_0^0([0, +\infty[)$ we have

$$\lim_{t\to 0^+} \mathcal{L}_t(f) = \mathcal{L}_0(f). \tag{11.4.7}$$

Let $g = \mathbf{1}_{[0,1]}$. Let f_+ and f_- be two continuous functions with compact support in $[0, +\infty[$ such that $0 \le f_\pm \le 1$, $f_- \le g \le f_+$ and

$$f_-(x) = \begin{cases} 1, & \text{if } 0 \le x \le 1-\varepsilon\,, \\ 0, & \text{if } x \ge 1 - \frac{\varepsilon}{2}\,, \end{cases} \qquad f_+(x) = \begin{cases} 1, & \text{if } 0 \le x \le 1\,, \\ 0, & \text{if } x \ge 1+\varepsilon\,. \end{cases}$$

We then have

$$\mathcal{L}_t(f_-) \le \mathcal{L}_t(g) \le \mathcal{L}_t(f_+). \tag{11.4.8}$$

According to (11.4.8) we have

$$\mathcal{L}_0(f_-) \le \liminf_{t\to 0^+} \mathcal{L}_t(g) \le \limsup_{t\to 0^+} \mathcal{L}_t(g) \le \mathcal{L}_0(f_+),$$

which implies

$$\frac{C_0}{\Gamma(\alpha)} \int_0^{1-\varepsilon} s^{\alpha-1}\, ds \le \liminf_{t\to 0^+} \mathcal{L}_t(g) \le \limsup_{t\to 0^+} \mathcal{L}_t(g) \le \frac{C_0}{\Gamma(\alpha)} \int_0^{1+\varepsilon} s^{\alpha-1}\, ds.$$

Letting ε tend to zero, we obtain

$$\lim_{t\to 0^+} \mathcal{L}_t(g) = \frac{C_0}{\alpha\,\Gamma(\alpha)}\,,$$

which can be written as

$$\lim_{t\to 0^+} t^{\alpha} \sum_{j\in\mathbf{N}^*} \mathbf{1}_{[0,1]}(\lambda_j t) = \frac{C_0}{\alpha\,\Gamma(\alpha)}\,.$$

This proves (11.4.3) and Lemma 11.4.1. □

Proof of Lemma 11.4.2 By making the change of variable $e^{-s} = x$, we are reduced to showing that if $g :]0, 1] \to \mathbf{R}$ is continuous and with compact support contained in $[\delta, 1]$ for some $\delta > 0$, there exists a sequence (P_n) of polynomials which satisfies $\sup_{]0,1]} \frac{1}{x}\,|g(x) - P_n(x)| \to 0$, as $n \to +\infty$. The function h defined by

$$h(x) = \begin{cases} \frac{1}{x}\, g(x) & \text{if } x \in]0, 1] \\ 0 & \text{if } x = 0, \end{cases}$$

is continuous on $[0, 1]$, because it is continuous on $]0, 1]$ and if $x_n \to 0$ we have $x_n \le \delta$ for $n \ge n_0$ and $h(x_n) = \frac{1}{x_n}\, g(x_n) = 0$ i.e. $h(x_n) \to 0$. The Stone-Weierstrass Theorem implies that there exists a sequence (Q_n) of polynomials which satisfies $\sup_{[0,1]} |h(x) - Q_n(x)| \to 0$, $n \to +\infty$. Then $\sup_{]0,1]} \frac{1}{x}\,|g(x) - x\,Q_n(x)| \to 0$. □

11.4.1 Continuation of the Proof of Theorem 11.3.1

Let (λ_n) be the sequence of eigenvalues of the Dirichlet problem for $-\Delta$. By virtue of Lemma 11.4.1, to prove Theorem 11.3.1, it suffices to have an asymptotic equivalent of $\sum_{n\ge 1} e^{-\lambda_n t}$ as $t \to 0^+$. But such an expression appears in the formula that gives the solution to the mixed problem for the heat equation. This observation allows us to clarify the strategy of the proof.

We introduce the following definition.

Definition 11.4.3 We call "heat kernel" the formal expression

$$p(t, x, y) = \sum_{n\ge 1} e^{-\lambda_n t}\, e_n(x)\, \overline{e_n(y)}\,, \quad t \ge 0\,,\ x, y \in \Omega\,.$$

This kernel allows us to solve the mixed problem for the heat equation. Since $\|e_n\|_{L^2(\Omega)} = 1$, we may hope to have

$$\int_{\Omega} p(t,x,x)\,dx = \sum_{n\geq 1} e^{-\lambda_n t}\,. \tag{11.4.9}$$

It would therefore suffice to have an asymptotic equivalent of the left-hand side of this equality. This equivalent is not easy to obtain, but if $\Omega = \mathbf{R}^d$ the heat kernel $k(t,x,y)$ can be computed by Fourier transform and we will be able to find an equivalent of $\int_\Omega k(t,x,x)\,dx$ as $t \to 0^+$. The end of the proof will then consist in comparing this last expression to $\int_\Omega p(t,x,x)\,dx$.

Study of the Heat Kernel

Theorem 11.4.4 (1) $p \in C^0([0,+\infty[, \mathcal{D}'(\Omega\times\Omega))$.

(2) $p(0,x,y) = \delta(x-y)$.

(3) *The series defining* p *and the derived series (in* (t,x,y)*) of all orders, converge uniformly on every compact subset of* $]0,+\infty[\times\Omega\times\Omega$.

(4) $p \in C^\infty(]0,+\infty[\times\Omega\times\Omega)$.

(5) *If* $g \in C_0^0(\Omega)$, *the solution to the mixed problem (10.5.1) is given by*

$$u(t,x) = \int_\Omega p(t,x,y)\,g(y)\,dy\,, \quad t>0\,,\ x\in\Omega.$$

Proof (1) Let us first show that the series defining p converges in $\mathcal{D}'(\Omega\times\Omega)$ for $t \geq 0$. Let $\Phi \in C_0^\infty(\Omega\times\Omega)$; set $b_n = \langle e_n \otimes \overline{e}_n, \Phi\rangle$. Since $\Phi \in C_0^\infty(\Omega\times\Omega)$ which is contained in $C_0^\infty(\mathbf{R}^d\times\mathbf{R}^d)$ we can write

$$\Phi(x,y) = (2\pi)^{-d}\int e^{i\,y\cdot\xi}\,\widehat{\Phi}^2(x,\xi)\,d\xi\,,$$

where $\widehat{\Phi}^2$ denotes the Fourier transform in y. Next, since supp Φ is contained in $K_1 \times K_2$, there exists $\theta \in C_0^\infty(\Omega)$ such that $\Phi(x,y) = \theta(y)\,\Phi(x,y)$ (take $\theta = 1$ on K_2). We then have, since $e_n \in C^\infty(\Omega)$,

$$\begin{aligned} b_n &= \int_\Omega\int_\Omega e_n(x)\,\overline{e_n(y)}\,\Phi(x,y)\,dx\,dy \\ &= (2\pi)^{-d}\int_\Omega\int_\Omega\int_{\mathbf{R}^d} e^{iy\cdot\xi}\,e_n(x)\,\overline{e_n(y)}\,\theta(y)\,\widehat{\Phi}^2(x,\xi)\,d\xi\,dx\,dy \end{aligned}$$

where the triple integral is absolutely convergent. Consequently,

$$b_n = (2\pi)^{-d}\int_{\mathbf{R}^d}\big(e_n, \overline{\widehat{\Phi}^2(\cdot,\xi)}\big)_{L^2}\,(\theta\, e^{i\langle\cdot,\xi\rangle}, e_n)_{L^2}\,d\xi.$$

We deduce that

$$\sum_{n=1}^{N}|b_n| \le (2\pi)^{-d}\int_{\mathbf{R}^d}\Big(\sum_{n=1}^{N}\big|(e_n, \overline{\widehat{\Phi}^2(\cdot,\xi)})_{L^2}\big|^2\Big)^{\frac{1}{2}}\Big(\sum_{n=1}^{N}\big|(e_n, \theta\, e^{i\langle\cdot,\xi\rangle})_{L^2}\big|^2\Big)^{\frac{1}{2}}\, d\xi.$$

On the other hand,

$$\sum_{n=1}^{N}\big|(e_n, \overline{\widehat{\Phi}^2(\cdot,\xi)})\big|^2 \le \sum_{n=1}^{+\infty}\big|(e_n, \overline{\widehat{\Phi}^2(\cdot,\xi)})\big|^2 = \|\widehat{\Phi}^2(\cdot,\xi)\|^2_{L^2}\,.$$

Similarly,

$$\sum_{n=1}^{N}\big|(e_n, \theta\, e^{i\langle\cdot,\xi\rangle})\big|^2 \le \|\theta\|^2_{L^2}\,.$$

It follows that

$$\sum_{n=1}^{N}|b_n| \le (2\pi)^{-d}\,\|\theta\|_{L^2}\int_{\mathbf{R}^d}\|\widehat{\Phi}^2(\cdot,\xi)\|_{L^2}\, d\xi\,.$$

We deduce that the series $\sum_{n=1}^{+\infty} e^{-\lambda_n t}\, b_n$ is convergent for $t \ge 0$ (since $e^{-\lambda_n t} \le 1$) and, by Lebesgue's theorem, that the function $t \mapsto \sum_{n\ge 1} e^{-\lambda_n t}\langle e_n \otimes \overline{e_n}, \Phi\rangle$ is continuous.

(2) It suffices to show that, for all $\varphi, \psi \in C_0^\infty(\Omega)$, we have

$$\langle p(0,\cdot,\cdot), \varphi\otimes\psi\rangle = \langle\delta(x-y), \varphi\otimes\psi\rangle = \int_\Omega \varphi(x)\,\psi(x)\, dx. \tag{11.4.10}$$

Let us set

$$\begin{aligned} a_N = \Big\langle\sum_{n=1}^{N} e_n\otimes\overline{e_n}, \varphi\otimes\psi\Big\rangle &= \sum_{n=1}^{N}\int_\Omega e_n(x)\,\varphi(x)\, dx\int_\Omega \overline{e_n(y)}\,\psi(y)\, dy \\ &= \Big\langle\sum_{n=1}^{N}(\psi, e_n)_{L^2}\, e_n, \varphi\Big\rangle. \end{aligned}$$

Since $\psi \in C_0^\infty(\Omega) \subset L^2(\Omega)$ we have $\lim_{N\to+\infty}\sum_{n=1}^{N}(\psi, e_n)_{L^2}\, e_n = \psi$ in $L^2(\Omega)$ hence also in $\mathcal{D}'(\Omega)$. We deduce that $\lim_{N\to+\infty} a_N = \langle\psi, \varphi\rangle = \int \psi(x)\,\varphi(x)\, dx$, whence (11.4.10).

(3) and (4). Let $p_N(t,x,y) = \sum_{n=0}^{N} e^{-\lambda_n t}\, e_n(x)\,\overline{e_n(y)}$. It is easily verified that

$$\Big(\frac{\partial}{\partial t} - \frac{1}{2}(\Delta_x + \Delta_y)\Big)\, p_N = 0.$$

On the other hand, it is known that this operator admits a fundamental solution given by the function

$$E(t,x,y) = \frac{H(t)}{(4\pi t)^n} \exp\Big[-\Big(\frac{|x|^2+|y|^2}{2t}\Big)\Big]$$

which is a C^∞ function outside the origin $t = x = y = 0$. Our results will be a consequence of the following lemma.

Lemma 11.4.5 *Let P be a differential operator with constant coefficients on $\mathbf{R}^k$, admitting a fundamental solution $E \in C^\infty(\mathbf{R}^k \setminus \{0\})$. Let X be an open subset of $\mathbf{R}^k$ and (u_j) a sequence of (C^∞) solutions of the equation $P\,u = 0$ in X. If (u_j) converges in $\mathcal{D}'(X)$ it also converges in $C^\infty(X)$.*

Let us first remark that this lemma implies assertions (3) and (4).

Proof of Lemma 11.4.5 Let K be a compact subset of X, $\omega \subset X$ a neighborhood of K, φ in $C_0^\infty(X)$, $\varphi = 1$ on a neighborhood of $\overline{\omega}$. Set

$$r_j = P(\varphi\, u_j) = \varphi\, P\, u_j + [P, \varphi]\, u_j = [P, \varphi]\, u_j.$$

We have $\operatorname{supp} r_j \subset X \setminus \omega$; thus $r_j \in C_0^\infty(X \setminus \omega)$ and there exists a compact $\widetilde{K}$ of $X \setminus \omega$ (independent of j) such that $\operatorname{supp} r_j \subset \widetilde{K}$, $\forall\, j$. Now, since $P\,E = \delta_0$ we have $\varphi\, u_j = E * r_j$. Let $\theta \in C^\infty(\mathbf{R}^k)$ such that there exists $\varepsilon > 0$ for which $\operatorname{supp}(1-\theta) \subset B(0,\varepsilon)$ (i.e. $\theta = 1$ outside $B(0,\varepsilon)$). We have $\operatorname{supp}[(1-\theta)\,E] * r_j \subset B(0,\varepsilon) + \widetilde{K}$. We choose ε small enough so that $[\widetilde{K} + B(0,\varepsilon)] \cap \omega = \emptyset$. Then for $x \in \omega$ we have $u_j = (\theta\, E) * r_j$ and by hypothesis $\theta\, E \in C^\infty(\mathbf{R}^k)$. Thus for $x \in K$,

$$\partial^\alpha u_j(x) = [\partial^\alpha(\theta\, E)] * r_j = \langle r_j, \partial^\alpha(\theta\, E)(x - \cdot)\rangle. \tag{11.4.11}$$

Since $r_j \to r$ in $\mathcal{D}'(X)$ this implies that the convergence is uniform on every compact of $C^\infty(X)$, that is, on every closed bounded subset of $C^\infty(X)$. Now if K is a compact subset of X and $\psi \in C^\infty(\mathbf{R}^k)$, the set $\{\psi_x\}_{x\in K}$, where ψ_x is the map $y \mapsto \psi(x-y)$, is a compact subset of $C^\infty(X)$. For this consider the map $\Phi : K \to C^\infty(X), x \mapsto \psi_x$. This map is continuous; indeed, if $x_n \to x$ in K and K_1 is a compact subset of X then,

$$\sup_{y\in K_1} |\partial^\alpha \psi(x_n - y) - \partial^\alpha \psi(x - y)| \le |x_n - x| \sup_{z\in K_2} |\partial^\alpha\, \partial\psi(z)|.$$

Thus, $\Phi(K) = \{\psi_x\}_{x\in K}$ is a compact subset of C^∞. The formula (11.4.11) shows that $(\partial^\alpha u_j)_{j\in\mathbf{N}}$ converges uniformly on K. □

End of the proof of Theorem 11.4.4. It remains to prove assertion (5). If $g \in C_0^0(\Omega)$, $\operatorname{supp} g \subset K$, we have

$$\int_{\Omega} p(t,x,y)\,g(y)\,dy = \int_{K} p(t,x,y)\,g(y)\,dy = \int \sum_{n\geq 1} e^{-\lambda_n t}\, e_n(x)\,\overline{e_n(y)}\,g(y)\,dy$$
$$= \sum_{n\geq 1} e^{-\lambda_n t}\, e_n(x)(g, e_n)_{L^2},$$

by uniform convergence of the series on the compact $\{t\} \times \{x\} \times K$. Thus (5) follows from (10.5.2). □

We deduce from Theorem 11.4.4 that

$$p(t,x,x) = \sum_{n\geq 1} e^{-\lambda_n t}\, |e_n(x)|^2 \in C^\infty(]0,+\infty[\times\Omega).$$

According to the Beppo Levi Theorem, we can write

$$\int_{\Omega} p(t,x,x)\,dx = \sum_{n\geq 1} e^{-\lambda_n t}\, \|e_n\|_{L^2}^2 = \sum_{n\geq 1} e^{-\lambda_n t} \leq +\infty. \tag{11.4.12}$$

By virtue of Lemma 11.4.1, it suffices to study the first term of (11.4.12). Since this study is not easy, we will begin by studying the heat kernel on the whole of $\mathbf{R}^d$. We shall denote by $H(t)$ the Heaviside function, $H(t) = 1$ for $t \geq 0$, $H(t) = 0$ for $t < 0$.

Proposition 11.4.6 *Let* $k(t,x) = (4\pi t)^{-\frac{d}{2}}\, H(t)\, e^{-\frac{|x|^2}{4t}}$. *We have*

(1) $k \in C^0([0,+\infty[, \mathcal{D}'(\mathbf{R}^d)) \cap C^\infty(]0,+\infty[\times\mathbf{R}^d)$.

(2) $k(0,\cdot) = \delta_0$ *(the Dirac measure at the origin).*

(3) $\left(\frac{\partial k}{\partial t} - \Delta k\right)(t,x) = 0$ *for* $(t,x) \in]0,+\infty[\times\mathbf{R}^d$.

(4) *Let* $g \in C_0^0(\mathbf{R}^d)$. *Set, for* $t \geq 0$, $x \in \mathbf{R}^d$, $u(t,x) = \big(k(t,\cdot) * g\big)(x)$. *Then* u *belongs to* $C^\infty(]0,+\infty[\times\mathbf{R}^d) \cap C^0([0,+\infty[\times\mathbf{R}^d)$ *and solves the Cauchy problem* $\frac{\partial u}{\partial t} - \Delta u = 0$ *in* $[0,+\infty[\times\mathbf{R}^d$, $u|_{t=0} = g$.

Proof (1) It is clear that $k \in C^\infty(]0,+\infty[\times\mathbf{R}^d)$. On the other hand, let φ in $C_0^\infty(\mathbf{R}^d)$; the change of variable $x = 2\sqrt{t}\,y$ allows us to write

$$\langle k(t,\cdot), \varphi\rangle = \pi^{-\frac{d}{2}} \int e^{-|y|^2}\, \varphi(2\sqrt{t}\,y)\,dy.$$

The dominated convergence theorem shows that the right-hand side is a continuous function on $[0,+\infty[$.

(2) The above formula implies that

$$\langle k(0,\cdot), \varphi\rangle = \pi^{-\frac{d}{2}} \int e^{-|y|^2}\, \varphi(0)\,dy = \varphi(0) = \langle \delta_0, \varphi\rangle.$$

(3) We can write, for fixed $t > 0$, $k(t,x) = (4\pi t)^{-\frac{d}{2}} F_t(|x|)$ where $F_t(r) = e^{-\frac{r^2}{4t}}$. Since F_t is radial we have

$$\Delta_x F_t(|x|) = F_t''(|x|) + \frac{d-1}{|x|} F_t'(|x|) = \left(\frac{|x|^2}{4t^2} - \frac{d}{2t}\right) e^{-\frac{|x|^2}{4t}} .$$

On the other hand, it is easy to see that

$$\frac{\partial k}{\partial t}(t,x) = (4\pi t)^{-\frac{d}{2}} \left(\frac{|x|^2}{4t^2} - \frac{d}{2t}\right) e^{-\frac{|x|^2}{4t}},$$

which proves (3).

(4) We can write, for $t > 0$,

$$u(t,x) = (4\pi t)^{-\frac{d}{2}} \int e^{-\frac{|x-y|^2}{4t}} g(y)\, dy = \pi^{-\frac{d}{2}} \int e^{-|z|^2} g(x - 2\sqrt{t}\, z)\, dz .$$

Using the first equality and the Lebesgue differentiation theorem, it is easy to show that the function $(t,x) \mapsto u(t,x)$ is C^∞ on $]0,+\infty[\times \mathbf{R}^d$; the second equality and the dominated convergence theorem show that $u \in C^0([0,+\infty[\times \mathbf{R}^d)$.

We have, by virtue of (2), $u_0 = \delta_0 * g = g$. On $]0,+\infty[\times \mathbf{R}^d$, u is the C^∞ function defined by $u(t,x) = k(t,\cdot) * g$. It follows from (3) that

$$\Delta u(t,x) = [(\Delta k(t,\cdot)) * g](x) = \left[\frac{\partial k}{\partial t}(t,\cdot) * g\right](x) = \frac{\partial}{\partial t}[k(t,\cdot) * g] = \frac{\partial u}{\partial t}(t,x).$$

□

The end of the proof of Theorem 11.3.1 consists in comparing p and k.

Comparison of p and k.

Here is the main result of this paragraph. Set $\theta(y) = d(y, \partial\Omega)$ for $y \in \Omega$, where $d(\cdot,\cdot)$ is the Euclidean distance.

Proposition 11.4.7 *For $(t,x,y) \in]0,+\infty[\times\Omega\times\Omega$ we have*

$$0 \le k(t, x-y) - p(t,x,y) \le \begin{cases} (4\pi t)^{-\frac{d}{2}} e^{-\frac{\theta(y)^2}{4t}}, & \text{if } 0 < t \le t_0(y), \\ (4\pi t_0(y))^{-\frac{d}{2}} e^{-\frac{\theta(y)^2}{4t_0(y)}}, & \text{if } t \ge t_0(y), \end{cases}$$

where $t_0(y) = \frac{\theta(y)^2}{2d}$ (d is the dimension here).

Proof Let $g \in C_0^\infty(\Omega)$ be positive. Let $\widetilde{g} \in C_0^\infty(\mathbf{R}^d)$ denote its extension by zero outside Ω. For $t \geq 0$, set $v = k(t, \cdot) * \widetilde{g}$. Then the distribution v is a solution of the problem

$$\frac{\partial v}{\partial t} - \Delta\, v = 0 \text{ in } \mathcal{D}'(]0, +\infty[\times \mathbf{R}^d), \quad v(0, \cdot) = \widetilde{g}, \tag{11.4.13}$$

and $v \in C^\infty(]0, +\infty[\times \mathbf{R}^d) \cap C^0([0, +\infty[\times \mathbf{R}^d)$ according to Proposition 11.4.6. In particular the function v belongs to $C^0(\overline{Q})$, where $\overline{Q} = [0, +\infty[\times \overline{\Omega}$. Moreover, for $t > 0$ we have

$$v(t, x) = (4\,\pi\, t)^{-\frac{d}{2}} \int_\Omega e^{-\frac{|x-y|^2}{4t}}\, g(y)\, dy, \tag{11.4.14}$$

since $\widetilde{g}$ is zero outside Ω.

If we set $\theta_0 = d(\operatorname{supp} g, \partial\Omega)$, we have $|x - y| \geq \theta_0$ for $x \in \partial\Omega$ and $y \in \operatorname{supp} g$. We deduce that

$$0 \leq v(t, x) \leq (4\,\pi\, t)^{-\frac{d}{2}}\, e^{-\frac{\theta_0^2}{4t}} \int_\Omega g(y)\, dy\,, \quad \forall (t, x) \in (0, +\infty) \times \partial\Omega. \tag{11.4.15}$$

Set for $t \in (0, +\infty)$,

$$h(t) = (4\,\pi\, t)^{-\frac{d}{2}}\, e^{-\frac{\theta_0^2}{4t}} \int_\Omega g(y)\, dy. \tag{11.4.16}$$

The function h is increasing for $t \leq t_0 = \frac{\theta_0^2}{2d}$ and decreasing for $t \geq t_0$. It follows from (11.4.15) that for $(t, x) \in (0, +\infty) \times \partial\Omega$ we have

$$0 \leq v(t, x) \leq H(t) = \begin{cases} h(t), & \text{if } t \leq t_0\,, \\ h(t_0), & \text{if } \ t > t_0. \end{cases} \tag{11.4.17}$$

Since the function H is continuous on $(0, +\infty)$ and C^1 for $0 < t \leq t_0$ and $t > t_0$ we have

$$H'(t) = \begin{cases} h'(t), & \text{if } t \leq t_0\,, \\ 0, & \text{if } \ t > t_0. \end{cases}$$

Therefore,

$$H'(t) \geq 0, \quad t \in (0, +\infty) \ \text{ and } \ \lim_{t \to 0} H(t) = 0. \tag{11.4.18}$$

Let $u \in C^0([0, +\infty[, L^2(\Omega)) \cap C^\infty(]0, +\infty[\times \Omega)$ be the solution, given by Theorem 10.5.1, of the problem

$$\frac{\partial u}{\partial t} - \Delta u = 0 \text{ in } \mathcal{D}'(]0, +\infty[\times\Omega), \quad u(0, \cdot) = g, \quad u(t, \cdot) \in H_0^1(\Omega), \quad t > 0. \tag{11.4.19}$$

It follows from Remark 10.5.3, that $u \in C^0(\overline{Q})$.

Set $w = v - u$. Then w is a solution of the problem

$$\begin{aligned} &(i) \ \ \frac{\partial w}{\partial t} - \Delta\, w = 0 \text{ in } \mathcal{D}'(]0, +\infty[\times\Omega), \\ &(ii) \ \ w(0, \cdot)|_{\Omega} = 0, \\ &(iii) \ \ w(t, \cdot)|_{\partial\Omega} = v(t, \cdot)|_{\partial\Omega}, \quad t > 0. \end{aligned} \tag{11.4.20}$$

We then have $w \in C^0(\overline{Q}) \cap C^2(Q)$, where $Q =]0, +\infty[\times\Omega$.

We claim that, with the notation in (11.4.17), we have

$$0 \le w(t, x) \le H(t), \quad \forall (t, x) \in Q. \tag{11.4.21}$$

For that we shall use Corollary 10.5.5.

Since $\partial Q = \{t = 0, x \in \Omega\} \cup \{t > 0, x \in \partial\Omega\}$, using (11.4.20) and the fact that $v \ge 0$ we see that

$$w \ge 0 \quad \text{on} \quad \partial Q. \tag{11.4.22}$$

We deduce from Corollary 10.5.5 (maximum principle) that

$$w \ge 0 \quad \text{on} \quad Q. \tag{11.4.23}$$

Now set for $(t, x) \in Q$, $\widetilde{w}(t, x) = w(t, x) - H(t)$. Then, using (11.4.20) and (11.4.18) we see that

$$\Big(\frac{\partial}{\partial t} - \Delta\Big)\widetilde{w}(t, x) = -H'(t) \le 0, \quad \forall (t, x) \in Q. \tag{11.4.24}$$

It follows from (11.4.17) and (11.4.20) (iii) that

$$w(t, x) \le H(t), \quad \forall (t, x) \in (0, +\infty) \times \partial\Omega. \tag{11.4.25}$$

Then, from (11.4.20), (11.4.18) and (11.4.25) we have

$$\widetilde{w}(0, x) = 0, \quad \forall x \in \Omega, \quad \widetilde{w}(t, x) \le 0 \ \ \forall (t, x) \in (0, +\infty) \times \partial\Omega,$$

that is,

$$\widetilde{w} \le 0 \text{ on } \partial Q. \tag{11.4.26}$$

By (11.4.24), (11.4.26) and Corollary 10.5.5 we have

$$\widetilde{w} \le 0 \text{ in } Q,$$

in other words,

$$w(t,x) \le H(t), \quad \forall (t,x) \in Q,$$

which, together with (11.4.22), proves (11.4.21).

Now, according to Theorem 11.4.4, (5) and (11.4.14), we have for $(t,x) \in Q$,

$$w(t,x) = \int_\Omega k(t, x-y)\, g(y)\, dy - \int_\Omega p(t,x,y)\, g(y)\, dy = \big\langle g, k(t, x-\cdot) - p(t,x,\cdot)\big\rangle.$$

where $\langle\ ,\ \rangle$ denotes the duality between $\mathcal{E}'(\Omega)$ and $C^\infty(\Omega)$, since $g \in \mathcal{E}'(\Omega)$ and for all $t > 0$, $x \in \Omega$, the function $y \mapsto k(t,x-y) - p(t,x,y)$ is C^∞ in Ω.

It follows frow (11.4.21) that

$$0 \le \big\langle g, k(t, x-\cdot) - p(t,x,\cdot)\big\rangle \le H(t) \quad \text{in } Q \tag{11.4.27}$$

Fix $y_0 \in \Omega$. Let $\psi \in C_0^\infty(\mathbf{R}^d)$ be such that

$$\psi \ge 0, \int \psi(y)\, dy = 1, \operatorname{supp} \psi \subset \{x : |x| \le 1\}.$$

Then we set $\psi_\varepsilon(y) = \varepsilon^{-n}\, \psi\big(\frac{y-y_0}{\varepsilon}\big)$. If ε is small enough, we have $\psi_\varepsilon \in C_0^\infty(\Omega)$, $\int_\Omega \psi_\varepsilon(x)\, dx = 1$ and $(\psi_\varepsilon) \to \delta_{y_0}$ in $\mathcal{E}'(\Omega)$. Moreover as the support of ψ_ε converges to y_0, $\theta_0 = d(\operatorname{supp} \psi_\varepsilon, \partial\Omega)$ converges to $\theta(y_0) = d(y_0, \partial\Omega)$.

Taking $g = \psi_\varepsilon$ in (11.4.27) and letting ε tend to zero, we obtain

$$0 \le k(t, x-y_0) - p(t,x,y_0) = \lim_{\varepsilon \to 0} \langle \psi_\varepsilon, k(t,x-\cdot) - p(t,x,\cdot)\rangle \le H(t),$$

which completes the proof of of Proposition 11.4.7. □

Corollary 11.4.8 *Let $(t,x) \in]0, +\infty[\times\Omega$; denote $\theta(x) = d(x, \partial\Omega)$. Then,*

(i) $0 \le (4\pi t)^{-\frac{d}{2}} - p(t,x,x) \le (4\pi t)^{-\frac{d}{2}}\, e^{-\frac{\theta(x)^2}{4t}}$, $t \in \big]0, \frac{\theta(x)^2}{2d}\big]$,

(ii) $(4\pi t)^{-\frac{d}{2}}\, |\Omega| - \int_\Omega p(t,x,x)\, dx \le C\, t^{-\frac{d}{2}+\frac{1}{2}}$, $t \to 0^+$.

Proof (i) follows from Proposition 11.4.7; it suffices to take $x = y \in \Omega$.

(ii) Let us set $g(t,x) = (4\pi t)^{-\frac{d}{2}} - p(t,x,x) \ge 0$. We will show that for $t > 0$ small enough, the function $x \to g(t,x,x)$ belongs to $L^1(\Omega)$ and

$$0 \le \int_\Omega g(t,x,x)\, dx \le C t^{-\frac{d}{2}+\frac{1}{2}}, \quad t \to 0^+. \tag{11.4.28}$$

Let us suppose this is proved for a moment. Since $p(t,x,x) = (4\pi t)^{-\frac{d}{2}} - g(t,x,x)$ and Ω is bounded, we deduce that $x \to p(t,x,x) \in L^1(\Omega)$ and

$$0 \leq (4\pi t)^{-\frac{d}{2}} |\Omega| - \int_\Omega p(t,x,x)\,dx = \int_\Omega g(t,x,x)\,dx \leq Ct^{-\frac{d}{2}+\frac{1}{2}},$$

which proves (ii).

It remains to prove (11.4.28). We write $\Omega = \Omega_1 \cup \Omega_2$ where

$$\Omega_1 = \{x \in \Omega : \theta(x) \leq \sqrt{2dt}\}, \quad \Omega_2 = \{x \in \Omega : \theta(x) > \sqrt{2dt}\},$$

and $I_j = \int_{\Omega_j} g(t,x,x)\,dx$, $j = 1, 2$.

To handle I_1 we will need the following lemma. For $\alpha > 0$ let us set

$$F(\alpha) = \int_{\theta(x)\leq\alpha} dx = \mu\{x \in \Omega : \theta(x) < \alpha\}.$$

Lemma 11.4.9 *There exists $C > 0$ such that $F(\alpha) \leq C\,\alpha$ for all $\alpha > 0$.*

Let us admit this lemma for a moment.

According to Definition 11.4.3, we have $p(t,x,x) \geq 0$, so that $g(t,x)$ is bounded above by $(4\pi t)^{-\frac{d}{2}}$. It follows from Lemma 11.4.9 that

$$I_1 \leq (4\pi t)^{-\frac{d}{2}} F(\sqrt{2dt}) \leq C_1 t^{-\frac{d}{2}+\frac{1}{2}}, \; t > 0. \tag{11.4.29}$$

Let us consider the term I_2.

We note that $\theta(x) > \sqrt{2dt}$ is equivalent to $t < \frac{\theta(x)^2}{2d}$; we can therefore use inequality (i) of Corollary 11.4.8. We obtain

$$I_2 \leq (4\pi t)^{-\frac{d}{2}} \int_{\theta(x)>\sqrt{2dt}} e^{-\frac{\theta(x)^2}{4t}}\,dx. \tag{11.4.30}$$

Lemma 11.4.10 *Let $s_0 = \sqrt{2dt}$ and $\psi(s) = e^{-\frac{s^2}{4t}}$ for $s > 0$. Then,*

$$\int_{\theta(x)>s_0} \psi(\theta(x))\,dx = -\int_{s_0}^{+\infty} \psi'(s)\,F(s)\,ds - \psi(s_0)\,F(s_0).$$

Let us also admit this lemma for a moment. We deduce from (11.4.30) and from Lemmas 11.4.9 and 11.4.10 that

$$I_2 \leq C\,(4\pi t)^{-\frac{d}{2}} \int_{\sqrt{2dt}}^{+\infty} \frac{s}{2t} e^{-\frac{s^2}{4t}}\,s\,ds\,, \; (\text{since } -\psi(s_0)\,F(s_0) \leq 0).$$

Set in the integral $s = \sqrt{t}\,x$. Then, $I_2 \leq \frac{1}{2}(4\pi t)^{-\frac{d}{2}+\frac{1}{2}} \int_0^{+\infty} x^2 e^{-\frac{x^2}{4}}\,dx$, which, added to (11.4.29), proves (11.4.28) and completes the proof of (ii) of Corollary 11.4.8. □

Proof of Lemma 11.4.9 Ω is a bounded open set so there exists $\rho \in C^1(\mathbf{R}^d)$ such that $\Omega = \{x \in \mathbf{R}^d : \rho(x) < 0\}$, $\partial\Omega = \{x \in \mathbf{R}^d : \rho(x) = 0\}$, $d\rho \neq 0$ on $\partial\Omega$.

Point 1: $\exists\, M > 0 : |\rho(x)| \leq M\,\theta(x), \forall\, x \in \Omega$.

There exists $R > 0$ such that $\overline{\Omega} \subset \{x \in \mathbf{R}^d : |x| < R\}$. Set $M = \sup_{|x| \leq R} \|\rho'(x)\|$. For $x \in \Omega$ and $y \in \partial\Omega$ we have $|\rho(x) - \rho(y)| \leq M\,|x - y|$. Since $\rho(y) = 0$ it follows $|\rho(x)| \leq M\,|x - y|$ and thus $|\rho(x)| \leq M \inf_{y \in \partial\Omega} |x - y|$ i.e. $|\rho(x)| \leq M\theta(x)$.

Set $O_\alpha = \{x \in \Omega : \rho(x) \leq M\,\alpha\}$; then,

$$F(\alpha) \leq \int_{O_\alpha} dx. \tag{11.4.31}$$

Point 2: for $x \in \partial\Omega$, there exists $i_x \in \{1, \ldots, n\}$, $\varepsilon_x > 0$, $c_x > 0$, with $\left|\frac{\partial\rho}{\partial i_x}(y)\right| \geq c_x$, for all $y \in B(x, \varepsilon_x)$ (open ball). Of course, we have $\partial\Omega \subset \bigcup_{x \in \partial\Omega} B(x, \varepsilon_x)$ and since $\partial\Omega$ is compact, there exist $N \geq 1$, $x_j \in \partial\Omega$, $\varepsilon_j > 0$, $c_j > 0$, $j = 1, \ldots, N$ such that $\partial\Omega \subset \bigcup_{j=1}^{N} B(x_j, \varepsilon_j)$ and, on $B(x_j, \varepsilon_j)$, we have $\left|\frac{\partial\rho}{\partial x_{\varphi(j)}}(y)\right| \geq c_j$, where $\varphi(j) \in \{1, \ldots, n\}$.

Point 3: there exists $\alpha_0 > 0$ such that $O_{\alpha_0} \subset \bigcup_{j=1}^{N} B(x_j, \varepsilon_j)$. We proceed by contradiction; otherwise, for every $k \geq 1$ there exists $y_k \in O_{\frac{1}{k}}$ such that $y_k \notin \bigcup_{j=1}^{N} B(x_j, \varepsilon_j)$. Since $\overline{\Omega}$ is compact, there exists a subsequence $(y_{\varphi(k)})$ that converges to $\overline{y} \in \overline{\Omega}$; moreover, $\rho(y_k) \leq \frac{M}{k}$ so $\rho(\overline{y}) = 0$, i.e. $\overline{y} \in \partial\Omega$, but $\overline{y} \notin \bigcup_{j=1}^{N} B(x_j, \varepsilon_j)$, which is a contradiction.

Consequence: for $\alpha \leq \alpha_0$ we have $O_\alpha \subset \bigcup_{j=1}^{N} (B(x_j, \varepsilon_j) \cap O_\alpha)$; thus,

$$F(\alpha) \leq \sum_{j=1}^{N} \int_{O_\alpha \cap B(x_j, \varepsilon_j)} dx. \tag{11.4.32}$$

Point 4: fix $j \in \{1, \ldots, N\}$. Rotating the coordinates we may assume that we have $\left|\frac{\partial\rho}{\partial x_d}\right| \geq c_d > 0$ on $B(x_j, \varepsilon_j)$. Let $\chi : O_\alpha \cap B(x_j, \varepsilon_j) \to \mathbf{R}^d$, $x \mapsto \chi(x) = y$ where $\chi_j(x) = x_j$, $j = 1, \ldots, d-1$, $\chi_d(x) = -\rho(x)$. The modulus of the determinant of the Jacobian of this transformation is $\left|\frac{\partial\rho}{\partial x_d}(x)\right|$. Thus, χ is a C^1 diffeomorphism onto its image. As in $O_\alpha \cap B(x_j, \varepsilon_j)$ we have $|x| \leq |x_j| + \varepsilon_j$ and $|\rho(x)| \leq M\,\alpha$, the image is contained in a set of the form $\{y = (y', y_d) : |y'| \leq A,\ y_d \in]0, M\,\alpha]\}$. By the change of variables theorem, we have

$$\begin{aligned}\int_{O_\alpha \cap B(x_j,\varepsilon_j)} dx &= \int_{\chi[O_\alpha \cap B]} \left|\frac{\partial \rho}{\partial x_d}(\chi^{-1}(y))\right|^{-1} dy \\ &\leq c_d^{-1} \int_{|y'|\leq A} \int_0^{M\alpha} dy_d\, dy' \leq K\,\alpha.\end{aligned}$$

It follows from (11.4.32) that $F(\alpha) \leq C\,\alpha$, for $\alpha \in]0, \alpha_0]$. For $\alpha > \alpha_0$ we have the inequality, $F(\alpha) \leq |\Omega| \leq \frac{|\Omega|}{\alpha_0}\,\alpha$. □

Proof of Lemma 11.4.10 This is a consequence of the Fubini-Tonelli Theorem. Indeed, let $\varphi = \psi'$ and $O = \{x \in \Omega : \theta(x) > s_0\}$. Set

$$I = \int_O \Big(\int_{\theta(x)}^{+\infty} \varphi(s)\, ds \Big)\, dx.$$

Then we have

$$\begin{aligned} I &= \int_O \int_{s_0}^{+\infty} \mathbf{1}_{\{s>\theta(x)\}}\, \varphi(s)\, ds\, dx = \int_{s_0}^{+\infty} \varphi(s) \Big(\int_O \mathbf{1}_{\{\theta(x)<s\}}\, dx \Big)\, ds \\ &= \int_{s_0}^{+\infty} \varphi(s) \Big(\int_{s_0<\theta(x)<s} dx \Big)\, ds = \int_{s_0}^{+\infty} \varphi(s)\, (F(s) - F(s_0))\, ds. \end{aligned}$$

As $\psi \to 0$ when $s \to +\infty$ it follows that

$$-\int_O \psi(\theta(x))\, dx = \int_{s_0}^{+\infty} \psi'(s)\, F(s)\, ds + \psi(s_0)\, F(s_0).$$

□

End of the Proof of Theorem 11.3.1.

We deduce from (11.4.12) and Corollary 11.4.8 that $\Sigma\, e^{-\lambda_n t}$ is convergent for all $t > 0$ and that $\sum_{n=1}^{+\infty} e^{-\lambda_n t}$ is equivalent to $\frac{|\Omega|}{(4\pi t)^{d/2}}$ as $t \to 0^+$. It follows from Lemma 11.4.1 that

$$N(\lambda) \sim \frac{|\Omega|}{(4\pi)^{d/2} \frac{d}{2} \Gamma(\frac{d}{2})}\, \lambda^{\frac{d}{2}} = (2\pi)^{-d}\, C_d\, |\Omega|\, \lambda^{\frac{d}{2}},$$

where $C_d = \int_{|x|<1} dx$.

Chapter 12
Almgren's Theory, Unique Continuation and Hausdorff Measure of Nodal Sets of Harmonic Functions

12.1 Prerequisites

Divergence theorem (see appendix for a reminder). Measure theory; Carathéodory's theorem. Holomorphic functions. Geometric integration formula.

12.2 The Problem

In this chapter we are interested in solutions to equations of the type

$$(\star) \quad Lu = 0 \quad \text{in an open set } \Omega \subset \mathbf{R}^d,$$

where L is an elliptic operator (the simplest model being the Laplacian) and more precisely in the places where such solutions can vanish. At first, we will study the propagation of their zeros and ask the following question: if such a solution vanishes somewhere is it identically zero on Ω?

Since the equations studied are linear, this question is of course equivalent to the following: if two solutions of the equation $Lu = f$ coincide somewhere do they coincide everywhere? This is why we speak of "unique continuation" in relation to this problem.

We will see that nontrivial solutions of $(\star)$ cannot vanish on an open subset (no matter how small), nor even on a set of strictly positive measure. This will lead us, in a second step, to try to estimate the possible size of the set of zeros using a finer notion of measure, namely the Hausdorff measure. To address these questions, several approaches are possible. Here, we have chosen to present a powerful and elegant tool known as "Almgren's theory," named after the American mathematician Frederick J. Almgren (1933–1997), who developed it.

C. Zuily, *Selected Topics in Partial Differential Equations*, Universitext,
https://doi.org/10.1007/978-3-032-24082-8_12

The results in this chapter are, for the most part, valid for solutions of elliptic operators L with real and Lipschitz coefficients. However, in order not to obscure our discussion with excessive technicalities, we will limit ourselves to the case where L is the Laplacian.

12.3 Almgren's Theory

12.3.1 Notation

For $t > 0$ and $x_0 \in \mathbf{R}^d$, we set

$$B(x_0, t) = \{x \in \mathbf{R}^d : |x - x_0| < t\} \quad \text{and} \quad S_t(x_0) = \{x \in \mathbf{R}^d : |x - x_0| = t\}.$$

For simplicity, we will write $B_r = B(0, r)$, $S_r = S_r(0)$.

We denote by $d\sigma_t$ the Lebesgue measure on S_t. Then, $d\sigma_t = t^{d-1} d\omega$ where $d\omega$ is the Lebesgue measure on the unit sphere.

We denote by ∇ the gradient, $x \cdot \nabla$ will denote the operator $\sum_{j=1}^d x_j \partial_j$, and div the divergence (of a vector field), i.e., if $F = (f_1, \dots, f_d)$, then $\operatorname{div} F = \sum_{j=1}^d \partial_j f_j$. Finally, $\Delta = \sum_{j=1}^d \partial_j^2$ will denote the Laplace operator.

We introduce the following quantities:

$$\begin{aligned} H(x_0, r) &= r^{1-d} \int_{S_r(x_0)} (u(x))^2 \, d\sigma_r, \quad I(x_0, r) = r^{1-d} \int_{B(x_0, r)} |\nabla u(x)|^2 \, dx, \\ N(x_0, r) &= \frac{r I(x_0, r)}{H(x_0, r)}, \end{aligned} \tag{12.3.1}$$

and we will denote by $H(r), I(r), N(r)$ the quantities corresponding to $x_0 = 0$. Finally, $N(x_0, r)$ is called the frequency function of u.

12.3.2 The "doubling"

The main result of this section is the following.

Theorem 12.3.1 (Doubling) *Let $R_0 > 0$ and $x_0 \in \mathbf{R}^d$. Let u be a real solution of $\Delta u = 0$ in $B(x_0, R_0)$. Then for $r \in (0, \frac{1}{2} R_0)$ we have*

$$\int_{B(x_0, 2r)} (u(x))^2 \, dx \le 2^d e^{(2 \ln 2) N(x_0, R_0)} \int_{B(x_0, r)} (u(x))^2 \, dx. \tag{12.3.2}$$

The remainder of this section is essentially devoted to the proof of this theorem. Note that the assumption that u is real was made solely to simplify the calculations that follow. Next, observe that, by replacing u with $u(\cdot + x_0)$, we may assume that $x_0 = 0$.

Proposition 12.3.2

1. *$H(r) = r^{-d} \int_{|x|<r} div\big(x(u(x))^2\big)\, dx$*
2. *$I(r) = r^{-d} \int_{|x|=r} u(x)(x \cdot \nabla u)(x)\, d\sigma_r$*
3. *$H'(r) = 2I(r), \quad N(r) = \frac{rH'(r)}{2H(r)}$*
4. *The function $r \mapsto N(r)$ is increasing.*

Proof 1. By the divergence theorem (see the appendix) and taking into account that the outward unit normal to the ball of radius r is $\nu = \frac{x}{r}$ we have

$$\int_{|x|<r} \operatorname{div}\big(x(u(x))^2\big)\, dx = \int_{|x|=r} \frac{|x|^2}{r}(u(x))^2\, d\sigma_r = r\int_{|x|=r} (u(x))^2\, d\sigma_r,$$

which proves point 1.

2. We have div $(u\nabla u) = |\nabla u|^2 + u\Delta u = |\nabla u|^2$ so that, again by the divergence theorem,

$$I(r) = r^{1-d} \int_{|x|<r} \operatorname{div}\,(u\nabla u)\, dx = r^{1-d}\int_{|x|=r} \frac{1}{r}(x\cdot\nabla)u(x)\, d\sigma_r,$$

which proves point 2.

3. Using point 1, we can write

$$\begin{aligned} H'(r) &= -dr^{-1}H(r) + r^{-d}\frac{d}{dr}\int_0^r \int_{|x|=t} \operatorname{div}\big(x(u(x))^2\big)\, dr\, d\sigma_t \\ &= -dr^{-1}H(r) + r^{-d}\int_{|x|=r} \operatorname{div}\big(x(u(x))^2\big)\, d\sigma_r. \end{aligned}$$

On the other hand, $\operatorname{div}\big(x(u(x))^2\big) = d(u(x))^2 + 2u(x)(x\cdot\nabla u)(x)$, so that

$$\begin{aligned} H'(r) &= -dr^{-1}H(r) + dr^{-d}\int_{|x|=r} (u(x))^2\, d\sigma_r + 2r^{-d}\int_{|x|=r} u(x)(x\cdot\nabla u)(x)\, d\sigma_r \\ &= -dr^{-1}H(r) + dr^{-1}H(r) + 2r^{-d}\int_{|x|=r} u(x)(x\cdot\nabla u)(x)\, d\sigma_r \\ &= 2I(r), \end{aligned}$$

by point 2. We immediately deduce that

$$N(r) = \frac{rI(r)}{H(r)} = \frac{rH'(r)}{2H(r)},$$

4. Using spherical coordinates, we can write

$$rI(r) = r^{2-d} \int_0^r \int_{|x|=t} |\nabla u(x)|^2 \, d\sigma_t \, dt.$$

Consequently,

$$\frac{d}{dr}\big(rI(r)\big) = (2-d)I(r) + r^{2-d} \int_{|x|=r} |\nabla u(x)|^2 \, d\sigma_r.$$

By the divergence theorem we have

$$\begin{aligned} r^{1-d} \int_{|x|<r} \operatorname{div}\big[|\nabla u(x)|^2 x\big] \, dx &= r^{1-d} \int_{|x|=r} |\nabla u(x)|^2 x \cdot \frac{x}{r} \, d\sigma_r \\ &= r^{2-d} \int_{|x|=r} |\nabla u(x)|^2 \, d\sigma_r, \end{aligned}$$

so that

$$\frac{d}{dr}\big(rI(r)\big) = (2-d)I(r) + r^{1-d} \int_{|x|<r} \operatorname{div}\big[|\nabla u(x)|^2 x\big] \, dx.$$

On the other hand,

$$\operatorname{div}\big[|\nabla u(x)|^2 x\big] = d|\nabla u(x)|^2 + (x \cdot \nabla)\big(|\nabla u(x)|^2\big).$$

We deduce that

$$\frac{d}{dr}\big(rI(r)\big) = 2I(r) + r^{1-d} \int_{|x|<r} (x \cdot \nabla)\big(|\nabla u(x)|^2\big) \, dx. \tag{12.3.3}$$

Let us denote $\operatorname{Hess}(u)(x) = (\partial_j \partial_k u(x))_{1 \le j,k \le d}$. We then have

$$\begin{aligned} (x \cdot \nabla)\big(|\nabla u(x)|^2\big) &= \sum_{j=1}^d x_j \partial_j \Big(\sum_{k=1}^d (\partial_k u(x))^2 \Big) = 2 \sum_{j=1}^d \Big(\sum_{k=1}^d (\partial_j \partial_k u(x)) \partial_k u(x) \Big) x_j, \\ &= 2 \sum_{j=1}^d \big(\operatorname{Hess}(u)(x) \nabla u(x)\big)_j x_j = 2\langle x, \operatorname{Hess}(u)(x) \nabla u(x) \rangle. \end{aligned}$$

We can write

$$\operatorname{div}\big((x \cdot \nabla u(x)) \nabla u(x)\big) = \sum_{k=1}^d \partial_k \Big(\sum_{j=1}^d x_j \partial_j u(x) \Big) \partial_k u(x).$$

Therefore

$$\operatorname{div}(x \cdot \nabla u(x))\nabla u(x) = \sum_{k=1}^{d}\sum_{j=1}^{d} \delta_{jk}\partial_j u(x)\partial_k u(x) + \sum_{k=1}^{d}\sum_{j=1}^{d} x_j(\partial_k\partial_j u(x))\partial_k u(x),$$

where δ_{jk} is the Kronecker symbol. We deduce that

$$2\operatorname{div}\big((x \cdot \nabla u(x))\nabla u(x)\big) = 2|\nabla u(x)|^2 + 2\langle x, \operatorname{Hess}(u)(x)\nabla u(x)\rangle.$$

Denoting by $'$ the derivative with respect to r and recalling that $B_r = \{x : |x| < r\}$ we deduce from (12.3.3) that

$$\begin{aligned}(rI(r))' &= 2I(r) + 2r^{1-d}\int_{B_r} \operatorname{div}\big((x \cdot \nabla u(x))\nabla u(x)\big)\,dx - 2r^{1-d}\int_{B_r} |\nabla u(x)|^2\,dx \\ &= 2r^{1-d}\int_{B_r} \operatorname{div}\big((x \cdot \nabla u(x))\nabla u(x)\big)\,dx,\end{aligned}$$

by the definition of $I(r)$. It follows from the divergence theorem that

$$\frac{d}{dr}(rI(r)) = 2r^{1-d}\int_{|x|=r} \big((x \cdot \nabla u(x)\big)^2\,d\sigma_r. \tag{12.3.4}$$

According to (12.3.1) we have

$$\begin{aligned}N'(r) &= \frac{1}{H(r)^2}\big((rI(r))'H(r) - rI(r)H'(r)\big) \\ &= \frac{1}{H(r)^2}\big((rI(r))'H(r) - 2rI(r)^2\big).\end{aligned} \tag{12.3.5}$$

Next, according to point 2. and the Cauchy-Schwarz inequality, we can write

$$2rI(r)^2 \le 2r^{1-d}\Big(\int_{|x|=r} (u(x))^2\,d\sigma_r\Big)\Big(\int_{|x|=r} \big((x \cdot \nabla u(x)\big)^2\,d\sigma_r\Big).$$

Using (12.3.4) and the definition of $H(r)$ we get

$$2rI(r)^2 \le (rI(r))'\,H(r).$$

Using (12.3.5) we obtain $N'(r) \ge 0$.

□

Remark 12.3.3 Let us compute $N(r)$ in the case where $u = P_m$ is a real, homogeneous, harmonic polynomial of degree m. By Euler's identity (see $(*)$ below) we have

$$(x \cdot \nabla)P_m = mP_m.$$

Using point 2. of Proposition 12.3.2 and the definition of $H(r)$ we obtain

$$I(r) = mr^{-d} \int_{|x|=r} (P_m(x))^2 \, d\sigma_r = \frac{m}{r} H(r),$$

so that $N(r) = \frac{rI(r)}{H(r)} = m$.

(∗) Since P_m is homogeneous of degree m we have, for $\lambda > 0$, $P_m(\lambda x) = \lambda^m P_m(x)$. Differentiating this equality with respect to λ and then taking $\lambda = 1$ we obtain $\sum_{j=1}^{d} x_j \partial_{x_j} P_m(x) = m P_m(x)$.

The equality $N(r) = m$ proved in the above Remark is a general fact as shown by the following result.

Proposition 12.3.4 *Let $R_0 > 0$ and u be a nontrivial solution of the equation $\Delta u = 0$ in $B(0, R_0)$. Then $\lim_{r\to 0} N(r)$ exists and is equal to the exact order of vanishing of u at zero.*

Proof Let us first note that by Proposition 12.3.8, u cannot vanish to infinite order at zero. The proof will be carried out in several steps. In that follows we shall set $B_r = B(0, r)$.

1. For $m \in \mathbf{N}$ let $\mathcal{P}_m$ be the set of homogeneous polynomials of degree m on $\mathbf{R}^d$; if $q_m \in \mathcal{P}_m$ then $q_m(tx) = t^m q_m(x)$ for all $t \in \mathbf{R}$ and all $x \in \mathbf{R}^d$. Set

$$\mathcal{H}_m = \{p_m \in \mathcal{P}_m : \Delta p_m = 0, \text{ in } \mathbf{R}^d\}.$$

Let $(q_m)_{m\in\mathbf{N}}$ be a sequence where $q_m \in \mathcal{P}_m$. If $\sum_{m=0}^{+\infty} q_m(x) = 0$ in B_ρ, for some $\rho > 0$, then

$$q_m(x) = 0, \quad \forall x \in B_\rho, \quad \forall m \in \mathbf{N}. \tag{12.3.6}$$

Indeed, suppose there exists $m \in \mathbf{N}$ and $x_0 \in B_\rho$ such that $q_m(x_0) \neq 0$. Let $m_0 \geq 0$ be the smallest such index, i.e., $q_j(x_0) = 0$ for $j < m_0$ and $q_{m_0}(x_0) \neq 0$. Let $t \in (0, 1)$; then $tx_0 \in B_\rho$. We can write

$$0 = \sum_{m=0}^{+\infty} q_m(tx_0) = \sum_{m=0}^{+\infty} t^m q_m(x_0) = t^{m_0} q_{m_0}(x_0) + O(t^{m_0+1}).$$

Dividing both sides by t^{m_0} and letting t tend to zero, we find that $q_{m_0}(x_0) = 0$, which is a contradiction.

Since a harmonic function is analytic, it is in a neighborhood of zero, the uniformly convergent sum of its Taylor series at zero. We can therefore write in B_ρ

$$u(x) = \sum_{\alpha\in\mathbf{N}^d} \frac{(\partial^\alpha u)(0)}{\alpha!} x^\alpha.$$

Set $p_m(x) = \sum_{\alpha=m} \frac{(\partial^\alpha u)(0)}{\alpha!} x^\alpha$. Then $p_m \in \mathcal{P}_m$ and $u(x) = \sum_{m=0}^{+\infty} p_m$.

In the sense of distributions, we have $\Delta u(x) = \sum_{m=0}^{+\infty} \Delta p_m$ and since both sides are continuous functions in $|x| < \rho$, the equality is also true pointwise. Moreover, since p_m is a polynomial of degree m, we have $\Delta p_0 = \Delta p_1 \equiv 0$ so that

$$0=\sum_{m=0}^{+\infty}\Delta p_m=\sum_{m=2}^{+\infty}\Delta p_m=\sum_{m=0}^{+\infty}\Delta p_{m+2}.$$

By applying (12.3.6) to $q_m=\Delta p_{m+2}\in\mathcal{P}_m$, we deduce that $\Delta p_j=0$ for all j, so that $p_j\in\mathcal{H}_j$.

2. Let $k\in\mathbf{N}$, and suppose that u vanishes exactly to order k at zero, i.e., either $u(0)\neq 0$ if $k=0$, or, if $k\geq 1$,

$$\partial_x^\alpha u(0)=0,\quad |\alpha|<k,\ \text{and}\ \ \exists\alpha_0\in\mathbf{N}^d,\,|\alpha_0|=k:\partial_x^{\alpha_0}u(0)\neq 0.$$

If $k=0$ we have $u(0)=p_0\neq 0$. If $k\geq 1$ then by definition $\partial_x^{\alpha_0}u(0)=0$ for $|\alpha|<k$. Since $p_m(x)=\sum_{|\alpha|=m}\frac{(\partial^\alpha u)(0)}{\alpha!}x^\alpha$ we have $p_m(x)\equiv 0$ if $m<k$. Moreover, $p_k\not\equiv 0$ otherwise we would have $(\partial^\alpha u)(0)=0$ for $|\alpha|=k$.

Consequently we can write

$$u(x)=p_k(x)+r_k(x),\qquad \text{where } r_k(x)=\sum_{m=k+1}^{+\infty}p_m(x).$$

Since $p_m\in\mathcal{P}_m$ we have $|p_m(x)|\leq C_m|x|^m$ and $\nabla p_m(x)\in\mathcal{P}_{m-1}$. Therefore we have $|\nabla p_m(x)|\leq C'_m|x|^{m-1}$. We deduce

$$|r_k(x)|\leq C_k|x|^{k+1},\quad |\nabla r_k(x)|\leq C'_k|x|^k.$$

3. We deduce from the previous step that for $|x|<\rho$,

$$\begin{aligned}(u(x))^2&=(p_k(x))^2+\psi_k(x),\quad |\psi_k(x)|\leq C_k|x|^{2k+1},\\ |\nabla u(x)|^2&=|\nabla p_k(x)|^2+\theta_k(x),\quad |\theta_k(x)|\leq C_k|x|^{2k-1}.\end{aligned}\tag{12.3.7}$$

We have

$$\begin{aligned}&\int_{|x|=r}(p_k(x))^2\,d\sigma_r=\int_{|\omega|=1}(p_k(r\omega))^2r^{d-1}\,d\omega=r^{2k+d-1}\int_{|\omega|=1}(p_k(r\omega))^2\,d\omega,\\ &\int_{|x|=r}(\psi_k(x))^2\,d\sigma_r\leq C(d,k)r^{2k+d},\\ &\int_{|x|<r}|\nabla p_k(x)|^2\,dx=\int_{|x|<1}|\nabla p_k(ry)|^2r^d\,dy=r^{2(k-1)+d}\int_{|x|<1}|\nabla p_k(y)|^2\,dy,\\ &\int_{|x|<r}|\theta_k(x)|^2\,dx\leq C'k\int_{|x|<r}|x|^{2k-1}\,dx<C'(d,k)r^{2k+d-1}.\end{aligned}$$

We deduce from (12.3.7) and from the above that

$$H(r)=r^{1-d}\int_{|x|=r}(u(x))^2\,d\sigma_r=r^{2k}\int_{|\omega|=1}(p_k(r\omega))^2\,d\omega+O(r^{2k+1}),$$

$$rI(r) = r^{2-d} \int_{|x|<r} |\nabla u(x)|^2 \, dx = r^{2k} \int_{|x|<1} |\nabla p_k(x)|^2 \, dx + O(r^{2k+1}).$$

We deduce that

$$N(r) = \frac{rI(r)}{H(r)} = \frac{\int_{|y|<1} |\nabla_x p_k(x)|^2 \, dx + O(r)}{\int_{|\omega|=1} |p_k(\omega)|^2 \, d\omega + O(r)}.$$

It follows that the limit of $N(r)$ as r tends to zero is the frequency function of the homogeneous harmonic polynomial p_k at the point $r = 1$. But as we saw in Remark 12.3.3, this is equal to k, the order of the polynomial, which proves Proposition 12.3.4. □

Proof of Theorem 12.3.1

Since N is increasing, we have $N(r) \leq N(R_0)$. Using point 3. of Proposition 12.3.2, it follows that $H'(r) \leq \frac{2}{r} N(R_0) H(r)$. Let $\rho \leq \frac{1}{2} R_0$. Integrating this inequality between ρ and 2ρ we obtain

$$H(2\rho) \leq e^{(2\ln 2)N(R_0)} H(\rho), \tag{12.3.8}$$

that is, taking into account (12.3.1),

$$2^{1-d} \int_{|x|=2\rho} (u(x))^2 \, d\sigma_{2\rho} \leq e^{(2\ln 2)N(R_0)} \int_{|x|=\rho} (u(x))^2 \, d\sigma_\rho.$$

Integrating this inequality between 0 and r and setting $2\rho = \rho'$ in the integral on the left-hand side we get

$$2^{-d} \int_0^{2r} \int_{|x|=\rho'} (u(x))^2 \, d\sigma_{\rho'} \, d\rho' \leq e^{(2\ln 2)N(R_0)} \int_0^r \int_{|x|=\rho} (u(x))^2 \, d\sigma_\rho \, d\rho,$$

which proves (12.3.2)

Remark 12.3.5 The inequality (12.3.2) is optimal. Indeed, for a real, homogeneous, harmonic polynomial P_m of degree m we have

$$\int_{|x|<2r} (P_m(x))^2 \, dx = 2^{2m+d} \int_{|x|<r} (P_m(x))^2 \, dx. \tag{12.3.9}$$

Indeed, write $P_m(x) = \sum_{|\alpha|=m} a_\alpha x^\alpha$ and set $J_t = \int_{|x|<t} (P_m(x))^2 \, dx$. In spherical coordinates $x = s\omega$, we can write

$$J_t = \sum_{\substack{|\alpha|=m \\ |\beta|=m}} a_\alpha a_\beta \int_{|x|<t} x^{\alpha+\beta}\, dx = \sum_{\substack{|\alpha|=m \\ |\beta|=m}} a_\alpha a_\beta \int_{s=0}^{t} \int_{S^{d-1}} s^{2m+d-1} \omega^{\alpha+\beta}\, ds\, d\omega$$

$$= \frac{t^{2m+d}}{2m+d} \sum_{\substack{|\alpha|=m \\ |\beta|=m}} a_\alpha a_\beta \int_{S^{d-1}} \omega^{\alpha+\beta} d\omega.$$

We deduce that $J_{2r} = 2^{2m+d} J_r$.

In Remark 12.3.3 we saw that in this case $N(r) = m$. The equality (12.3.9) thus reads

$$\int_{|x|<2r} (P_m(x))^2\, dx = 2^d e^{(2\ln 2)N(r)} \int_{|x|<r} (P_m(x))^2\, dx,$$

which shows that the inequality (12.3.2) cannot be improved.

Remark 12.3.6 Let $R_0 > 0$, $k \in \mathbf{N}$ and $r > 0$ such that $2^k r \leq R_0$. By iterating the inequality (12.3.2) we obtain

$$\int_{|x|<2^k r} (u(x))^2\, dx \leq (2^d 4^{N(R_0)})^k \int_{|x|<r} (u(x))^2\, dx. \tag{12.3.10}$$

12.3.3 *Three Spheres and Three Balls Inequalities*

Proposition 12.3.7 *Let $x_0 \in \mathbf{R}^d$, and u a harmonic function in $B(x_0, R_0)$. Then for all r in $(0, \frac{1}{10}R_0)$ we have the following inequalities.*

$$\int_{|x-x_0|=2r} \big(u(x)\big)^2\, d\sigma_{2r} \leq \Big(\int_{|x-x_0|=r} \big(u(x)\big)^2 d\,\sigma_r\Big)^{\frac{1}{2}} \Big(\int_{|x-x_0|=4r} \big(u(x)\big)^2 d\,\sigma_{4r}\Big)^{\frac{1}{2}}. \tag{12.3.11}$$

$$\int_{|x-x_0|<2r} \big(u(x)\big)^2\, dx \leq \Big(\int_{|x-x_0|<r} \big(u(x)\big)^2 dx\Big)^{\frac{1}{2}} \Big(\int_{|x-x_0|<4r} \big(u(x)\big)^2 dx\Big)^{\frac{1}{2}}. \tag{12.3.12}$$

Proof We can assume that $x_0 = 0$. According to points 3. and 4. of Proposition 12.3.2, we can write

$$\frac{\rho H'(\rho)}{H(\rho)} = N(\rho) \leq N(2\rho) = \frac{2\rho H'(2\rho)}{H(2\rho)}.$$

Integrating this inequality between r and $2r$ gives

$$\ln\big(H(2r)\big) - \ln\big(H(r)\big) \leq \ln\big(H(4r)\big) - \ln\big(H(2r)\big),$$

where ln is the natural logarithm, so that

$$\ln\big(H(2r)\big) \leq \ln\big(H(r)^{\frac{1}{2}}\big) + \ln\big(H(4r)^{\frac{1}{2}}\big) = \ln\big(H(r)^{\frac{1}{2}} H(4r)^{\frac{1}{2}}\big),$$

which, taking into account the definition of $H(r)$, shows (12.3.11).

To obtain (12.3.12) we have just to integrate (12.3.11) between 0 and r and use the Cauchy-Schwarz inequality. □

12.3.4 Some Applications

It is known that nontrivial distributional solutions of the equation $\Delta u = 0$ are analytic functions and therefore cannot vanish at a point to infinite order. Here is a result of this type, which uses the previous arguments.

Proposition 12.3.8 *Let $u \in L^2_{loc}(\mathbf{R}^d)$ be a solution of the equation $\Delta u = 0$ in the set $B_{R_0} = \{x \in \mathbf{R}^d : |x| < R_0\}$. If for every $N \in \mathbf{N}$ there exists $C_N > 0$ such that*

$$F(r) := \int_{|x|<r} \big(u(x)\big)^2 \, dx \leq C_N r^N \quad \forall r \in (0, R_0), \tag{12.3.13}$$

then $u \equiv 0$.

Proof Suppose that u is not identically zero. Note that

$$F(r) = \int_0^r \int_{|x|=\rho} \big(u(x)\big)^2 \, d\sigma_\rho \, d\rho = \int_0^r \rho^{d-1} H(\rho) \, d\rho.$$

By Proposition 12.3.2, we can write for $r \in (0, R_0)$ and $s \in (0, r)$,

$$\frac{H'(s)}{H(s)} = \frac{2}{s} N(s) \leq \frac{2}{s} N(R_0).$$

Integrating this inequality between ρ and R_0 we obtain

$$\ln\big(H(R_0)\big) - \ln\big(H(\rho)\big) \leq 2 \ln\big(\frac{R_0}{\rho}\big) N(R_0),$$

from which we deduce

$$H(\rho) \geq H(R_0) \big(\frac{\rho}{R_0}\big)^{2N(R_0)}.$$

Consequently,

$$F(r) \geq \frac{H(R_0)}{R_0^{2N(R_0)}} \int_0^r \rho^{d-1+2N(R_0)} \, d\rho = \frac{H(R_0)}{R_0^{2N(R_0)} (d + 2N(R_0))} r^{d+2N(R_0)},$$

which contradicts (12.3.13). Therefore $u \equiv 0$. □

The advantage of this type of method lies in the fact that it extends to the case where the Laplacian is replaced by an elliptic operator with real and Lipschitz coefficients.

12.3.5 Unique Continuation from an Open Set

Theorem 12.3.9 *Let Ω be a connected open subset of $\mathbf{R}^d$. Let $u \in C^\infty(\Omega)$ be a solution of $\Delta u = 0$ in Ω. Suppose there exist $x_0 \in \Omega$ and $r_0 > 0$ such that $u = 0$ in $B((x_0, r_0)$. Then u is identically zero in Ω.*

Remark 12.3.10 1. The fact that u is of class C^∞ in Ω is not an assumption since we know, by the hypoellipticity of the Laplacian, that any harmonic distribution is infinitely differentiable.

2. This result is not surprising since we also know that any harmonic distribution is analytic and that an analytic function which vanishes on a small open set is identically zero.

3. The advantage of the method used to prove Theorem 12.3.9 is that it adapts to the case where the Laplacian is replaced by an elliptic operator with real Lipschitz coefficients, for which we can no longer use the analyticity argument.

Proof Since the open set Ω is connected, it is arcwise connected. Let $K \subset \Omega$ be a compact set and $x \in K$. There exists a path γ of finite length joining x_0 to x. We can find $r \in (0, r_0)$ small enough and a finite sequence $(x_j)_{0 \le j \le J}$ such that $x_j \in \gamma$, $|x_j - x_{j+1}| = r$, $\overline{B}(x_j, 4r) \subset \Omega$ and $x \in B(x_J, r)$. We will show by induction that

$$u = 0 \quad \text{in} \quad B(x_j, r), \quad 0 \le j \le J. \tag{12.3.14}$$

By assumption, (12.3.14) is true for $j = 0$. Suppose it is true at order j. According to inequality (12.3.12) applied at the point x_j, we have that $u = 0$ in $B(x_j, 2r)$. But $B(x_{j+1}, r) \subset B(x_j, 2r)$ since $|x - x_j| \le |x - x_{j+1}| + |x_j - x_{j+1}| < 2r$. It follows that $u = 0$ in $B(x_{j+1}, r)$, which completes the induction. Using (12.3.14) with $j = J$ and using the fact that $x \in B(x_J, r)$, we deduce that there exists $r_1 > 0$ such that $u = 0$ in $B(x, r_1)$. Since K can be covered by a finite number of such balls, we deduce that $u = 0$ in K, for any compact K in Ω, and thus $u = 0$ in Ω. □

Here is a corollary of Theorem 12.3.9 which applies to eigenfunctions of the Laplacian.

Theorem 12.3.11 *Let Ω be a connected open subset of $\mathbf{R}^d$, $\lambda > 0$, and $u \in C^\infty(\Omega)$ a solution of $\Delta u + \lambda u = 0$ in Ω. Suppose there exist $x_0 \in \Omega$ and $r_0 > 0$ such that $u = 0$ in $B(x_0, r_0)$. Then u is identically zero in Ω.*

Proof Let $T > 0$ and $\widetilde{\Omega} = (-T, T) \times \Omega$. Set $v(t, x) = e^{t\sqrt{\lambda}} u(x)$ for $(t, x) \in \widetilde{\Omega}$. Then we have $(\partial_t^2 + \Delta)v = 0$ in $\widetilde{\Omega}$. Moreover, if $0 < \delta < T$, we have $v = 0$ in $(-\delta, \delta) \times B(x_0, r_0)$, which is a ball contained in $\widetilde{\Omega}$. We deduce from Theorem 12.3.9 applied to the Laplacian $\partial_t^2 + \Delta$ in $\mathbf{R}^{d+1}$ that $v = 0$ in $\widetilde{\Omega}$. In particular, we will have $v(0, x) = u(x) = 0$ in Ω. □

12.4 Unique Continuation from Sets of Positive Measure

The main result of this section is the following. It extends Theorem 12.3.9.

Theorem 12.4.1 *Let u be a harmonic function in an open set Ω of $\mathbf{R}^d$. If u vanishes on a subset $E \subset \Omega$ of nonzero measure, then u is identically zero in Ω.*

Remark 12.4.2 This result can be applied, for example, to holomorphic functions in an open set $\mathcal{O}$ of $\mathbf{C}$. If such a function vanishes on a set $E \subset \mathcal{O}$ of positive measure, then it is identically zero. Indeed, since the function f is holomorphic, it satisfies the equation $\overline{\partial} f = 0$ in $\mathcal{O}$. Then $\partial\overline{\partial} f = 0$ in $\mathcal{O}$. It suffices to note that $\partial\overline{\partial} = \frac{1}{4}\Delta$ and to apply Theorem 12.4.1.

The remainder of this section is devoted to the proof of this theorem. We begin with some preliminary results.

12.4.1 Density Points

For $x \in \mathbf{R}^d$ and $r > 0$, $B(x, r)$ will denote the ball centered at x with radius r.

Proposition 12.4.3 *Let μ be the Lebesgue measure on $\mathbf{R}^d$. Let $E \subset \mathbf{R}^d$ be a measurable set. Then,*

$$\lim_{r \to 0} \frac{\mu(B(x, r) \cap E)}{\mu(B(x, r))} = 1, \quad \text{for } \mu\text{-almost every } x \in E.$$

This result is nontrivial only if E has strictly positive measure. It is an immediate corollary of the following Lebesgue differentiation theorem.

Theorem 12.4.4 (Lebesgue) *Let $f \in L^1_{loc}(\mathbf{R}^d)$. Then, for almost every $x \in \mathbf{R}^d$,*

$$\lim_{r \to 0} \frac{1}{\mu(B(x, r))} \int_{B(x,r)} f(t)\, d\mu(t) = f(x).$$

Proposition 12.4.3 is obtained by applying Theorem 12.4.4 to $f = \chi_E$, the characteristic function of E.

Proof of Theorem 12.4.4 It suffices to show that, for almost every $x \in \mathbf{R}^d$,

$$\lim_{r \to 0} \frac{1}{\mu(B(x, r))} \int_{B(x,r)} |f(t) - f(x)|\, d\mu(t) = 0. \tag{12.4.1}$$

We will use the following lemma.

Lemma 12.4.5 (Hardy-Littlewood maximal inequality) *Let $f \in L^1(\mathbf{R}^d)$. Set*

$$Mf(x) = \sup_{r>0} \frac{1}{\mu(B(x,r))} \int_{B(x,r)} |f(t)|\, d\mu(t).$$

There exists $C > 0$ depending only on d such that for all $\alpha > 0$,

$$\mu\{x \in \mathbf{R}^d : Mf(x) > \alpha\} \le C \frac{\|f\|_{L^1}}{\alpha}$$

Let us show how the lemma implies (12.4.1). Set

$$T_r(f)(x) = \frac{1}{\mu(B(x,r))} \int_{B(x,r)} |f(t) - f(x)|\, d\mu(t)$$

and $T(f)(x) = \overline{\lim}_{r\to 0} T_r(f)(x)$. We will show that the set $\{x \in \mathbf{R}^d : T(f)(x) > \alpha\}$ is negligible, which will imply that

$$\{x \in \mathbf{R}^d : T(f)(x) > 0\} = \cup_{k\ge 1}\{x \in \mathbf{R}^d : T(f)(x)\} > \frac{1}{k}\}$$

is also negligible, i.e., $T(f)(x) = 0$, μ-a.e.

Let $n \ge 1$. There exists $g \in C^0(\mathbf{R}^d)$ such that $\|f - g\|_{L^1} \le \frac{1}{n}$. Set $h = f - g$ then

$$T_r(h)(x) \le \frac{1}{\mu(B(x,r))} \int_{B(x,r)} |h(t)|\, d\mu(t) + |h(x)| \le M(h)(x) + |h(x)|.$$

Since g is continuous, the dominated convergence theorem shows that, for all $x \in \mathbf{R}^d$,

$$\lim_{r\to 0} T_r(g)(x) = \lim_{r\to 0} \frac{1}{\mu(B(x,r))} \int_{B(x,r)} |g(t) - g(x)|\, d\mu(t) = 0,$$

i.e., $T(g)(x) = 0$. Since $f = g + h$ we have

$$T(f)(x) \le T(g)(x) + T(h)(x) \le Mh(x) + |h(x)|,$$

thus,

$$\{x : T(f)(x) > \alpha\} \subset \{x : Mh(x) > \alpha/2\} \cup \{x : |h(x)| > \alpha/2\}. \tag{12.4.2}$$

According to Lemma 12.4.5 we have

$$\mu\{x : Mh(x) > \alpha\} \le C \frac{2\|h\|_{L^1}}{\alpha} \le \frac{2C}{\alpha n}. \tag{12.4.3}$$

Next, since $\int_{\mathbf{R}^d} |h(t)|\, d\mu(t) \leq \frac{1}{n}$ we have

$$\frac{1}{n} \geq \int_{\{t:|h(t)|>\alpha/2\}} |h(t)|\, d\mu(t) \geq \frac{\alpha}{2}\mu\{t : |h(t)| > \alpha/2\},$$

whence

$$\mu\{t : |h(t)| > \alpha/2\} \leq \frac{2}{\alpha n}. \tag{12.4.4}$$

Using (12.4.2), (12.4.3), (12.4.4), we finally obtain

$$\mu\{x : T(f)(x) > \alpha\} \leq \frac{C_1}{n},$$

for all $n \geq 1$ whence $\mu\{x : T(f)(x) > \alpha\} = 0$. □

Proof of Lemma 12.4.5 By inner regularity of the measure, it suffices to show that for any compact K contained in the set

$$\left\{x : \sup_{r>0} \frac{1}{\mu(B(x,r))} \int_{B(x,r)} |f(t)|\, d\mu(t) > \alpha\right\},$$

we have $\mu(K) \leq \frac{C}{\alpha}\|f\|_{L^1}$.

By hypothesis, for every $x \in K$ there exists r_x such that

$$\frac{1}{\mu(B(x,r_x))} \int_{B(x,r_x)} |f(t)|\, d\mu(t) > \alpha.$$

By compactness, K is covered by a finite number of balls $B(x, r_x)$. The Vitali lemma (see Lemma 12.6.7 in the appendix) implies that

$$\exists \big(B(x_j, r_{x_j})\big)_{j=1\dots,N} \text{ disjoint} : K \subset \cup_{j=1}^{N} B(x_j, 3r_{x_j}).$$

Then,

$$\begin{aligned}\mu(K) \leq \sum_{j=1}^{N} \mu(B(x_j, 3r_{x_j})) \leq 3^d \sum_{j=1}^{N} \mu(B(x_j, r_{x_j})) &\leq \frac{3^d}{\alpha} \sum_{j=1}^{N} \int_{B(x_j,r_{x_j})} |f(t)|\, d\mu(t),\\ &\leq \frac{3^d}{\alpha} \int_{\mathbf{R}^d} |f(t)|\, d\mu(t),\end{aligned}$$

since the balls $B(x_j, r_{x_j})$ are disjoint. □

12.4.2 An Inequality

In what follows, "measurable" will mean measurable with respect to the Lebesgue measure, which we will denote by μ.

Lemma 12.4.6 *Let $x_0 \in \mathbf{R}^d$ and $r > 0$. Let F be a measurable subset of $B(x_0, r)$ such that $\mu(F) > 0$. Then for any $v \in H^1(B(x_0, r))$ and for any measurable set $A \subset B(x_0, r)$ we have*

$$\mu(F) \int_A |v(x)|^2 \, dx \le 2\mu(A) \int_F |v(x)|^2 \, dx + C\, r^{d+1} \, \mu(A)^{\frac{1}{d}} \int_{B(x_0,r)} |\nabla_x v(x)|^2 \, dx,$$

where $C = \frac{2^{d+3}}{d+1} \left(\mu(S^{d-1})\right)^{1-\frac{1}{d}}$.

Proof It suffices to prove this lemma for $x_0 = 0$ and for $v \in C^\infty(\overline{B_r})$ where B_r denotes $B(0, r)$. Let $y \in F$ and $x \in A$, $x \neq y$. We have

$$\begin{aligned} v(y) - v(x) &= \int_0^{|y-x|} \frac{d}{d\rho} \left(v(x + \rho \frac{y-x}{|y-x|} \right) d\rho, \\ &= \int_0^{|y-x|} \nabla_x v\left(x + \rho \frac{y-x}{|y-x|}\right) \cdot \frac{y-x}{|y-x|} \, d\rho. \end{aligned}$$

Using the Cauchy-Schwarz inequality we obtain

$$|v(x)|^2 \le 2|v(y)|^2 + 2|y-x| \int_0^{|y-x|} \left| \nabla_x v\left(x + \rho \frac{y-x}{|y-x|}\right) \right|^2 d\rho.$$

Integrating this inequality for $y \in F$ we get

$$\mu(F)|v(x)|^2 \le 2 \int_F |v(y)|^2 \, dy + \int_{B_r} |y-x| \int_0^{|y-x|} \left| \nabla_x v\left(x + \rho \frac{y-x}{|y-x|}\right) \right|^2 d\rho \, dy. \tag{12.4.5}$$

We set, in the second integral on the right-hand side, $y = x + t\omega$, $t > 0$, $\omega \in S^{d-1}$. We first show that for $x \in B_r$,

$$y = x + t\omega \in B_r, t > 0, \omega \in S^{d-1} \Longleftrightarrow y = x + t\omega, 0 < t < t^*(x, \omega), \tag{12.4.6}$$

where $0 < t^* \le 2r$. Indeed, $y \in B_r$ is equivalent to $|x + t\omega|^2 < r^2$, therefore to $t^2 + 2(x \cdot \omega)t + |x|^2 - r^2 < 0$. The reduced discriminant of this quadratic in t is equal to $\Delta = (x \cdot \omega)^2 + r^2 - |x|^2$, which is strictly positive. Therefore, this quadratic has two real roots, and since their product is $|x|^2 - r^2 < 0$, it has one positive root t^* and one negative root t_*. Consequently, this polynomial is strictly negative if and only if $0 < t < t^*$. Moreover, $t^* = -x \cdot \omega + \sqrt{(x \cdot \omega)^2 + r^2 - |x|^2}$, therefore we have $0 < t^* \le |x| + r \le 2r$.

Making the change of variables $y = x + t\omega$ in the second integral on the right-hand side of (12.4.5), we obtain

$$\mu(F)|v(x)|^2 \le 2\int_F |v(y)|^2\,dy + 2\int_{S^{d-1}}\int_{t=0}^{t^*} t\int_{\rho=0}^{t} |\nabla_x v(x+\rho\omega)|^2 t^{d-1}\,d\rho\,dt\,d\omega.$$

By Fubini's Theorem, we can write

$$\mu(F)|v(x)|^2 \le 2\int_F |v(y)|^2\,dy + \int_{S^{d-1}}\int_{\rho=0}^{t^*}\Big(\int_{t=\rho}^{t^*} t^d\,dt\Big)|\nabla_x v(x+\rho\omega)|^2\,d\rho\,d\omega.$$

Since $t^* < 2r$, we obtain

$$\mu(F)|v(x)|^2 \le 2\int_F |v(y)|^2\,dy + \frac{2^{d+2}}{d+1}r^{d+1}\int_{S^{d-1}}\int_{\rho=0}^{t^*}\frac{|\nabla_x v(x+\rho\omega)|^2}{\rho^{d-1}}\rho^{d-1}d\rho\,d\omega.$$

Setting $z = x + \rho\omega$ and using (12.4.6) again, we obtain

$$\mu(F)|v(x)|^2 \le 2\int_F |v(y)|^2\,dy + \frac{2^{d+2}}{d+1}r^{d+1}\int_{B_r}\frac{|\nabla_x v(z)|^2}{|z-x|^{d-1}}\,dz.$$

Integrating both sides of this inequality with respect to x over A, we obtain

$$\mu(F)\int_A |v(x)|^2\,dx \le 2\mu(A)\int_F |v(y)|^2\,dy + \frac{2^{d+2}}{d+1}r^{d+1}\int_{B_r}|\nabla_x v(z)|^2\,g(z)\,dz, \tag{12.4.7}$$

where $g(z) = \int_A \frac{1}{|z-x|^{d-1}}\,dx$.

To compute g let $\delta > 0$ and set

$$A_1 = \{x \in A : |z-x| < \delta\},\quad A_2 = \{x \in A : |z-x| > \delta\}.$$

We have

$$\int_{A_1}\frac{1}{|z-x|^{d-1}}\,dx \le \int_{|z-x|<\delta}\frac{1}{|z-x|^{d-1}}\,dx \le \mu(S^{d-1})\,\delta,$$
$$\int_{A_2}\frac{1}{|z-x|^{d-1}}\,dx \le \delta^{-(d-1)}\mu(A).$$

We choose δ such that $\mu(S^{d-1})\,\delta = \delta^{-(d-1)}\mu(A)$, that is, $\delta = \left(\frac{\mu(A)}{\mu(S^{d-1})}\right)^{\frac{1}{d}}$ and we obtain the estimate

$$g(z) \le 2\big(\mu(S^{d-1})\big)^{1-\frac{1}{d}}\mu(A)^{\frac{1}{d}}.$$

Using (12.4.7), we finally obtain

$$\mu(F)\int_A |v(x)|^2\,dx \le 2\mu(A)\int_F |v(y)|^2\,dy$$
$$+\frac{2^{d+3}}{d+1}\big(\mu(S^{d-1})\big)^{1-\frac{1}{d}} r^{d+1}\mu(A)^{\frac{1}{d}}\int_{B_r} |\nabla_x v(z)|^2\,dz.$$

□

12.4.3 First Proof of Theorem 12.4.1

We know from Proposition 12.4.3 that almost every point $x_0 \in E$ is a density point. Denoting $|A| = \mu(A)$ we have

$$\lim_{r\to 0}\frac{|E\cap B(x_0,r)|}{|B(x_0,r)|} = 1.$$

Let $\varepsilon \in (0,\frac{1}{2})$. Then there exists $r_0 > 0$ such that for $r \le r_0$, with r_0 small enough,

$$|E\cap B(x_0,r)| \ge (1-\varepsilon)|B(x_0,r)|, \quad |E^c\cap B(x_0,r)| \le \varepsilon|B(x_0,r)|. \tag{12.4.8}$$

We will apply Lemma 12.4.6 with $F = E\cap B(x_0,r)$ and $A = E^c\cap B(x_0,r)$. Since $u = 0$ on E and $B(x_0,r) = (E\cap B(x_0,r))\cup(E^c\cap B(x_0,r))$, it follows that

$$|E\cap B(x_0,r)|\int_{B(x_0,r)} (u(x))^2\,dx \le C(d)r^{d+1}|E^c\cap B(x_0,r)|^{\frac{1}{d}}\int_{B(x_0,r)} |\nabla u(x)|^2\,dx.$$

Using (12.4.8) and the fact that $|B(x_0,r)|^{\frac{1}{d}-1} \le C_1(d)r^{1-d}$, we obtain

$$\int_{B(x_0,r)} (u(x))^2\,dx \le C_2(d)\frac{\varepsilon^{\frac{1}{d}}}{1-\varepsilon}r^2\int_{B(x_0,r)} |\nabla u(x)|^2\,dx. \tag{12.4.9}$$

We will show that there exists $C_3(d) > 0$ such that

$$\int_{B(x_0,r)} |\nabla u(x)|^2\,dx \le \frac{C_3(d)}{r^2}\int_{B(x_0,2r)} (u(x))^2\,dx. \tag{12.4.10}$$

Indeed, let $\chi \in C_0^\infty(\mathbf{R}^d)$, $\chi(y) = 1$ if $|y| \le 1$, $\chi(y) = 0$ if $|y| \ge 2$. Set $\chi_r(x) = \chi(\frac{x-x_0}{r})$. We can write, with $L^2 = L^2(\Omega)$,

$$0 = -(\Delta u, \chi_r^2 u)_{L^2} = (\nabla u, \chi_r^2\nabla u)_{L^2} + 2(\nabla u, \chi_r(\nabla\chi_r)u)_{L^2}.$$

We deduce that

$$\|\chi_r\nabla u\|_{L^2}^2 \le 2\|\chi_r\nabla u\|_{L^2}\|(\nabla\chi_r)u\|_{L^2}^2.$$

As $\chi_r = 1$ on $B(x_0, r)$ and $|\nabla\chi_r(x)| \le \frac{C}{r}$, we obtain (12.4.10). Using (12.4.9), (12.4.10), it follows that

$$\int_{B(x_0,r)} \big(u(x)\big)^2\, dx \le C_4(d)\varepsilon^{\frac{1}{d}} \int_{B(x_0,2r)} \big(u(x)\big)^2\, dx. \tag{12.4.11}$$

We then use Theorem 12.3.1. We obtain

$$\int_{B(x_0,r)} \big(u(x)\big)^2\, dx \le C_5(d)\varepsilon^{\frac{1}{d}} 4^{N(p,r_0)} \int_{B(x_0,r)} \big(u(x)\big)^2\, dx.$$

Taking ε small enough (depending on u), we deduce that $u \equiv 0$ in $B(x_0, r)$. But, according to Theorem 12.3.9, this implies that $u \equiv 0$ in Ω.

12.4.4 Second Proof of Theorem 12.4.1

Here is another proof of the theorem in question that does not use Theorem 12.3.9. We begin with the following lemma.

Lemma 12.4.7 *Let $f : \mathbf{R}^+ \to \mathbf{R}^+$ be an increasing function such that*

$$\forall\delta > 0 \;\; \exists r_\delta \in (0, 1) : f(r) \le \delta f(2r), \quad \forall r \le r_\delta.$$

Then for every $N \in \mathbf{N}^*$ there exist C_N, $r_N > 0$ such that

$$f(r) \le C_N r^N, \quad \forall r \le r_N.$$

Proof Fix $m \in \mathbf{N}^*$ such that $2^{m-1} < \frac{r_\delta}{r} \le 2^m$. Since $2^{m-2}r < r_\delta$, using the hypotheses we can write

$$f(1) \ge f(r_\delta) \ge f(2^{m-1}r) \ge \delta^{-1} f(2^{m-2}r) \ge \cdots \ge \delta^{-(m-1)} f(r).$$

We have $(m-1)\ln 2 \le \ln\big(\frac{r_\delta}{r}\big) \le m\ln 2$. Therefore setting $\alpha = \frac{1}{\ln 2}$, it follows that $m \ge \ln[\big(\frac{r_\delta}{r}\big)^\alpha]$ so that, for $0 < \delta \le 1$,

$$\delta^m \le \delta^{\ln\left[\left(\frac{r_\delta}{r}\right)^\alpha\right]} = e^{(\ln\delta)\ln\left[\left(\frac{r_\delta}{r}\right)^\alpha\right]} = e^{\ln\left[\left(\frac{r}{r_\delta}\right)^{\alpha\ln\frac{1}{\delta}}\right]} = \Big(\frac{r}{r_\delta}\Big)^{\alpha\ln\frac{1}{\delta}}.$$

We deduce that

$$f(r) \le \delta^{-1}\Big(\frac{r}{r_\delta}\Big)^{\alpha\ln\frac{1}{\delta}} f(1).$$

Let $N \in \mathbf{N}^*$. There exists $\delta_N \in (0, 1)$ such that $C_1 N \le \alpha\ln\frac{1}{\delta_N} \le C_2 N$. We associate to it by the hypothesis r_N. Then $\delta_N^{-1} \le e^{\frac{C_2}{\alpha}N}$ and for $r \le r_N$ we have

$$f(r) \leq e^{\frac{C_2}{\alpha}N}\Big(\frac{r}{r_N}\Big)^{\frac{C_1}{\alpha}N} f(1).$$

Setting $C_N = e^{\frac{C_2}{\alpha}N} r_N^{-\frac{C_1}{\alpha}N}$ we obtain $f(r) \leq C_N r^N$ for $r \leq r_N$. □

Let us return to the proof of Theorem 12.4.1. We set $f(r) = \int_{B(x_0,r)} \big(u(x)\big)^2\, dx$. From the inequality (12.4.11) we see that the function f satisfies the hypothesis of Lemma 12.4.7. We deduce from Proposition 12.3.8 that $u \equiv 0$.

12.5 Estimate of the Hausdorff Measure of the Nodal Set of Harmonic Functions

Recall that the nodal set of a continuous function is the set of points where it vanishes. When u is a nontrivial solution of the equation $\Delta u = 0$, by virtue of Theorems 12.3.9 and 12.4.1, this set can be neither open nor of positive Lebesgue measure. To measure its size, one must then use a finer notion of measure, namely the Hausdorff measure.

12.5.1 The Hausdorff Measure and Its Main Properties

Let $A \subset \mathbf{R}^d$, $s \in [0, +\infty)$. For $\delta > 0$ we set

$$\mathcal{H}^s_\delta(A) = \inf\Big\{\sum_{j=0}^{+\infty} \alpha(s)\Big(\frac{\operatorname{diam} U_j}{2}\Big)^s : A \subset \cup_{j=1}^{+\infty} U_j,\ \operatorname{diam} U_j \leq \delta\Big\},$$

where $\alpha(s) = \frac{\pi^{\frac{s}{2}}}{\Gamma(\frac{s}{2}+1)}$, Γ denoting the Gamma function.

Here $\operatorname{diam} U_j = \sup_{x,y\in U_j} d(x, y)$, denotes the diameter of the set U_j.

If $\delta_1 < \delta_2$, since $\operatorname{diam} U_j \leq \delta_1 \Rightarrow \operatorname{diam} U_j \leq \delta_2$ we have $\mathcal{H}^s_{\delta_2}(A) \leq \mathcal{H}^s_{\delta_1}(A)$, in other words the map $\delta \mapsto \mathcal{H}^s_\delta(A)$ is decreasing.

Definition 12.5.1 For A, s as above, we set

$$\mathcal{H}^s(A) = \lim_{\delta\to 0} \mathcal{H}^s_\delta(A) \in [0, +\infty].$$

Note that we also have

$$\mathcal{H}^s(A) = \sup_{\delta>0} \mathcal{H}^s_\delta(A).$$

Remark 12.5.2 (i) The quantity $\alpha(s)$ is introduced so that, when $s = d$, $\mathcal{H}^d$ coincides exactly with the Lebesgue measure $\mathcal{L}^d$.

Here are some elementary properties.

Theorem 12.5.3 *(i)* $\mathcal{H}^s(\emptyset) = 0$.
(ii) If $A \subset B$, *then* $\mathcal{H}^s(A) \leq \mathcal{H}^s(B)$.
(iii) $\mathcal{H}^s(\cup_{j=1}^{+\infty}(A_j)) \leq \sum_{j=1}^{+\infty} \mathcal{H}^s(A_j) \leq +\infty$.

Thus, $\mathcal{H}^s$ is an outer measure.

Definition 12.5.4 Let $A \subset \mathbf{R}^d$. We say that A is $\mathcal{H}^s$-measurable if for all $E \subset \mathbf{R}^d$ we have

$$\mathcal{H}^s(E) = \mathcal{H}^s(E \cap A) + \mathcal{H}^s(E^c \cap A).$$

By Carathéodory's theorem, the $\mathcal{H}^s$-measurable sets form a σ-algebra.

Definition 12.5.5 The restriction of $\mathcal{H}^s$ to the σ-algebra of measurable sets is called **the s-dimensional Hausdorff measure**. Note that this σ-algebra contains the Borel sets.

Theorem 12.5.6 *1. Let* $(A_j)_{j \in \mathbf{N}^*}$ *be a countable family of pairwise disjoint measurable sets. Then,*

$$\mathcal{H}^s(\cup_{j=1}^{+\infty} A_j) = \sum_{j=1}^{+\infty} \mathcal{H}^s(A_j).$$

2. $\mathcal{H}^0$ *is the counting measure, i.e., if* A *is a finite set,* $\mathcal{H}^0(A) = cardinality(A)$, *and if* A *is infinite* $\mathcal{H}^0(A) = +\infty$.
3. $\mathcal{H}^1 = \mathcal{L}^1$ *(the Lebesgue measure) on* $\mathbf{R}$.
4. $\mathcal{H}^s \equiv 0$ *on* $\mathbf{R}^d$ *for all* $s > d$.
5. $\mathcal{H}^s(\lambda A) = \lambda^s \mathcal{H}^s(A)$ *for all* $\lambda > 0$ *and all* $A \subset \mathbf{R}^d$.
6. $\mathcal{H}^s(L(A)) = \mathcal{H}^s(A)$ *for any affine isometry* $L : \mathbf{R}^d \to \mathbf{R}^d$ *and any* $A \subset \mathbf{R}^d$.

Example 12.5.7 Let D be a countable set; for all $s > 0$ we have $\mathcal{H}^s(D) = 0$. Indeed, if we write $D = \{x_1, x_2, \dots x_k, \dots\}$ we have $D \subset \cup_{j=1}^{+\infty}\{x_j\}$ and since the diameter of a point is zero, we have $\mathcal{H}^s_\delta(D) = 0$ for all $\delta > 0$ and thus $\mathcal{H}^s(D) = 0$.

Proposition 12.5.8 *Let* $f : \mathbf{R}^d \to \mathbf{R}^d$ *be an* α*-Lipschitz function where* $\alpha > 0$, *i.e.,*

$$|f(x) - f(y)| \leq \alpha |x - y|, \quad \forall x, y \in \mathbf{R}^d.$$

Let $s \geq 0$. *If* A *is* $\mathcal{H}^s$ *measurable, so is* $f(A)$, *and*

$$\mathcal{H}^s(f(A)) \leq \alpha^s \mathcal{H}^s(A).$$

Remark 12.5.9 As a consequence of this result, we obtain that if A and B are two C^1-diffeomorphic sets, then

$$\mathcal{H}^s(A) < +\infty \iff \mathcal{H}^s(B) < +\infty, \quad \mathcal{H}^s(A) = 0 \iff \mathcal{H}^s(B) = 0.$$

Proposition 12.5.10 *Let $A \subset \mathbf{R}^d$ and $0 \le s < t < +\infty$.*
(*i*) *If $\mathcal{H}^s(A) < +\infty$ then $\mathcal{H}^t(A) = 0$.*
(*ii*) *If $\mathcal{H}^t(A) > 0$ then $\mathcal{H}^s(A) = +\infty$.*

Proof Assertion (ii) is the contrapositive of (i). Let us prove (i). Suppose $\mathcal{H}^s(A)$ is finite and let $\delta > 0$. There exist sets $(U_j)_{j\in\mathbf{N}^*}$ such that $A \subset \cup_{j=1}^{+\infty} U_j$, diam $U_j \le \delta$ and

$$\sum_{j=0}^{+\infty} \alpha(s)\Big(\frac{\text{diam } U_j}{2}\Big)^s \le \mathcal{H}^s_\delta(A) + 1 \le \mathcal{H}^s(A) + 1.$$

Then,

$$\begin{aligned}\mathcal{H}^t_\delta(A) &\le \sum_{j=0}^{+\infty} \alpha(t)\Big(\frac{\text{diam } U_j}{2}\Big)^t = \frac{\alpha(t)}{\alpha(s)} 2^{s-t} \sum_{j=0}^{+\infty} \alpha(s)\Big(\frac{\text{diam } U_j}{2}\Big)^s (\text{diam } U_j)^{t-s} \\ &\le \frac{\alpha(t)}{\alpha(s)} 2^{s-t}\delta^{t-s}\big(\mathcal{H}^s(A) + 1\big).\end{aligned}$$

As δ tends to zero, the right-hand side tends to zero since $t - s > 0$. □

Definition 12.5.11 The Hausdorff dimension of a set $A \subset \mathbf{R}^d$ is defined by

$$\dim_{\mathcal{H}}(A) = \begin{cases} (i)\ \inf\{s \in [0, +\infty) : \mathcal{H}^s(A) = 0\} \\ (ii)\ \sup\{t \in [0, +\infty[: \mathcal{H}^t(A) = +\infty\}. \end{cases}$$

Note that

$$\begin{cases} \mathcal{H}^a(A) < +\infty \Rightarrow \dim_{\mathcal{H}}(A) \le a, \\ \mathcal{H}^b(A) > 0 \Rightarrow \dim_{\mathcal{H}}(A) \ge b. \end{cases}$$

Indeed, suppose $d = \dim_{\mathcal{H}}(A) > a$. According to (i) of the definition, for $a < t < d$ we have $\mathcal{H}^t(A) > 0$. From (ii) of Proposition 12.5.10 we have $\mathcal{H}^a(A) = +\infty$, which is a contradiction. Similarly, suppose $d = \dim_{\mathcal{H}}(A) < b$. According to (ii) of the definition, for $d < s < b$ we have $\mathcal{H}^s(A) < +\infty$. According to (i) of Proposition 12.5.10 we have $\mathcal{H}^b(A) = 0$, which is a contradiction.

In summary,

$$\dim_{\mathcal{H}}(A) = a \iff 0 < \mathcal{H}^a(A) < +\infty.$$

The Hausdorff measure jumps from $+\infty$ to zero, and the place where this jump occurs is precisely the Hausdorff dimension.

12.5.2 m-Rectifiable Sets

Definition 12.5.12 Let $0 \le m \le d$. A $\mathcal{H}^m$-measurable set $E \subset \mathbf{R}^d$ is said to be m-rectifiable if,

$$\mathcal{H}^m(E) < +\infty \quad \text{and if} \quad E \subset E_0 \cup \left(\cup_{j=1}^{+\infty} f_j(\mathbf{R}^m) \right),$$

where $\mathcal{H}^m(E_0) = 0$ and $f_j : \mathbf{R}^m \to \mathbf{R}^d$ is a Lipschitz function for $j = 1, 2, \dots$.

Remark 12.5.13 Since $\theta(A \cup B) \subset \theta(A) \cup \theta(B)$, it follows from the definition and from Remark 12.5.9 that rectifiability is invariant under C^1 diffeomorphisms.

Example 12.5.14 (i) Let $d \geq 2$ and $1 \leq p \leq d-1$. The set

$$E = \{x \in B(0,1) \subset \mathbf{R}^d : x_1 = \cdots = x_p = 0\},$$

is $(d-p)$-rectifiable. Indeed,

$$E \subset Q = \{|x_i| \leq 1, p+1 \leq i \leq d\},$$

and the cube Q of side 2 and dimension $d-p$ can be covered by $O_d\left(\frac{1}{\delta^{d-p}}\right)$ cubes of side δ. Each of these cubes has diameter $O_d(\delta)$, so we deduce that

$$\mathcal{H}_\delta^{d-p}(Q) \leq C(d)\left(\frac{1}{\delta^{d-p}}\right)\delta^{d-p} \leq C'(d).$$

Writing $x' = (x_{p+1}, \dots, x_d)$ we have $E \subset \{(x', 0) : x' \in B(0,1) \cap \mathbf{R}^{d-p}\}$. It suffices to set $f(x') = (x', 0)$.

(ii) By diffeomorphism the set

$$E = \{x \in B(0,1) \subset \mathbf{R}^d : g_1(x) = \cdots = g_p(x) = 0\},$$

where the g_j are C^2 functions whose differentials $dg_1(x), \dots, dg_p(x)$ are linearly independent at every point of E, is $(d-p)$-rectifiable.

(iii) Let u be a nontrivial harmonic function in $\mathbf{R}^d$, and set

$$S = \{x \in B(0,1) : u(x) = 0\}.$$

Then S is $(d-1)$-rectifiable.

Indeed, we write $S = \cup_{k=1}^{+\infty} S_k$ where

$$S_k = \{x \in B(0,1) : \partial^\alpha u(x) = 0, \forall |\alpha| \leq k-1, \left(\partial^\alpha u\right)_{|\alpha|=k} \not\equiv 0\}.$$

Note that the above union is finite because u is analytic. Indeed, if there exists a point where all derivatives of u vanish, then u is identically zero.

First of all, the set $S_1(u)$ is a submanifold of dimension $n-1$. According to (ii), $S_1(u)$ is $(d-1)$-rectifiable. We will see that for $k \geq 2$, $\mathcal{H}^{d-1}(S_k(u)) = 0$ and thus $\mathcal{H}^{d-1}(\cup_{k\geq 2} S_k(u)) = 0$. For this, we show that $S_k(u)$ is contained in a smooth manifold of dimension $d-2$ and thus, by (ii), $\mathcal{H}^{d-2+\varepsilon}(S_k(u)) = 0$ for $\varepsilon > 0$.

For every $x \in S_k(u)$ there exists $|\beta| = k-2$ such that $(\partial_j \partial_k v(x))_{1\leq j,k\leq d} \neq 0$ where $v = \partial^\beta u$. Moreover, $\Delta v = 0$. The nonzero matrix $M = (\partial_j \partial_k v(x))_{1\leq j,k\leq d}$ is real and symmetric so there exists an orthogonal matrix O such that $O^{-1}MO = \mathrm{diag}(\lambda_j) := D$ with $D \neq 0$. Thus there exists at least one eigenvalue, say λ_1, that is nonzero. Since

$$0 = \Delta v = \mathrm{Tr} M = \mathrm{Tr}(O^{-1}MO) = \sum_{j=1}^{d} \lambda_j,$$

there exists another eigenvalue, say λ_2, that is nonzero. Set $w(y) = v(Oy)$. Then

$$\nabla_y w(y) = O^{-1}\nabla_x v(x), \quad \text{and} \quad \mathrm{Hess}_y(w)(y) = O^{-1}\mathrm{Hess}_x(v)(x)O = D.$$

We have

$$\begin{aligned} &\partial_1^2 w(y) = \lambda_1, \quad \partial_j\partial_1 w(y) = 0, \quad 2 \leq j \leq d,\\ &\partial_2^2 w(y) = \lambda_2, \quad \partial_j\partial_2 w(y) = 0, \quad 1 \leq j \leq d, j \neq 2. \end{aligned}$$

It follows that

$$\nabla_y \partial_1 w(y) = (\lambda_1, 0, \ldots, 0), \quad \nabla_y \partial_2 w(y) = (0, \lambda_2, 0, \ldots, 0).$$

These are therefore two linearly independent vectors. Thus,

$$\Sigma = \{y \in V(0) : \partial_1 w(y) = 0, \partial_2 w(y) = 0\},$$

is a smooth submanifold of dimension $d-2$ and $S_k(u) \subset \Sigma$.

12.5.3 The Geometric Integration Formula

The Formula

Let m, d be two integers such that $m \leq d$. We set

$$\Lambda(m, d) = \{\lambda = (\lambda_1, \lambda_2 \ldots, \lambda_m) \in \mathbf{N}^m : 1 \leq \lambda_1 < \lambda_2 < \cdots < \lambda_m \leq d\}.$$

For $\lambda \in \Lambda$ we denote by $P_\lambda : \mathbf{R}^d \to \mathbf{R}^m$ the map

$$P_\lambda(x_1, x_2, \ldots, x_d) = (x_{\lambda_1}, x_{\lambda_2}, \ldots, x_{\lambda_m}).$$

We also denote by $d\mathcal{L}^m$ the Lebesgue measure on $\mathbf{R}^m$.

Theorem 12.5.15 (Federer [11], Theorem 3.2.27) *Let $m, d \in \mathbf{N}^*$, $m \leq d$. Let E be an m-countably rectifiable subset of $\mathbf{R}^d$. Set, for $\lambda \in \Lambda(m, d)$,*

$$a_\lambda = \int_{\mathbf{R}^m} \mathcal{H}^0(E \cap P_\lambda^{-1}(y))d\mathcal{L}^m(y) = \int_{\mathbf{R}^m} \mathcal{H}^0(\{x \in E : P_\lambda(x) = y\})d\mathcal{L}^m(y).$$

Then,

$$\Big(\sum_{\lambda\in\Lambda(m,d)} a_\lambda^2\Big)^{\frac{1}{2}} \leq \mathcal{H}^m(E) \leq \sum_{\lambda\in\Lambda(m,d)} a_\lambda.$$

For example, take $d = 2$ and $m = 1$. In this case $\Lambda(1, 2) = \{\lambda = 1 \text{ or } \lambda = 2\}$. Then for $y \in \mathbf{R}$, $P_1^{-1}(y) = (y, x_2)$ and $P_2^{-1}(y) = (x_1, y)$. Thus $E \cap P_1^{-1}(y)$ is the intersection of E with the vertical line $x_1 = y$ and $E \cap P_2^{-1}(y)$ is the intersection of E with the horizontal line $x_2 = y$. The formula of the theorem then says that to compute the 1-dimensional Hausdorff measure of E it suffices to know the number of points of its intersection with the lines parallel to the axes. Suppose that E is the circle centered at the origin with radius R. If $y < -R$ or $y > R$ $P_j^{-1}(y) = \emptyset$. If $-R < y < R$ there are two points of intersection of the lines $x_1 = y$ or $x_2 = y$ with E. Thus $a_1 = a_2 = \int_{-R}^{R} 2\,dy = 4R$. We thus have

$$4\sqrt{2}R \leq \mathcal{H}^1(E) = 2\pi R \leq 8R.$$

12.5.4 Reminders on Holomorphic Functions

Let us first recall Jensen's formula.

Theorem 12.5.16 *Let $R > 0$ and f a holomorphic function in $B(0, R)$ such that $f(0) \neq 0$. For $r \in (0, R)$ denote by $\alpha_1, \ldots, \alpha_{n(r)}$ the zeros of f in $\overline{B}(0, r)$. Then*

$$|f(0)| \prod_{n=1}^{n(r)} \frac{r}{|\alpha_n|} = exp\Big(\frac{1}{2\pi} \int_{-\pi}^{\pi} ln\, |f(re^{i\theta})|\, d\theta\Big).$$

Corollary 12.5.17 *Let f be a holomorphic function in $B(0, R)$ such that $|f(0)| \neq 0$. For r such that $0 < 2r < R$ let $M(r) = \sup_{|z|=r} |f(z)|$, and $n(r)$ the number of zeros of f in $\overline{B}(0, r)$. Then,*

$$n(r) \leq \frac{1}{ln\, 2} ln\, M(2r) - \frac{1}{ln\, 2} ln\, |f(0)|. \tag{12.5.1}$$

Proof First, we have

$$\begin{aligned} \exp\Big(\frac{1}{2\pi} \int_{-\pi}^{\pi} \ln |f(2re^{i\theta})|\, d\theta\Big) &\leq \exp\Big(\frac{1}{2\pi} \sup_{|z|=2r} \ln |f(z)| \int_{-\pi}^{\pi} d\theta\Big) \\ &\leq \exp\big(\ln M(2r)\big) \leq M(2r). \end{aligned} \tag{12.5.2}$$

Let $\alpha_1, \ldots, \alpha_{n(2r)}$ be the zeros of f in $\overline{B}(0, 2r)$ such that

$$|\alpha_1| \leq |\alpha_2| \leq \cdots \leq |\alpha_{n(2r)}|.$$

We have $n(r) \leq n(2r)$ and

$$|\alpha_n| \leq r \quad \text{if } 1 \leq n \leq n(r). \tag{12.5.3}$$

Apply Theorem 12.5.16 with $2r$ instead of r, and inequality (12.5.2). Since $\frac{2r}{|\alpha_n|} \geq 1$ it follows, using (12.5.3),

$$M(2r) \geq |f(0)| \prod_{n=1}^{n(2r)} \frac{2r}{|\alpha_n|} \geq |f(0)| \prod_{n=1}^{n(r)} \frac{2r}{|\alpha_n|} \geq |f(0)| 2^{n(r)}.$$

We deduce that

$$\ln M(2r) \geq n(r) \ln 2 + \ln |f(0)|,$$

which proves (12.5.1). □

Remark 12.5.18 In Corollary 12.5.17 one can replace the point 0 by any point p. It suffices to apply it to the function $u(\cdot + p)$.

Here is the analogue of Corollary 12.5.17 in the case of rectangles in **C**.

Lemma 12.5.19 *Let* $a > 0$, $\rho_0 > 0$ *such that* $\rho_0 \leq \frac{1}{10}a$. *Let*

$$Q = \{z \in \mathbf{C} : |Re\, z| < a, |Im\, z| < \rho_0\}.$$

There exists $C > 0$ *independent of* a, ρ_0 *such that for any holomorphic function* u *in a neighborhood of* $3Q$ *we have*

$$\#\{z \in Q : u(z) = 0\} \leq C \frac{a}{\rho_0} ln \big(\sup_{z \in 3Q} |u(z)| \big).$$

Proof In the interval $I = [-a, a]$, the function u has a finite set of zeros which we denote by $\mathcal{P} = \{p_1, \ldots, p_K\}$.

Consider, for $0 \leq j \leq J := \left[\frac{a}{\rho_0}\right]$, the points $\alpha_j = -a + (2j+1)\rho_0$ and set, for small $\varepsilon > 0$,

$$Q_j = \{z \in \mathbf{C} : |\mathrm{Re}\, z - \alpha_j| < \rho_0 + \varepsilon, \quad |\mathrm{Im}\, z| < \rho_0 + \varepsilon\}.$$

We have $\overline{Q} \subset \cup_{j=0}^{J} Q_j$.

If $\alpha_j \in \mathcal{P}$, we consider a nearby point β_j ($\beta_j = \alpha_j + \varepsilon$) such that $\beta_j \notin \mathcal{P}$ and denote by Q'_j the analogue of Q_j where α_j is replaced by β_j. We still have $\overline{Q} \subset \cup_{j=0}^{J} Q'_j$.

Finally set

$$B_j = \{z \in \mathbf{C} : |z - (\beta_j, 0)| < \sqrt{2}(\rho_0 + \varepsilon)\}.$$

Since $Q'_j \subset B_j$, we have $\overline{Q} \subset \cup_{j=0}^{J} B_j$.

Since $\rho_0 \leq \frac{1}{10}a$, if ε is small enough, we have

$$\overline{Q} \subset \cup_{j=0}^{J} B_j \subset \frac{3}{2}Q, \quad J = [\frac{a}{\rho_0}]. \tag{12.5.4}$$

It follows that

$$\#\{z \in Q : u(z) = 0\} \leq \sum_{j=0}^{J} \#\{z \in B_j : u(z) = 0\}. \tag{12.5.5}$$

According to Corollary 12.5.17, Remark 12.5.18, we have denoting $q_j = (\beta_j, 0)$,

$$\begin{aligned}\#\{z \in B_j : u(z) = 0\} &\leq \frac{1}{\ln 2}\ln\big(\sup_{\partial(2B_j)} |u(z)|\big) - \frac{1}{\ln 2}\ln|u(q_j)| \\ &\leq \frac{2}{\ln 2}\ln\big(\sup_{z\in 3Q} |u(z)|\big),\end{aligned}$$

by (12.5.4).

It follows from (12.5.5) that

$$\#\{z \in Q : u(z) = 0\} \leq \frac{4}{\ln 2}\frac{a}{\rho_0}\ln\big(\sup_{z\in 3Q} |u(z)|\big).$$

□

Holomorphic Extension of Harmonic Functions

Proposition 12.5.20 *For any $\delta \in (0, 1)$ there exists $\eta_1 > 0$ such that any C^∞ solution of the equation $\Delta u = 0$ in $\overline{B}(0, 1)$ extends to a holomorphic function $\widetilde{u}$ in the set $Q = \{z \in \mathbf{C}^d : |Rez| < \delta, Imz| < \eta_1\}$ which satisfies*

$$\sup_{Q} |\widetilde{u}| \leq C \sup_{B(0,1)} |u|.$$

The proof of this result will use the following lemma.

Lemma 12.5.21 *Let $\delta < 1$. For any $\varepsilon > 0$ and any $p \in \mathbf{R}^d$ such that $B(p, \varepsilon)$ is contained in $B(0, 1)$ and for any $k \in \mathbf{N}$ we have*

$$\sup_{|\alpha|=k} |\partial^\alpha u(p)| \leq \Big(\frac{2dk}{\varepsilon}\Big)^k \sup_{B(p,\varepsilon)} |u|. \tag{12.5.6}$$

Proof The case $k = 0$ is trivial since $|u(p)| \leq \sup_{B(p,\varepsilon)} |u|$.

When $k = 1$, define $v_i = \partial_{x_i} u$ for $1 \leq i \leq d$ and use the fact that v_i is harmonic. We apply the mean value formula (see Proposition 12.6.6) to v_i and then the divergence theorem (see Theorem 12.6.4). We obtain

$$\partial_{x_i} u(p) = \frac{1}{|B(p, \frac{\varepsilon}{2})|} \int_{B(p,\frac{\varepsilon}{2})} \partial_{x_i} u(x)\, dx = \frac{1}{|B(p, \frac{\varepsilon}{2})|} \int_{\partial B(p,\frac{\varepsilon}{2})} \nu_i(x) u(x)\, d\sigma,$$

where ν is the unit outward normal. We get

$$|\partial_{x_i} u(p)| \leq \frac{1}{|B(p, \frac{\varepsilon}{2})|} \int_{\partial B(p,\frac{\varepsilon}{2})} |u(x)|\, d\sigma \leq \frac{|\partial B(p, \frac{\varepsilon}{2})|}{|B(p, \frac{\varepsilon}{2})|} \sup_{\partial B(p,\frac{\varepsilon}{2})} |u| \leq \frac{2d}{\varepsilon} \sup_{\partial B(p,\frac{\varepsilon}{2})} |u|.$$

If $q \in \partial B(p, \frac{\varepsilon}{2})$, we have $B(q, \frac{\varepsilon}{2}) \subset B(p, \varepsilon)$ and thus

$$|u(q)| \leq \sup_{B(q,\frac{\varepsilon}{2})} |u| \leq \sup_{B(p,\varepsilon)} |u|.$$

It follows that

$$|\partial_{x_i} u(p)| \leq \frac{2d}{\varepsilon} \sup_{B(p,\varepsilon)} |u|,$$

which proves (12.5.6) for $k = 1$.

We then proceed by induction, assuming the inequality holds at order k. Let $\alpha \in \mathbf{N}^d$ such that $|\alpha| = k + 1$. We have $\alpha = \beta + e_i$ with $|\beta| = k$. Set $w = \partial^\beta u$ and $v = \partial^\alpha u = \partial_{x_i} w$. Since v is harmonic, we can write, using Proposition 12.6.6 in the Appendix and denoting by ν the unit outward normal, for $\rho \in]0, \varepsilon]$,

$$\begin{aligned} \partial^\alpha u(p) = v(p) &= \frac{1}{|B(p,\rho)|} \int_{B(p,\rho)} v(x)\, dx = \frac{1}{|B(p,\rho)|} \int_{B(p,\rho)} \partial_{x_i} w(x)\, dx, \\ &= \frac{1}{|B(p,\rho)|} \int_{\partial B(p,\rho)} \nu_i(x) w(x)\, d\sigma = \frac{1}{|B(p,\rho)|} \int_{\partial B(p,\rho)} \nu_i(x) \partial^\beta u(x)\, d\sigma. \end{aligned}$$

We deduce that

$$|\partial^\alpha u(p)| \leq \frac{|\partial B(p,\rho)|}{|B(p,\rho)|} \sup_{\partial B(p,\rho)} |\partial^\beta u| \leq \frac{d}{\rho} \sup_{\partial B(p,\rho)} |\partial^\beta u|. \tag{12.5.7}$$

We take $\rho = \frac{\varepsilon}{k+1}$ and $r = k\rho$. If $q \in \partial B(p,\rho)$ we have $B(q,r) \subset B(p,\varepsilon) \subset B(0,1)$. Then, by induction,

$$|\partial^\beta u(q)| \leq \left(\frac{2dk}{r}\right)^k \sup_{B(q,r)} |u| \leq \left(\frac{2dk}{r}\right)^k \sup_{B(p,\varepsilon)} |u|.$$

Since $\frac{d}{\rho} = \frac{d(k+1)}{\varepsilon}$ and $\frac{1}{r} = \frac{k+1}{k\varepsilon}$, we deduce from (12.5.7) that

$$|\partial^\alpha u(p)| \le \frac{d(k+1)}{\varepsilon}\left(\frac{2d(k+1)}{\varepsilon}\right)^k \sup_{B(p,\varepsilon)} |u| \le \left(\frac{2d(k+1)}{\varepsilon}\right)^{k+1} \sup_{B(p,\varepsilon)} |u|,$$

which proves (12.5.6) at order $k+1$. □

In what follows, we will take $\varepsilon > 0$ such that $\varepsilon + \delta < 1$. Note that if $p \in B_\delta(0)$, we have $B(p, \varepsilon) \subset B_{\delta+\varepsilon}(0)$.

Let us then recall the following estimates for $\alpha \in \mathbf{N}^d$:

$$|\alpha|! \le C(d)^{|\alpha|}\alpha!, \quad |\alpha|^{|\alpha|} \le e^{|\alpha|}|\alpha|!, \quad \sum_{|\alpha|=k} 1 \le C(d)k^{d-1} \le C'(d)2^k. \tag{12.5.8}$$

It follows from (12.5.6) and (12.5.8) that there exists $C_1(d) > 0$ such that for all $\alpha \in \mathbf{N}^d$,

$$|\partial^\alpha u(p)| \le C_1(d)^{|\alpha|+1}\varepsilon^{-|\alpha|}\alpha! \sup_{B(0,\delta+\varepsilon)} |u|. \tag{12.5.9}$$

Proof of Proposition 12.5.20 For $z \in \mathbf{C}$ set

$$A_k = \sum_{|\alpha|=k} \frac{|\partial^\alpha u(p)|}{\alpha!}|z-p|^{|\alpha|}.$$

According to (12.5.9), (12.5.8) we have

$$|A_k| \le C_2(d)^{k+1}\varepsilon^{-k}|z-p|^k \sup_{B(0,\delta+\varepsilon)} |u|. \tag{12.5.10}$$

The series $\sum_{k=0}^{+\infty} A_k$ is uniformly convergent for $|z-p| \le C_3(d)\varepsilon$, $C_3(d)$ sufficiently small.

For $p \in B(0,\delta)$ and $|z-p| \le C_3(d)\varepsilon$ set

$$\widetilde{u}(z) = \sum_{k=0}^{+\infty} \frac{\partial^\alpha u(p)}{\alpha!}(z-p)^\alpha$$

Then $\widetilde{u}$ is holomorphic in $\{z \in \mathbf{C} : |z-p| < C_3(d)\varepsilon\}$ for all $p \in B_\delta(0)$ and by gluing, a holomorphic function in $Q = \{z \in \mathbf{C} : |\mathrm{Re}\, z| < \delta, |\mathrm{Im}\, z| \le C_3(d)\varepsilon\}$. Moreover, according to Taylor's formula, this function coincides with u when $\mathrm{Im}\, z = 0$. Finally, from (12.5.10) we have

$$\sup_{z\in Q} |\widetilde{u}(z)| \le C_4(d) \sup_{B(0,\delta+\varepsilon)} |u|.$$

□

12.5.5 Estimate of the Measure of Nodal Sets of Harmonic Functions

Recall that the nodal set of u is the set where u vanishes. The aim of this paragraph is to study the measure of such sets when u is a harmonic function.

Example 12.5.22 Consider in $\mathbf{R}^2$ the homogeneous polynomial of degree $k \in \mathbf{N}^*$, $u_k(x, y) = \operatorname{Re}(z^k)$. It is harmonic in $\mathbf{R}^2$ because $\Delta = 4\partial\overline{\partial}$ and thus $\Delta(z^k) = 0$. We have

$$u_k(x, y) = \operatorname{Re}(z^k) = r^k \cos(k\theta),$$

so that

$$\{(x, y) \in \mathbf{R}^2 : u_k(x, y) = 0\} = \{z = re^{i\theta} : r > 0, \quad \theta = \frac{\pi}{2k} + \frac{j\pi}{k}, 0 \leq j \leq 2k - 1\}.$$

This set therefore consists of k lines passing through the origin. Then, for $R > 0$,

$$\mathcal{H}^1(\{x \in B(0, R) : u_k(x) = 0\}) = 2Rk, \tag{12.5.11}$$

since the diameter of $B(0, R)$ is equal to $2R$ and the set consists of k diameters.

The aim of this paragraph is to generalize (12.5.11) to the case of arbitrary solutions of the equation $\Delta u = 0$.

We will use, in what follows, the following result and recall that the frequency function $N(x_0, r)$ was defined in (12.3.1).

Lemma 12.5.23 *Let u be a solution of $\Delta u = 0$ in $B(0, 1) \subset \mathbf{R}^d$ and $x_0 \in B(0, \frac{1}{4})$.*

1. *There exist constants $c_1(d), c_2(d)$ depending only on d such that*

$$N(x_0, \frac{1}{2}) \leq c_1(d)N(0, 1) + c_2(d). \tag{12.5.12}$$

2. *There exists $\varepsilon(d)$ depending only on d such that if $N(0, 1) \leq \varepsilon(d)$ then $u(x) \neq 0$ in $B(0, \frac{1}{2})$.*
3. *If $\{x \in B(0, \frac{1}{2}) : u(x) = 0\} \neq \emptyset$ then there exists $c(d) > 0$ such that for x_0 in $B(0, \frac{1}{4})$,*

$$N(x_0, \frac{1}{2}) \leq c(d)N(0, 1).$$

Proof 1. First, we have $B(x_0,\frac{3}{4})\subset B(0,1)$ and $B(0,\frac{1}{4})\subset B(x_0,\frac{1}{2})$. It follows from Remark 12.3.6 that

$$\int_{B(x_0,\frac{3}{4})}\big(u(x)\big)^2\,dx\le\int_{B(0,2^2\frac{1}{4})}\big(u(x)\big)^2\,dx\le(2^d4^{N(1,0)})^2\int_{B(0,\frac{1}{4})}\big(u(x)\big)^2\,dx,$$

from which

$$\int_{B(x_0,\frac{3}{4})}\big(u(x)\big)^2\,dx\le 2^{2d}4^{2N(1,0)}\int_{B(x_0,\frac{1}{2})}\big(u(x)\big)^2\,dx. \tag{12.5.13}$$

According to (12.3.8) we have

$$\int_{|x-x_0|=\frac{5}{8}}\big(u(x)\big)^2\,d\sigma\le C_1(d)4^{2N(1,0)}\int_{|x-x_0|=\frac{1}{2}}\big(u(x)\big)^2\,d\sigma. \tag{12.5.14}$$

Since the function $r\mapsto\int_{|x-x_0|=r}\big(u(x)\big)^2\,d\sigma$ is increasing, we can write

$$\begin{aligned}\int_{B(x_0,\frac{3}{4})}\big(u(x)\big)^2\,dx&\ge\int_{B(x_0,\frac{3}{4})\backslash B(x_0,\frac{5}{8})}\big(u(x)\big)^2\,dx=\int_{\frac{5}{8}}^{\frac{3}{4}}r^{d-1}\int_{|x-x_0|=r}\big(u(x)\big)^2\,d\sigma\,dr\\&\ge C_2(d)\int_{|x-x_0|=\frac{5}{8}}\big(u(x)\big)^2\,dx,\end{aligned}$$

and

$$\int_{B(x_0,\frac{1}{2})}\big(u(x)\big)^2\,dx=\int_0^{\frac{1}{2}}r^{d-1}\int_{|x-x_0|=r}\big(u(x)\big)^2\,d\sigma\,dr\le C_3(d)\int_{|x-x_0|=\frac{1}{2}}\big(u(x)\big)^2\,d\sigma.$$

Recall that if we set $\widetilde{H}(x_0,r)=\int_{|x-x_0|=r}\big(u(x)\big)^2\,d\sigma$ we have

$$\frac{d}{dr}\big(\ln\widetilde{H}(x_0,r)\big)=\frac{\widetilde{H}'(x_0,r)}{\widetilde{H}(x_0,r)}=\frac{2N(x_0,r)}{r}-\frac{d-1}{r}.$$

Since the function $r\mapsto N(x_0,r)$ is increasing, we can write

$$\int_{\frac{5}{8}}^{\frac{1}{2}}\frac{d}{dr}(\ln\widetilde{H}(x_0,r))\,dr=\int_{\frac{5}{8}}^{\frac{1}{2}}\Big(\frac{2N(x_0,r)}{r}-\frac{d-1}{r}\Big)\,dr\ge C_4(d)N(x_0,\frac{1}{2})+1),$$

since $\int_{\frac{5}{8}}^{\frac{1}{2}}\frac{dr}{r}=\ln\frac{5}{4}$. We deduce that

$$\ln\Big(\int_{|x-x_0|=\frac{5}{8}}\big(u(x)\big)^2\,d\sigma\Big)-\ln\Big(\int_{|x-x_0|=\frac{1}{2}}\big(u(x)\big)^2\,d\sigma\Big)\ge C(d)\big(N(x_0,\frac{1}{2})+1\big).$$

Using (12.5.13) it follows that

$$C_4(d)\big(N(x_0,\frac{1}{2})+1\big) \le \ln C_1(d) + 4(\ln 2)N(0,1),$$

which proves (12.5.12).

2. We can assume that $\int_{|x|=1} \big(u(x)\big)^2\, d\sigma = 1$. From the definition of $N(0,1)$ we deduce that

$$\int_{B(0,1)} |\nabla u(x)|^2\, dx \le N(0,1).$$

Suppose that $N(0,1) \le \varepsilon(d)$ where $\varepsilon(d)$ is to be chosen. There then exists $C_1(d) > 0$ such that

$$\sup_{B(0,\frac{1}{2})} |\nabla u(x)| \le C_1(d)\Big(\int_{B(0,1)} |\nabla u(x)|^2\, dx\Big)^{\frac{1}{2}} \le C_1(d)\sqrt{\varepsilon(d)}. \tag{12.5.15}$$

Next,

$$1 = \int_{|x|=1} \big(u(x)\big)^2\, d\sigma \le 2^d 4^{N(0,1)} \int_{|x|=\frac{1}{2}} \big(u(x)\big)^2\, d\sigma \le 2^d 4^{\varepsilon(d)} \int_{|x|=\frac{1}{2}} \big(u(x)\big)^2\, d\sigma.$$

So there exists a point p_0 such that $|p_0| = \frac{1}{2}$ and $|u(p_0)| \ge \Big(\frac{2^{-d}4^{-\varepsilon(d)}}{\mu\{|x|=\frac{1}{2}\}}\Big)^{\frac{1}{2}} = C_2(d)$. Using (12.5.15) we have, for $x_0 \in B_{\frac{1}{2}}(0)$,

$$|u(x_0) - u(p_0)| \le |x_0 - p_0| \sup_{B_{\frac{1}{2}}(0)} |\nabla u| \le C_3(d)\sqrt{\varepsilon(d)},$$

and thus

$$|u(x_0)| \ge |u(p_0)| - C_3(d)\sqrt{\varepsilon(d)} \ge C_2(d) - C_3(d)\sqrt{\varepsilon(d)} > 0$$

if $\varepsilon(d)$ is small enough.

3. If this set is non-empty, we have, according to point 2, $N(0,1) \ge \varepsilon(d)$. Then, according to point 1, it follows that

$$N(x_0,\frac{1}{2}) \le c_1(d)N(0,1) + \frac{c_2(d)}{\varepsilon(d)}\varepsilon(d) \le \big(c_1(d) + \frac{c_2(d)}{\varepsilon(d)}\big)N(0,1).$$

□

The Main Theorem

As before, in what follows, $B(x_0, r)$ will denote the open ball in $\mathbf{R}^d$ centered at x_0 with radius r and $S_r(x_0) = \partial B(x_0, r)$ the sphere $\{x : |x - x_0| = r\}$. We will then denote by $|B(x_0, r)|$ (resp. $|\partial B(x_0, r)|$) the d-dimensional (resp. $(d-1)$-dimensional) Lebesgue measure of this set. Note that $\frac{|\partial B(x_0,r)|}{|B(x_0,r)|} = \frac{d}{r}$.

The purpose of this paragraph is to prove the following result.

Theorem 12.5.24 *Let u be a harmonic function in $\overline{B}(0, 1) \subset \mathbf{R}^d$. Then there exists a positive constant $C = C(d)$ depending only on the dimension such that*

$$\mathcal{H}^{d-1}(\{x \in B(0, \frac{1}{2}) : u(x) = 0\}) \leq C(d)N(0, 1),$$

where $N(0, 1)$ is the frequency function of u at the point $r = 1$, i.e.

$$N(0, 1) = \frac{\int_{B(0,1)} |\nabla u(x)|^2 \, dx}{\int_{S_1(0)} (u(x))^2 \, d\omega}.$$

Remark 12.5.25 (i) Let us note that this result is consistent with the result in Example 12.5.22, which concerns the homogeneous harmonic polynomial of degree k, $u_k(x, y) = \mathrm{Re}\,(z^k)$ (for which we have seen that the 1-Hausdorff measure of the zeros in $B(0, 1)$ is equal to $2k$), since for this polynomial we have $N(0, 1) = k$.

(ii) More generally, let P_m be any homogeneous harmonic polynomial in $\mathbf{R}^d$. Let us set $E = \{x \in B(0, \frac{1}{2}) : P_m(x) = 0\}$. To estimate $\mathcal{H}^{d-1}(E)$ we use Theorem 12.5.15. The set $\Lambda(d-1, d)$ has d elements $\lambda_1 = (2, \ldots, d)$, $\lambda_2 = (1, 3, \ldots, d), \ldots \lambda_d = (1, 2, \ldots, d-1)$ and we have

$$\mathcal{H}^{d-1}(E) \leq \sum_{j=1}^{d} a_{\lambda_j}.$$

All the a_{λ_j} are estimated in the same way. For example,

$$a_{\lambda_1} = \int_{\mathbf{R}^{d-1}} \mathcal{H}^0\big\{(t, y') \in B(0, \frac{1}{2}) : P_m(t, y') = 0\big\}\, dy'.$$

Now, for fixed y', $t \mapsto P_m(t, y')$ is a polynomial of degree m which therefore has at most m zeros. Thus,

$$a_{\lambda_1} \leq \int_{|y'| \leq \frac{1}{2}} m \, dy' \leq c(d)m.$$

We saw in Remark 12.3.2 that for a homogeneous polynomial of degree m we have $N(0, 1) = m$. It follows that

$$\mathcal{H}^{d-1}\{x \in B(0, \frac{1}{2}) : P_m(x) = 0\} \leq c(d)N(0, 1),$$

which is exactly the conclusion of Theorem 12.5.24.

The proof of Theorem 12.5.24 will use the following lemma.

The $L^\infty - L^2$ Estimate

Lemma 12.5.26 *There exists* $C > 0$ *depending only on the dimension* d *such that for all* δ *in* $(0, 1)$ *and any solution* u *of* $\Delta u = 0$ *in* $\overline{B}(0, 1)$ *we have*

$$\|u\|_{L^\infty(B_\delta(0))} \leq \frac{C(d)}{(1-\delta)^{\frac{1}{2}}} \|u\|_{L^2(B(0,1))}.$$

Proof Since u is harmonic in $B(0, 1)$, for all $y \in B(0, 1)$ and $R > 0$ such that $\overline{B}(y, R)$ is included in $B(0, 1)$ we have, by Proposition 12.6.6,

$$u(y) = \frac{1}{|B(y, R)|} \int_{B(y,R)} u(x)\,dx,$$

where $|B(y, R)|$ is the Lebesgue measure of the ball $B(y, R)$.

Let $\delta \in (0, 1)$. For $|y| < \delta$ we have $\overline{B}(y, 1-\delta) \subset B(0, 1)$ so, by the formula above and the Cauchy-Schwarz inequality,

$$|u(y)|^2 \leq \frac{1}{|B(y, 1-\delta)|} \int_{B(y,1-\delta)} |u(x)|^2\,dx,$$

from which we deduce that

$$\sup_{|y|<\delta} |u(y)| \leq \frac{C(d)}{(1-\delta)^{\frac{1}{2}}} \|u\|_{L^2(B(0,1)}.$$

□

Proof of Theorem 12.5.24 It will be done in several steps.

Recall that from point 3 of Lemma 12.5.23, if $\{x \in B(0, \frac{1}{2}) : u(x) = 0\} \neq \emptyset$, then

$$\begin{aligned} &(i)\ \text{there exists } \varepsilon(d) > 0 \text{ such that} N(0,1) \geq \varepsilon(d), \\ &(ii)\ \text{there exists } C(d) > 0 \text{ such that } N(p, \frac{1}{2}) \leq C(d) N(0,1), \quad \forall p \in \overline{B}(0, \frac{1}{4}). \end{aligned} \tag{12.5.16}$$

Step 1. There exist $J \geq 1$ and $p_1, \dots, p_J \in S_{\frac{1}{4}}(0)$ such that

$$\overline{B}(0, \frac{1}{2}) \subset \bigcup_{j=1}^{J} B(p_j, \frac{5}{16}). \tag{12.5.17}$$

First, for any $x \in \overline{B}(0, \frac{1}{2})$ there exists $p \in S_{\frac{1}{4}}(0)$ such that $|x - p| < \frac{5}{16}$. Indeed, if $x \neq 0$ set $p = \frac{x}{4|x|}$. Then $|p| = \frac{1}{4}$ and $x - p = x\left(1 - \frac{1}{4|x|}\right)$ so $|x - p| = \left||x| - \frac{1}{4}\right|$. If $|x| \geq \frac{1}{4}$ we have $|x - p| = |x| - \frac{1}{4} \leq \frac{1}{2} - \frac{1}{4} = \frac{1}{4} < \frac{5}{16}$, and if $|x| < \frac{1}{4}$ we have $|x - p| = \frac{1}{4} - |x| \leq \frac{1}{4} < \frac{5}{16}$. Next, $0 \in B_{\frac{5}{16}}(p)$ for all $p \in S_{\frac{1}{4}}(0)$, since $|p| = \frac{1}{4} < \frac{5}{16}$.

We deduce that $\overline{B}(0, \frac{1}{2}) \subset \bigcup_{p \in S_{\frac{1}{4}}(0)} B(p, \frac{5}{16})$. The compactness of $\overline{B}(0, \frac{1}{2})$ implies (12.5.17).

Step 2. We can suppose that

$$\int_{B(0,1)} (u(x))^2 \, dx = 1. \tag{12.5.18}$$

According to Remark 12.3.6 with $k = 2$ and $r = \frac{1}{4}$ we have

$$1 \leq 2^{2d} 4^{2N(0,1)} \int_{B(0,\frac{1}{4})} (u(x))^2 \, dx.$$

If $p \in S_{\frac{1}{4}}(0)$, then $B(0, \frac{1}{4}) \subset B(p, \frac{1}{2})$, since $|x - p| \leq |x| + |p| < \frac{1}{4} + \frac{1}{4} = \frac{1}{2}$. Therefore,

$$1 \leq 2^{2d} 4^{2N(0,1)} \int_{B(p,\frac{1}{2})} (u(x))^2 \, dx. \tag{12.5.19}$$

Using (12.3.10) with $k = 4, r = \frac{1}{32}, R_0 = \frac{1}{2}$ we obtain

$$\int_{B(p,\frac{1}{2})} (u(x))^2 \, dx \leq (2^d 4^{N(p,\frac{1}{2})})^4 \int_{B(p,\frac{1}{32})} (u(x))^2 \, dx. \tag{12.5.20}$$

Using (12.5.12), it follows from (12.5.19) and (12.5.20) that there exist positive constants, $C_1(d)$, $c_1(d)$ such that

$$1 \leq C_1(d) 4^{c_1(d) N(0,1)} \int_{B(p,\frac{1}{32})} (u(x))^2 \, dx.$$

Therefore, there exists $x_p \in B(p, \frac{1}{32})$ such that

$$|u(x_p)| \geq C_2(d) 2^{-c_1(d) N(0,1)}. \tag{12.5.21}$$

Note that $|x_p| \leq |x_p - p| + |p| \leq \frac{1}{32} + \frac{1}{4} = \frac{9}{32}$.

Step 3. We know that u is holomorphic and bounded in the set

$$O = \{Z \in \mathbf{C}^d : |\text{Re}\, Z| < \frac{31}{32}, \quad |\text{Im}\, Z| < \eta_1\}.$$

Let

$$Q = \{z \in \mathbf{C} : |\mathrm{Re}\, z| < \frac{11}{32}, \quad |\mathrm{Im}\, Z| < \eta_0\},$$

where η_0 is much smaller than $\min(1, \eta_1)$. Set $Z = x_p + z\omega$ where $\omega \in \mathbf{S}^{d-1}$.

If $z \in 2Q$ then $Z \in \mathcal{O}$. Indeed, $|\mathrm{Re}\, Z| \le |x_p| + |\mathrm{Re}\, z| < \frac{9}{32} + \frac{22}{32} = \frac{31}{32}$ and $|\mathrm{Im}\, Z| < 2\eta_0 < \eta_1$. We can then define the function

$$g(z, \omega) = 2^{c_1(d)N(0,1)} u(x_p + z\omega)$$

as a holomorphic function in $2Q$. Moreover, according to (12.5.21),

$$|g(0, \omega)| = 2^{c_1(d)N(0,1)} |u(x_p)| \ge C_2(d),$$

$$\sup_{z \in 2Q} |g(z, \omega)| \le C_2(d) 2^{c_1(d)N(0,1)} \sup_{\mathcal{O}} |u| \le C_2'(d) 2^{c_1(d)N(0,1)} \sup_{B(0,\rho)} |u|, \quad \rho = \frac{63}{64}.$$

It follows from Lemma 12.5.26 and from the fact that $\int_{B(0,1)} (u(x))^2\, dx = 1$ that

$$|g(0, \omega)| \ge C_2(d), \quad \sup_{z \in 2Q} |g(z, \omega)| \le C_2''(d) 2^{c_1(d)N(0,1)}.$$

According to Lemma 12.5.19 we have

$$\#\{z \in Q : u(x_p + z\omega) = 0\} \le \frac{C}{\eta_0} C_3 + C_4(d) N(0, 1),$$

in particular, with $I = (-\frac{11}{32}, \frac{11}{32})$,

$$\#\{t \in I : u(x_p + t\omega) = 0\} \le \frac{C}{\eta_0} C_3 + C_4(d) N(0, 1).$$

Since every point of $B(x_p, \frac{11}{32})$ can be written as $x_p + t\omega, t \in I, \omega \in \mathbf{S}^{d-1}$ we have

$$\#\{x \in B(x_p, \frac{11}{32}) : u(x) = 0\} \le \frac{C}{\eta_0} C_3 + C_4(d) N(0, 1).$$

Next, we have $B(p, \frac{5}{16}) \subset B(x_p, \frac{11}{32})$ which follows from the fact that

$$|x - x_p| \le |x - p| + |p - x_p| < \frac{5}{16} + \frac{1}{32} = \frac{11}{32}.$$

We deduce that

$$\#\{x \in B(p, \frac{5}{16}) : u(x) = 0\} \le \frac{C}{\rho_0} C_3 + C_4(d) N(0, 1).$$

It follows from Step 1. that

$$\#\{x \in B(0, \frac{1}{2}) : u(x) = 0\Big\} \leq C_5 + C_6(d)N(0, 1)$$

and from point (i) of (12.5.16) that

$$\#\{x \in B(0, \frac{1}{2}) : u(x) = 0\} \leq C_7(d)N(0, 1).$$

We then apply the geometric integration formula described in Theorem 12.5.15.

Let $E = \{x \in B_{\frac{1}{2}}(0) : u(x) = 0\}$. The set $\Lambda(d-1, d)$ is finite. It consists of the d elements $(2, \ldots, d), (1, 3, \ldots, d), \ldots, (1, 2, \ldots, d-1)$. If λ is one of these elements we have

$$a_\lambda = \int_{\mathbf{R}^{d-1}} \mathcal{H}^0(\{x \in E : P_\lambda(x) = y\}dy.$$

Since $|y| = |P_\lambda(x)| \leq |x|$ and $x \in B(0, \frac{1}{2})$, we have $|y| \leq \frac{1}{2}$ in the above integral. Next,

$$\mathcal{H}^0(\{x \in E : P_\lambda(x) = y\} \leq \mathcal{H}^0(E) = \#\{x \in B(0, \frac{1}{2}) : u(x) = 0\} \leq C_7(d)N(0, 1).$$

We deduce from Theorem 12.5.15 that

$$\mathcal{H}^{d-1}\{x \in B(0, \frac{1}{2}) : u(x) = 0\} \leq C_8(d)N(0, 1),$$

which completes the proof of Theorem 12.5.24. □

12.6 Appendix

12.6.1 Divergence Theorem

Definition 12.6.1 Let Ω be an open subset of $\mathbf{R}^d$ and $k \in \mathbf{N}^* \cup \{+\infty\}$. We say that Ω is of class C^k if there exists a function $\rho \in C^k(\mathbf{R}^d, \mathbf{R})$ such that

$$\begin{cases} \Omega = \left\{x \in \mathbf{R}^d : \rho(x) < 0\right\}, \\ \partial\Omega = \left\{x \in \mathbf{R}^d : \rho(x) = 0\right\}, \ \text{grad}\, \rho(x)) \neq 0, \ \forall x \in \partial\Omega, \end{cases} \tag{12.6.1}$$

where $\partial\Omega$ denotes the boundary of Ω, and grad $\rho(x)$ is the vector $\left(\frac{\partial \rho}{\partial x_i}(x)\right)_{i=1,\ldots,d}$.

Example 12.6.2 (i) The set $\Omega = \{x \in \mathbf{R}^d : |x| < R\}$ is an open set with C^∞ boundary. To see that it suffices to take $\rho(x) = |x|^2 - R^2$. If $\Omega = \{x \in \mathbf{R}^d : |x| > R\}$ we take $\rho(x) = R^2 - |x|^2$.

(ii) Let $\psi : \mathbf{R}^{n-1} \to \mathbf{R}$ be a function of class C^k, where $k \in \mathbf{N}^*$. Set

$$\Omega = \{x \in \mathbf{R}^d : x_d > \psi(x_1, \dots, x_{d-1})\}.$$

Then Ω is an open set of class C^k with $\rho(x) = \psi(x_1, \dots, x_{d-1}) - x_d$.

For an open set of class C^1, the direction of the vector grad $\rho(x)$ for $x \in \partial\Omega$, does not depend on the function ρ. Indeed, if Ω is defined by another function $\widetilde{\rho} \in C^1$ which vanishes on $\partial\Omega$, one shows that for any point $x_0 \in \partial\Omega$, there exists a neighborhood V_{x_0} and a function α belonging to $C^0(V_{x_0}) \cap C^1(V_{x_0} \setminus \partial\Omega)$ such that $\alpha(x) > 0$ for $x \in V_{x_0}$, $\widetilde{\rho}(x) = \alpha(x)\rho(x)$ in V_{x_0} and $\lim_{\substack{x \to x_0 \\ x \notin \partial\Omega}} \big(\frac{\partial\alpha}{\partial x_i}\, \rho(x)\big) = 0$ for $i = 1, \dots, n$. Then $\frac{\partial\tilde{\rho}}{\partial x_i}(x_0) = \alpha(x_0)\, \frac{\partial\rho}{\partial x_i}(x_0)$ and thus the vectors grad $\tilde{\rho}(x_0)$ and grad $\rho(x_0)$ have the same direction.

Definition 12.6.3 (i) For $x \in \partial\Omega$, the vector grad $\rho(x)$ is called the **outward normal** to Ω at the point x. The vector $n(x) = \frac{\text{grad}\, \rho(x)}{\|\text{grad}\, \rho(x)\|}$ is the **unit outward normal** to Ω at the point $x \in \partial\Omega$.

$(ii) = \big\langle n(x), \frac{\partial}{\partial x}\big\rangle = \sum_{i=1}^n n_i(x)\, \frac{\partial}{\partial x_i}$ is called the **outward normal derivative** to Ω.

Note that grad $\rho(x)$ is the outward normal when $\Omega = \{x \in \mathbf{R}^d : \rho(x) < 0\}$.

If $F = (f_1, \dots, f_d)$ is a mapping from Ω into $\mathbf{C}^d$ we denote

$$\text{div}\, F = \sum_{i=1}^d \frac{\partial f_i}{\partial x_i}, \tag{12.6.2}$$

which is called the **divergence** of F.

We can then state the Divergence Theorem, which gives an integration by parts formula in any dimension.

Theorem 12.6.4 *Let $\Omega \subset \mathbf{R}^d$ be an open set of class C^1. There exists a positive measure $d\sigma$ on $\partial\Omega$ such that for all $\varphi \in C_0^1(\mathbf{R}^d)$ and all $f_1, \dots, f_d \in C^1(\mathbf{R}^d)$, we have*

$$\int_\Omega \mathit{div}\, F(x)\, \varphi(x)\, dx = -\int_\Omega F(x) \cdot \mathit{grad}\, \varphi(x)\, dx + \int_{\partial\Omega} \varphi(x)\, F(x) \cdot n(x)\, d\sigma \tag{12.6.3}$$

where $F(x) = (f_1(x), \dots, f_d(x))$, $X \cdot Y = \sum_{i=1}^d X_i\, Y_i$ and $n(x)$ is the unit outward normal to Ω at $x \in \partial\Omega$. The same result holds for $\varphi \in C^1(\mathbf{R}^d)$ and $f_j \in C_0^1(\mathbf{R}^d)$, $j = 1, \dots, d$.

Remark 12.6.5 1. This result is also known as the "Gauss-Green Theorem" or the "Ostrogradsky Theorem".

2. If Ω is a bounded open set with C^1 boundary, it suffices to assume that φ and f_j belong to $C^1(\overline{\Omega})$.

12.6.2 *Mean Value Formula for Harmonic Functions*

Proposition 12.6.6 *Let Ω be an open subset of $\mathbf{R}^d$. Let $u \in C^2(\Omega)$ such that for all $x \in \Omega$, $\Delta u(x) \leq 0$ (resp. ≥ 0), (resp. $= 0$).*

Let $x_0 \in \Omega$ and $r > 0$ such that $\overline{B}(x_0, r) \subset \Omega$. Denoting by $\mu(A)$ the Lebesgue measure of the set A, we then have

$$u(x_0) \geq (\textit{resp.} \ \leq)(\textit{resp.} \ =)\frac{1}{\mu(B(x_0, r))}\int_{B(x_0,r)} u(x)\,dx.$$

Proof Setting $\widetilde{u}(x) = u(x + x_0)$ we may assume that $x_0 = 0 \in \Omega$. It also suffices to treat the case where $\Delta u \leq 0$, the other cases following by replacing u with $-u$. As $\Delta u = \mathrm{div}\nabla u \leq 0$, we can apply (according to Remark 12.6.5) formula (12.6.3) with $F = \nabla u$ and $\varphi = 1$ in the open set $B(0, \rho)$ for $\rho \in]0, r[$. The outward unit normal to this open set is $n(x) = \frac{x}{\rho}$, so, switching to spherical coordinates $x = \rho\omega$, $\omega \in \mathbf{S}^{d-1}$,

$$0 \geq \int_{B(0,\rho)} \mathrm{div}\nabla u(x)\,dx = \int_{|x|=\rho} \frac{x}{\rho} \cdot \nabla u(x)\,d\sigma_\rho = \int_{\mathbf{S}^{d-1}} \sum_{j=1}^{d} \omega_j \frac{\partial u}{\partial x_j}(\rho\omega)\rho^{d-1}\,d\omega,$$

since $d\sigma_\rho = \rho^{d-1} d\omega$.

Now, for $\omega \in \mathbf{S}^{d-1}$ we have $\frac{\partial}{\partial\rho}[u(\rho\omega)] = \sum_{j=1}^{d} \omega_j \frac{\partial u}{\partial x_j}(\rho\omega)$. We deduce that

$$0 \geq \rho^{d-1}\frac{\partial}{\partial\rho}\int_{\mathbf{S}^{d-1}} u(\rho\omega)\,d\omega,$$

and thus

$$0 \geq \frac{\partial}{\partial\rho}\int_{\mathbf{S}^{d-1}} u(\rho\omega)\,d\omega = \frac{\partial}{\partial\rho}\Big(\rho^{1-d}\int_{|x|=\rho} u(x)\,d\sigma_\rho(x)\Big),$$

so that for all (ρ, t) such that $0 < \rho < t < r$, we have

$$\int_{\mathbf{S}^{d-1}} u(\rho\omega)\,d\omega = \rho^{1-d}\int_{|x|=\rho} u(x)\,d\sigma_\rho(x) \geq t^{1-d}\int_{|x|=t} u(x)\,d\sigma_t(x).$$

Since $\lim_{\rho\to 0}\int_{\mathbf{S}^{d-1}} u(\rho\omega)\,d\omega = \mu(\mathbf{S}^{d-1})u(0)$ we deduce that for all $t \in (0, r)$ we have

$$\int_{|x|=t} u(x)\, d\sigma_t(x) \le t^{d-1}\mu(\mathbf{S}^{d-1})u(0).$$

Integrating this equality between 0 and r gives

$$\int_{|x|<r} u(x)\, dx \le \int_0^r \int_{|x|=t} u(x)\, d\sigma_t(x)\, dt = \frac{r^d}{d}\mu(\mathbf{S}^{d-1})u(0).$$

To conclude, it suffices to note that $\frac{r^d}{d}\mu(\mathbf{S}^{d-1}) = \mu(B(0,r))$. □

12.6.3 The Vitali Lemma

We shall denote by $B(a,\rho)$ a ball of center a and radius ρ in a metric space. We begin by a simple observation.

$(*)$ Set for $i = 1, 2,\ B_i = B_i(a_i, \rho_i)$ with $\rho_2 \le \rho_1$. If $B_2 \cap B_1 \ne \emptyset$ then $B_2 \subset 3B_1$

where $3B_1$ denotes the ball of same center with radius 3-times the radius of B_1.

Indeed let $y \in B_2 \cap B_1$. Then for $x \in B_2$ we can write

$$d(x,a_1) \le d(x,a_2) + d(a_2,y) + d(y,a_1) \le \rho_2 + \rho_2 + \rho_1 \le 3\rho_1.$$

Lemma 12.6.7 *Let $B_1, \dots, B_N$ a finite collection of balls in a metric space. Then there exists a subcollection $B_{j_1}, \dots, B_{j_m}$ of these balls which are* **disjoint** *and satisfy*

$$\cup_{i=1}^N B_i \subset \cup_{k=1}^m 3B_{j_k},$$

Proof Let us denote by r_j the radius of B_j. We may assume (changing the order) that we have $r_1 \ge r_2 \ge \cdots \ge r_N$. Now we take $B_{j_1} = B_1$. If all the other balls $B_2, \dots, B_N$ meet B_1 by $(*)$ we would have $\cup_{j=2}^N B_j \subset 3B_1 = 3B_{i_1}$ and the lemma is proved since we have also $B_1 \subset 3B_1$.

Otherwise let B_{j_2} be the ball of greatest radius that does not meet B_{j_1}. Therefore $B_{j_1} \cap B_{j_2} = \emptyset$. By definition the balls B_i with $2 \le i < j_2$ meet B_{i_1}. By $(*)$ their union is contained in $3B_1$. If all the balls B_i with $j_2 < i \le N$ would meet B_{j_2} by $(*)$ their union would be contained in $3B_{j_2}$ so we would have $\cup_{j=1}^N B_j \subset 3B_{j_1} \cup 3B_{j_2}$.

If not, call B_{j_3}, $(j_3 > j_2)$, the ball of greatest radius which does not meet B_{j_1} and B_{j_2}. By definition, the balls B_i with $j_2 + 1 \le i < j_3$ meet B_{j_1} or B_{j_2}. Therefore by $(*)$ their union is contained in $3B_{j_1} \cup 3B_{j_2}$. If all the balls B_i with $j_3 < i \le N$ would meet B_{j_3} then by $(*)$ their union would be contained in $3B_{j_3}$ so we would have $\cup_{j=1}^N B_j \subset 3B_{j_1} \cup 3B_{j_2} \cup 3B_{j_3}$. Otherwise we call B_{j_4}, $(j_4 > j_3)$, the ball of greatest radius which does not meet B_{j_1}, B_{j_2} and B_{j_3}. And we continue by the same argument. The procedure has to stop since we have a finite number of balls. □

12.7 Comments

The estimation of the Hausdorff measure of the nodal sets of eigenvectors of the Dirichlet problem for the Laplacian in a bounded open set is a subject still under study today.

In 1990, H. Donnelly and C. Fefferman showed (in particular) that if Ω is a bounded open set in $\mathbf{R}^d$ with real-analytic boundary and if we denote by $e_\lambda \in H_0^1(\Omega)$ an eigenvector of the Laplacian associated to the eigenvalue λ that is, $-\Delta e_\lambda = \lambda e_\lambda$, there exist two positive constants c_1 and c_2 independent of λ such that, for λ large enough,

$$c_1\sqrt{\lambda} \le \mathcal{H}^{d-1}\{x \in \Omega : e_\lambda(x) = 0\} \le c_2\sqrt{\lambda}. \tag{12.7.1}$$

It was conjectured by S. T. Yau that this result would still hold for elliptic operators with C^∞ coefficients and for domains with C^∞ boundary. Despite recent advances, the full conjecture is still open today.

For an example in dimension 1 see paragraph 13.4.1 in the next Chapter.

Chapter 13
Nodal Domains of Eigenfunctions of the Laplacian

13.1 Prerequisites

This chapter follows Chap. 10, which deals with the spectral theory of the Dirichlet problem for the Laplacian in a bounded open set of $\mathbf{R}^d$.

13.2 The Problem

Let us first recall the main results of Chap. 10. Let Ω be a bounded open set in $\mathbf{R}^d$. The spectrum of the Dirichlet problem in Ω for the operator $-\Delta$ consists of a sequence $(\lambda_n)_{n\in\mathbf{N}}$ of positive real eigenvalues tending to $+\infty$, which we order as,

$$0 < \lambda_1 \leq \lambda_2 \leq \ldots \leq \lambda_n \ldots$$

some of the λ_j may be equal according to the multiplicity of the eigenvalue (which is finite).

There then exists an orthonormal basis $(e_n)_{n\in\mathbf{N}}$ of $L^2(\Omega)$ consisting of eigenvectors of $-\Delta$ that is, satisfying

$$-\Delta e_n = \lambda_n e_n, \quad e_n \in H_0^1(\Omega). \tag{13.2.1}$$

Note that thanks to equation (13.2.1) we have $e_n \in \cap_{k\in\mathbf{N}} H^k(\Omega) \subset C^\infty(\Omega)$.

We then introduce the following definition.

Definition 13.2.1 A nodal domain of e_n is a connected component of the set $\{x \in \Omega : e_n(x) \neq 0\}$.

The aim of this chapter is then to prove the following theorem due to R. Courant (see the Comments).

C. Zuily, *Selected Topics in Partial Differential Equations*, Universitext,
https://doi.org/10.1007/978-3-032-24082-8_13

Theorem 13.2.2 *The eigenvector e_n associated to the eigenvalue λ_n has at most n nodal domains.*

13.3 Proof of Theorem 13.2.2

In what follows, we denote by $(\cdot,\cdot)_{L^2}$ the inner product of $L^2(\Omega)$ and by $\langle\cdot,\cdot\rangle$ the duality between the spaces $H_0^1(\Omega)$ and $H^{-1}(\Omega)$ as well as that of distributions.

Proposition 13.3.1 *Let $w \in H^1(\Omega)$ be a nonzero element, orthogonal in $L^2(\Omega)$ to the vectors $e_1, \dots, e_{n-1}$ for $n \geq 2$. Then,*

1. $\|\nabla w\|_{L^2}^2 \geq \lambda_n \|w\|_{L^2}^2$.
2. *We have $\|\nabla w\|_{L^2}^2 = \lambda_n \|w\|_{L^2}^2$ if and only if w is an eigenvector of $-\Delta$ for the eigenvalue λ_n.*

Proof Let $\alpha_j = (w, e_j)_{L^2}$, $j \geq 1$. Then by assumption $\alpha_1 = \cdots = \alpha_{n-1} = 0$. Since the family $(e_j)_{j\in\mathbf{N}}$ is orthonormal, we have

$$w = \sum_{j=n}^{+\infty} \alpha_j e_j, \qquad \|w\|_{L^2}^2 = \sum_{j=n}^{+\infty} |\alpha_j|^2. \tag{13.3.1}$$

We will show that

$$(i)\ \ (\nabla w, \nabla e_j)_{L^2} = \lambda_j \alpha_j, \qquad (ii)\ \ (\nabla e_j, \nabla e_k)_{L^2} = \lambda_k \delta_{jk}, \tag{13.3.2}$$

where δ_{jk} denotes the Kronecker symbol, $\delta_{jk} = 0$ if $j \neq k$, $\delta_{jj} = 1$.

To do this, we first show that for $w \in H^1$ we have

$$(\nabla w, \nabla e_j)_{L^2} = -(w, \Delta e_j)_{L^2}, \tag{13.3.3}$$

which makes sense since $e_j \in H^2$. Indeed, since $e_j \in H_0^1$ there exists a sequence $(\varphi_k) \subset C_0^\infty(\Omega)$ which converges to e_j in H^1. We have

$$(\nabla w, \nabla \varphi_k)_{L^2} = \langle \nabla w, \nabla \overline{\varphi}_k \rangle = -\langle w, \Delta \overline{\varphi}_k \rangle.$$

The left-hand side converges to $(\nabla w, \nabla e_j)_{L^2}$. On the other hand, $\Delta \overline{e}_j \in H^{-1}(\Omega)$ and

$$\begin{aligned} |\langle w, \Delta \overline{\varphi}_k \rangle - \langle w, \Delta \overline{e}_j \rangle| &\leq \|w\|_{H^1} \|\Delta(\overline{\varphi}_k - \overline{e}_j)\|_{H^{-1}}, \\ &\leq \|w\|_{H^1} \|\overline{\varphi}_k - \overline{e}_j\|_{H^1} \to 0, \quad k \to +\infty, \end{aligned}$$

so that $\langle w, \Delta \overline{\varphi}_k \rangle$ converges to $\langle w, \Delta \overline{e}_j \rangle = (w, \Delta e_j)_{L^2}$, which proves (13.3.3).

Assertion (i) of (13.3.2) follows from (13.3.3) since $-\Delta e_j = \lambda_j e_j$ where λ_j is real and from the definition of α_j. Assertion (ii) of (13.3.2) follows from (13.3.3) and from the fact that we have $(e_j, e_k)_{L^2} = \delta_{jk}$.

Let us proceed to the proof of point 1. of the proposition. Let $N \geq n$. We can write

$$\left\| \nabla\Big(w - \sum_{j=n}^{N} \alpha_j e_j\Big) \right\|_{L^2}^2 = \|\nabla w\|_{L^2}^2 - 2\mathrm{Re}\Big(\sum_{j=n}^{N} \overline{\alpha}_j (\nabla w, \nabla e_j)_{L^2}\Big) + \sum_{j=n}^{N}\sum_{k=n}^{N} \alpha_j \overline{\alpha}_k (\nabla e_j, \nabla e_k)_{L^2}.$$

Using (13.3.2) we obtain

$$\left\| \nabla\Big(w - \sum_{j=n}^{N} \alpha_j e_j\Big) \right\|_{L^2}^2 = \|\nabla w\|_{L^2}^2 - 2\sum_{j=n}^{N} \lambda_j |\alpha_j|^2 + \sum_{j=n}^{N} \lambda_j |\alpha_j|^2 = \|\nabla w\|_{L^2}^2 - \sum_{j=n}^{N} \lambda_j |\alpha_j|^2.$$

The first term on the left-hand side being positive and the λ_j being increasing, we obtain

$$\|\nabla w\|_{L^2}^2 \geq \sum_{j=n}^{N} \lambda_j |\alpha_j|^2 \geq \lambda_n \sum_{j=n}^{N} |\alpha_j|^2. \tag{13.3.4}$$

To prove point 1. of the proposition, it suffices to let N tend to $+\infty$ and use (13.3.1).

Let us move on to point 2. of the proposition. First, if w is an eigenvector of $-\Delta$ corresponding to the eigenvalue λ_n we have by definition $w \in H_0^1(\Omega) \cap H^2(\Omega)$ and $-\Delta w = \lambda_n w$. Then,

$$(-\Delta w, w)_{L^2} = \lambda_n (w, w)_{L^2} = \lambda_n \|w\|_{L^2}^2.$$

To conclude, it suffices to note that for $w \in H_0^1(\Omega) \cap H^2(\Omega)$ we have the equality $(-\Delta w, w)_{L^2} = \|\nabla w\|_{L^2}^2$.

Conversely, if we have $\|\nabla w\|_{L^2}^2 = \lambda_n \|w\|_{L^2}^2$, by the inequalities (13.3.4) with N goes to $+\infty$, we have

$$\sum_{j=n}^{+\infty} \lambda_j |\alpha_j|^2 = \lambda_n \sum_{j=n}^{+\infty} |\alpha_j|^2,$$

which implies that $\sum_{j=n+1}^{+\infty} \lambda_j |\alpha_j|^2 = 0$. Since the λ_j are all strictly positive, this implies that $\alpha_j = 0$ for $j \geq n+1$. It follows from (13.3.1) that $w = \alpha_n e_n$. Point 2 then follows from (13.2.1). □

Proposition 13.3.2 *Let $u \in H^2(\Omega) \cap C^\infty(\Omega)$ and ω a nodal domain of u. Let us set $v = 1_\omega u$. Then,*

$$\begin{aligned} &(i) \quad \partial_k v = 1_\omega \partial_k u, \\ &(ii) - (\Delta u, u)_{L^2(\omega)} = -(\Delta v, v)_{L^2} = \|\nabla v\|_{L^2}^2. \end{aligned} \tag{13.3.5}$$

Proof The idea of the proofs is to use the divergence formula, but since we have no information about the regularity of the boundary of ω we cannot do so. We therefore need to use an indirect approach, which we will now present. First, let us recall that the function u belongs to the space $C^\infty(\Omega) \cap H^2(\Omega)$. Let us also recall some definitions.

A critical point of u is a point x such that $\nabla u(x) = 0$. The images by u of the critical points are called the critical values of u. The complement in $\mathbf{R}$ of the set of critical values forms the set of regular values of u. We will use the following lemma, which we will admit here (see [14] page 16, for a proof).

Lemma 13.3.3 (Sard's Lemma) *Let $f \in C^\infty(\Omega, \mathbf{R})$. Then the set of critical values of f has Lebesgue measure zero.*

Note that the hypothesis $f \in C^\infty$ can be weakened.

By definition of the nodal domain, the continuous function u does not vanish on the connected open set ω. It is therefore of constant sign. By possibly considering the function $-u$, we may thus assume that $u(x) > 0$ for all $x \in \omega$.

Next, using Lemma 13.3.3, we can find a sequence $(\varepsilon_i) \subset \mathbf{R}^+$ of regular values of u that tends to zero as i tends to $+\infty$. Set

$$\omega_i = \{x \in \omega : u(x) > \varepsilon_i\}, \quad u_i = u - \varepsilon_i. \tag{13.3.6}$$

The boundary of ω_i is the set $\Gamma_i = \{x \in \omega : u(x) = \varepsilon_i\}$. Since ε_i is a regular value of u, we have $\nabla u(x) \neq 0$ for all $x \in \Gamma_i$. This shows that Γ_i is a C^1 hypersurface. Furthermore, since $\overline{\omega}_i \subset \omega \subset \Omega$, the function u is C^1 on $\overline{\omega}_i$. We can therefore apply the divergence formula to u_i on ω_i (see Theorem 12.6.4 of Chap. 12).

Letting μ denote the Lebesgue measure, we first have

$$\lim_{i \to +\infty} \mu(\omega \setminus \omega_i) = 0. \tag{13.3.7}$$

Indeed,

$$\mu(\omega \setminus \omega_i) = \int_\omega 1_{\{0<u\leq\varepsilon_i\}}(x)\, dx.$$

For fixed x in ω, we have $u(x) > 0$, so if i is large enough that $\varepsilon_i < u(x)$, then $1_{\{0<u\leq\varepsilon_i\}}(x) = 0$. Since $1_{\{0<u\leq\varepsilon_i\}}(x) \leq 1$ and ω is bounded (since Ω is), the dominated convergence theorem yields the conclusion.

13.3.1 Proof of Point (i) of Proposition 13.3.2

With the notations of (13.3.6) we have

$$\lim_{i\to+\infty} 1_{\omega_i} u_i = 1_\omega u, \tag{13.3.8}$$

in $\mathcal{D}'(\Omega)$. Indeed, let $\varphi \in C_0^\infty(\Omega)$. We have

$$\int_\Omega \big[(1_\omega u)(x) - (1_{\omega_i} u_i)(x)\big]\varphi(x)\,dx = \int_{\omega\setminus\omega_i} u(x)\varphi(x)\,dx + \varepsilon_i \int_{\omega_i} \varphi(x)\,dx = (1) + (2).$$

We have

$$|(1)| \le \mu(\omega \setminus \omega_i)^{\frac{1}{2}} \|u\|_{L^2} \|\varphi\|_{L^\infty}, \quad |(2)| \le \varepsilon_i \int_\Omega |\varphi(x)|\,dx$$

and we use (13.3.7) as well as the fact that $\varepsilon_i \to 0$.

We deduce from (13.3.8) that

$$\lim_{i\to+\infty} \nabla 1_{\omega_i} u_i = \nabla 1_\omega u \quad \text{in } \mathcal{D}'(\Omega). \tag{13.3.9}$$

We will show that, in $\mathcal{D}'(\Omega)$, we have

$$\nabla 1_{\omega_i} u_i = 1_{\omega_i} \nabla u. \tag{13.3.10}$$

Indeed, let $\varphi \in C_0^\infty(\Omega)$ and $k \in \{1, \dots, d\}$. By the divergence theorem (Theorem 12.6.4), letting $n(x)$ denote the normal at the point $x \in \Gamma_i$, we can write

$$\begin{aligned}\langle \partial_k 1_{\omega_i} u_i, \varphi\rangle &= -\langle 1_{\omega_i} u_i, \partial_k \varphi\rangle = -\int_{\omega_i} u_i(x)\partial_k\varphi(x)\,dx,\\ &= \int_{\omega_i} \partial_k u_i(x)\varphi(x)\,dx - \int_{\Gamma_i} u_i(x) n_k(x)\varphi(x)\,dx = \langle 1_{\omega_i}\partial_k u, \varphi\rangle,\end{aligned}$$

since $\partial_k u_i = \partial_k u$ and $u_i = 0$ on Γ_i, which proves (13.3.10).

Finally, analogously to (13.3.8), we have

$$\lim_{i\to+\infty} 1_{\omega_i} \nabla u = 1_\omega \nabla u. \tag{13.3.11}$$

Point (i) of Proposition 13.3.5 follows from (13.3.9), (13.3.10), (13.3.11).

13.3.2 Proof of Point (ii) of Proposition 13.3.2

Using the divergence theorem, we can write

$$-\int_{\omega_i} \Delta u_i(x) u_i(x)\, dx = \int_{\omega_i} |\nabla u_i(x)|^2\, dx + \int_{\Gamma_i} n_i(x) \cdot \nabla u_i(x) u_i(x)\, d\sigma_i,$$

where $n_i(x) = -\frac{\nabla u_i(x)}{|\nabla u_i(x)|}$. The second integral on the right-hand side is zero since $u_i = 0$ on Γ_i. Next, $\Delta u_i = \Delta u, \quad \nabla u_i = \nabla u$, so the above equality can be written as

$$-\int_{\omega_i} \Delta u(x) u(x)\, dx + \varepsilon_i \int_{\omega_i} \Delta u(x)\, dx = \int_{\omega_i} |\nabla u(x)|^2\, dx. \tag{13.3.12}$$

We need to take the limit as $\varepsilon_i \to 0$ in the above equality. For this, we first note that using the Cauchy-Schwarz inequality and the fact that $\omega_i \subset \Omega$, we have

$$\varepsilon_i \left| \int_{\omega_i} \Delta u(x)\, dx \right| \leq \varepsilon_i\, \mu(\Omega)^{\frac{1}{2}} \left(\int_{\Omega} |\Delta u(x)|^2\, dx \right)^{\frac{1}{2}} \leq \varepsilon_i\, \mu(\Omega)^{\frac{1}{2}} \|u\|_{H^2(\Omega)},$$

so that

$$\lim_{I \to +\infty} \varepsilon_i \int_{\omega_i} \Delta u(x)\, dx = 0. \tag{13.3.13}$$

We then have

$$\begin{aligned} \left| \int_{\omega} \Delta u(x) u(x)\, dx - \int_{\omega_i} \Delta u(x) u(x)\, dx \right| &= \left| \int_{\omega \backslash \omega_i} \Delta u(x) u(x)\, dx \right| \\ &\leq \left(\int_{\omega \backslash \omega_i} |u(x)|^2\, dx \right)^{\frac{1}{2}} \left(\int_{\omega \backslash \omega_i} |\Delta u(x)|^2\, dx \right)^{\frac{1}{2}}, \end{aligned} \tag{13.3.14}$$

and

$$\left| \int_{\omega} |\nabla u(x)|^2\, dx - \int_{\omega_i} |\nabla u(x)|^2, dx \right| = \int_{\omega \backslash \omega_i} |\nabla u(x)|^2\, dx. \tag{13.3.15}$$

Now, since $u \in H^2(\Omega)$, it follows from (13.3.7), (13.3.14) and (13.3.15) that

$$\lim_{i \to +\infty} \int_{\omega_i} \Delta u(x) u(x)\, dx = \int_{\omega} \Delta u(x) u(x)\, dx,$$

$$\lim_{i \to +\infty} \int_{\omega} |\nabla u(x)|^2\, dx = \int_{\omega_i} |\nabla u(x)|^2\, dx.$$

Using these equalities, (13.3.13) and (13.3.12), we obtain (13.3.5) since $\Delta u = \Delta v$ and $\nabla u = \nabla v$ on ω. $\square$

13.3.3 End of the Proof of Theorem 13.2.2

Let $u \in H_0^1(\Omega)$ such that $-\Delta u = \lambda_n u$, and suppose that u has at least $n+1$ nodal domains, $\omega_1, \ldots, \omega_{n+1}$. Note that the ω_j are open sets and that $\omega_j \cap \omega_k = \emptyset$ if $j \neq k$. Set

$$v_j = 1_{\omega_j} u.$$

According to Proposition 13.3.5 we have $\nabla v_j = 1_{\omega_j} \nabla u$.

For $j \neq k$ the supports of v_j and v_k as well as those of ∇v_j and ∇v_k are disjoint (since $\omega_j \cap \omega_k = \emptyset$). It follows that, in $L^2(\Omega)$, v_j and v_k are orthogonal as well as ∇v_j and ∇v_k. Set

$$w = \sum_{j=1}^{n} \mu_j v_j, \quad \mu_j \in \mathbf{R} \setminus 0.$$

Then,

$$\begin{aligned} \|w\|_{L^2}^2 &= \sum_{j=1}^{n} \mu_j^2 \|v_j\|_{L^2}^2 = \sum_{j=1}^{n} \mu_j^2 \|u\|_{L^2(\omega_j)}^2 > 0, \\ \|\nabla w\|_{L^2}^2 &= \sum_{j=1}^{n} \mu_j^2 \|\nabla v_j\|_{L^2}^2. \end{aligned} \tag{13.3.16}$$

According to (13.3.16) and Proposition 13.3.2 we have

$$\begin{aligned} \|\nabla w\|_{L^2}^2 &= \sum_{j=1}^{n} \mu_j^2 \|\nabla v_j\|_{L^2}^2 = \sum_{j=1}^{n} \mu_j^2 (-\Delta v_j, v_j) = \sum_{j=1}^{n} \mu_j^2 (-\Delta u, u)_{L^2(\omega_j)} \\ &= \lambda_n \sum_{j=1}^{n} \mu_j^2 \|u\|_{L^2(\omega_j)}^2 = \lambda_n \sum_{j=1}^{n} \mu_j^2 \|v_j\|_{L^2}^2 = \lambda_n \|w\|_{L^2}^2. \end{aligned}$$

Let us choose the μ_j so that w is orthogonal in L^2 to the vectors $e_1, \ldots, e_{n-1}$ that is,

$$\sum_{j=1}^{n} \mu_j (v_j, e_k)_{L^2} = 0, \quad k = 1, \ldots, n-1.$$

This is possible because we have $n-1$ linear equations in n unknowns $\mu_1, \ldots, \mu_n$. This system gives rise to a solution $w \not\equiv 0$ to which we can apply point 2. of Proposition 13.3.1. It follows that w is an eigenvector of $-\Delta$, corresponding to the eigenvalue λ_n. On the other hand, by construction $w = 0$ on $\omega_{n+1} \subset \Omega$. According to Theorem 12.3.11 of Chap. 12, the function w is identically zero in Ω, which is absurd.

Theorem 13.2.2 is thus proved.

13.4 Nodal Set of the First Eigenvalue

In this paragraph we aim to show the following result.

Theorem 13.4.1 *The first eigenvalue λ_1 admits as its only nodal set the open set Ω. It is simple.*

Proof Let $u \in H_0^1(\Omega) \cap H^2(\Omega)$ such that $-\Delta u = \lambda_1 u$ in $\mathcal{D}'(\Omega)$. Recall that we have $u \in C^\infty(\Omega)$.

Point 1. We have $\nabla|u| = \text{sgn} u\, \nabla u \in L^2(\Omega)$, where sgn denotes the sign.

Indeed, for $\varepsilon > 0$ set $u_\varepsilon(x) = \sqrt{u(x)^2 + \varepsilon^2}$. Then $\lim_{\varepsilon \to 0} u_\varepsilon = |u|$ in $\mathcal{D}'(\Omega)$. It follows that for $k = 1, \dots d$,

$$\lim_{\varepsilon \to 0} \partial_k u_\varepsilon = \partial_k |u|, \quad \text{in } \mathcal{D}'(\Omega). \tag{13.4.1}$$

On the other hand $\partial_k u_\varepsilon = \frac{u\, \partial_k u}{\sqrt{u^2+\varepsilon^2}}$. Let us set $\widetilde{\Omega} = \{x \in \Omega : u(x) \neq 0\}$. For $\varphi \in C_0^\infty(\Omega)$ set

$$I_\varepsilon = \int_\Omega \frac{u(x)\, \partial_k u(x)}{\sqrt{u^2(x) + \varepsilon^2}} \varphi(x)\, dx = \int_{\widetilde{\Omega}} \frac{u(x)\, \partial_k u(x)}{\sqrt{u^2(x) + \varepsilon^2}} \varphi(x)\, dx.$$

We have

$$\lim_{\varepsilon \to 0} \frac{u(x) \partial_k u(x)}{\sqrt{u^2(x) + \varepsilon^2}} \varphi(x) = \text{sgn} u(x) \partial_k u(x) \varphi(x), \text{ for fixed } x \text{ in } \widetilde{\Omega}, \quad \text{and}$$
$$\left| \frac{u\, \partial_k u}{\sqrt{u^2 + \varepsilon^2}} \varphi \right| \leq |\partial_k u| |\varphi| \in L^1(\Omega).$$

The dominated convergence theorem shows that $\lim_{\varepsilon \to 0} \partial_k u_\varepsilon = \text{sgn} u\, \partial_k u$ in $\mathcal{D}'(\Omega)$. The Point 1 then follows from (13.4.1). We deduce that $|u| \in H^1(\Omega)$.

Point 2. Set $w = |u| \in H^1(\Omega)$. We have

$$\|\nabla w\|_{L^2}^2 = \|\nabla u\|_{L^2}^2 = \lambda_1 \|u\|_{L^2}^2 = \lambda_1 \|w\|_{L^2}^2. \tag{13.4.2}$$

Indeed, the first equality results from Point 1. The second equality is a consequence of the fact that u is an eigenvector associated to the eigenvalue λ_1 and the third equality is trivial.

Point 3. $|u|$ is an eigenvector of $-\Delta$ associated to the eigenvalue λ_1.

This follows from Point 2 of Proposition 13.3.1.

Point 4. Let $u_+ = \frac{1}{2}(|u| + u)$ and $u_- = \frac{1}{2}(|u| - u)$. Then $u_\pm$ is an eigenvector of $-\Delta$ associated to the eigenvalue λ_1.

This follows immediately from Point 3 above and the definition of u.

Point 5. End of the proof of Theorem 13.4.1. Being an eigenvector, u is not identically zero. Suppose there exist x_1, x_2 in Ω such that $u(x_1) < 0, u(x_2) > 0$.

Since the function u is continuous in Ω, there exist two open neighborhoods V_{x_1} and V_{x_2} such that $u < 0$ in V_{x_1} and $u > 0$ in V_{x_2}. We deduce that $u_+ = 0$ in V_{x_1} and $u_- = 0$ in V_{x_2}. Since these are eigenvectors of $-\Delta$, Theorem 12.3.11 of Chap. 12 states that $u_+ = u_- = 0$ in Ω, which implies that $u = u_+ - u_- = 0$ in Ω and leads to a contradiction. Therefore, u has constant sign in Ω, i.e., $u \geq 0$ (or $u \leq 0$). We deduce that $\Delta u = -\lambda_1 u \leq 0$ (or $\Delta u \geq 0$) in Ω. The maximum principle (see Theorem 13.5.1 in the appendix) then implies that $u > 0$ (or $u < 0$) in Ω.

Finally, let us assume that there exist two non-collinear eigenvectors u_1, u_2 associated with the eigenvalue λ_1. From the above we have $u_i(x) \neq 0, i = 1, 2$ for every $x \in \Omega$. Let $x_0 \in \Omega$ be fixed and set $v = u_2(x_0)u_1 - u_1(x_0)u_2$. Then v is not identically zero, otherwise we would have $u_1(x) = \frac{u_1(x_0)}{u_2(x_0)} u_2(x)$ for all $x \in \Omega$. On the other hand, v is an eigenfunction of $-\Delta$ associated with λ_1 which satisfies $v(x_0) = 0$. This contradicts the fact that any eigenfunction is strictly positive (or strictly negative). Therefore, the eigenspace associated with the eigenvalue λ_1 is of dimension 1 and λ_1 is simple. This completes the proof of Theorem 13.4.1. □

13.4.1 An Example in Dimension $d = 1$

Let us consider the interval $I = (0, 1)$. The functions in $H_0^1(I)$ are continuous functions on the closed interval $[0, 1]$. We must therefore solve the problem

$$u \in H^1(I), \qquad -u'' = \lambda u, \quad u(0) = u(1) = 0. \tag{13.4.3}$$

If $\lambda \leq 0$ the only solution to problem (13.4.3) is $u \equiv 0$. Indeed, if $\lambda = 0$ the solution of the equation is $u(x) = ax + b$ and the fact that $u(0) = u(1) = 0$ implies that $u \equiv 0$. If $\lambda < 0$ we must have $u(x) = C_1 e^{\sqrt{-\lambda}x} + C_2 e^{-\sqrt{-\lambda}x}$. The boundary conditions imply that $C_1 + C_2 = 0$ and $C_1 e^{\sqrt{-\lambda}} + C_2 e^{-\sqrt{-\lambda}} = 0$. Thus $C_2 = -C_1$ and $C_1(e^{\sqrt{-\lambda}} - e^{-\sqrt{-\lambda}}) = 0$. If $C_1 = 0$ we have $u \equiv 0$ and if $C_1 \neq 0$ we must have $e^{\sqrt{-\lambda}} = e^{-\sqrt{-\lambda}}$ so $e^{2\sqrt{-\lambda}} = 1$ that is, $\lambda = 0$ and we are brought back to the first case.

If $\lambda > 0$ the solution of the equation is $u(x) = C_1 e^{i\sqrt{\lambda}x} + C_2 e^{-i\sqrt{\lambda}x}$ and the boundary conditions give $C_2 = -C_1$ and $C_1(e^{i\sqrt{\lambda}} - e^{-i\sqrt{\lambda}}) = 2C_1 \sin(\sqrt{\lambda}) = 0$. We must have $C_1 \neq 0$ (otherwise $u \equiv 0$) and $\sin(\sqrt{\lambda}) = 0$ which implies $\lambda = k^2\pi^2$, $k \in \mathbf{N} \setminus 0$.

The solution to problem (13.4.3) with $\lambda_k = k^2\pi^2, k \geq 1$, with $L^2(I)$ norm equal to 1, is therefore $e_{\lambda_k}(x) = \sqrt{2}\sin(k\pi x)$. The nodal domains of this eigenfunction are the intervals $I_i = (\frac{i}{k}\pi, \frac{i+1}{k}\pi), i = 0, \ldots, k-1$. There are thus exactly k of them.

The cases of the Dirichlet problem in a square or in a disk in $\mathbf{R}^2$ are treated, for example, in the book [9].

Back to Formula (12.7.1)

This same example allows us to test formula (12.7.1). Indeed, if $d = 1$ we have $\mathcal{H}^{d-1}(E) = \mathcal{H}^0(E)$, the number of points in the set E. Let $e_{\lambda_k}(x) = \sqrt{2}\sin(k\pi x)$ be an eigenvector associated with the eigenvalue $\lambda_k = k^2\pi^2$. The set $\{x \in (0,1) : e_{\lambda_k}(x) = 0\}$ is, for $k \geq 2$, the set $\{\frac{1}{k}, \frac{2}{k}, \ldots, \frac{k-1}{k}\}$. It contains $k-1$ points so that

$$\frac{1}{2\pi}\sqrt{\lambda_k} \leq \mathcal{H}^0\{x \in (0,1) : e_{\lambda_k}(x) = 0\} \leq \frac{1}{\pi}\sqrt{\lambda_k}.$$

13.5 Appendix

13.5.1 The Strong Maximum Principle

Theorem 13.5.1 *Let Ω be a connected open subset of $\mathbf{R}^d$ and $u \in C^2(\Omega)$ such that $\Delta u \leq 0$ and $u \geq 0$ in Ω. Then we have the following alternative in Ω: either $u > 0$, or $u \equiv 0$.*

Proof Set $\Omega_0 = \{x \in \Omega : u(x) = 0\}$. If $\Omega_0 = \emptyset$ we have $u(x) > 0$ for all $x \in \Omega$. We can therefore assume that Ω_0 is non-empty. Since u is continuous, the set Ω_0 is closed in Ω. Let us show that it is also open. Let $y \in \Omega_0$. We can apply Proposition 12.6.6 (mean value formula) to the function u in a ball $B(y,r)$ whose closure is contained in Ω. It follows, since $u \geq 0$ in Ω,

$$0 = u(y) \geq \frac{1}{\mu(B(y,r))}\int_{B(y,r)} u(z)\,dz \geq 0.$$

We deduce that $u(z) = 0$ in $B(y,r)$, in other words $B(y,r) \subset \Omega_0$. Thus, Ω_0 is open. By the connectedness of Ω we have $\Omega_0 = \Omega$ so, $u \equiv 0$ in Ω. □

13.6 Comments

Richard Courant (1888–1972) was a European mathematician who emigrated to the United States in 1933. He founded a mathematics institute in New York, which is today called the "Courant Institute" and is part of New York University (NYU). He is the author, in collaboration with the great German mathematician David Hilbert (his thesis advisor), of a two-volume treatise on analysis and partial differential equations, entitled "Methods of Mathematical Physics".

References

For Prerequisites

1. Hubbard, J., West, B.: Differential Equations: A Dynamical System Approach. Texts in Applied Mathematics, vol. 5. Springer (1991)
2. Krantz, S.G.: Function Theory of Several Complex Variables. AMS Chelsea Publishing, American Math Society (2001)
3. Zuily, C.: Elements de distributions et d'équations aux dérivées partielles. Dunod, Sciences sup (2002)
4. Zuily, C.: Problèmes de distributions et d'équations aux dérivées partielles. Cassini (2011)

For Further Study

5. Alazard, T.: Analysis and Partial Differential Equations. Universitext, Springer (2024)
6. Alazard, T., Zuily, C.: Tools and Problems in Partial Differential Equations. Universitext, Springer (2020)
7. Bressan, A.: Hyperbolic Systems of Conservation Laws. Oxford University Press (2000)
8. Cheverry, C., Raymond, N.: A guide to spectral theory. In: Advanced Texts. Birkhauser (2021)
9. Courant, R., Hilbert, D.: Methods of Mathematical Physics. Wiley-Vch Verlag (2004)
10. Evans, L.C.: Partial Differential Equations Graduate Studies, vol. 19. American Mathematical Society, Providence, Rhode Island (2010)
11. Federer, H.: Geometric measure theory. In: Classics in Mathematics. Springer (1996)
12. John, F.: Partial Differential equations. In: Applied Mathematical Sciences, vol. 1. Springer (1986)
13. Lewin, M.: Théorie spectrale et mécanique quantique. Course at Ecole Polytechnique (2021). Personal page of M. Lewin
14. Milnor, J.: Topology from the Differentiable Viewpoint. Princeton Landmarks in Mathematics (1997)
15. Rauch, J.: Partial differential equations. In: Graduate Text. Springer (1991)
16. Strauss, W.: Partial Differential Equations: An Introduction. Wiley (2008)

C. Zuily, *Selected Topics in Partial Differential Equations*, Universitext,
https://doi.org/10.1007/978-3-032-24082-8

Index

C. Zuily, *Selected Topics in Partial Differential Equations*, Universitext,
https://doi.org/10.1007/978-3-032-24082-8

Zeitfracht Medien GmbH
Ferdinand-Jühlke-Straße 7
99095 Erfurt, Deutschland
produktsicherheit@kolibri360.de